Solar Thermal Systems: Thermal Analysis and its Application

Edited by

Manoj Kumar Gaur

Madhav Institute of Technology & Science
Gwalior
India

Brian Norton

Tyndall National Institute
University College Cork
Cork
Ireland

&

Gopal Nath Tiwari

Jan Nayak Chandrashekhar University
Ballia
India

Solar Thermal Systems: Thermal Analysis and its Application

Editors: Manoj Kumar Gaur, Brian Norton & Gopal Tiwari

ISBN (Online): 978-981-5050-95-0

ISBN (Print): 978-981-5050-96-7

ISBN (Paperback): 978-981-5050-97-4

© 2022, Bentham Books imprint.

Published by Bentham Science Publishers Pte. Ltd. Singapore. All Rights Reserved.

First published in 2022.

need for a court order if at any point you breach any terms of this License Agreement. In no event will any delay or failure by Bentham Science Publishers in enforcing your compliance with this License Agreement constitute a waiver of any of its rights.

3. You acknowledge that you have read this License Agreement, and agree to be bound by its terms and conditions. To the extent that any other terms and conditions presented on any website of Bentham Science Publishers conflict with, or are inconsistent with, the terms and conditions set out in this License Agreement, you acknowledge that the terms and conditions set out in this License Agreement shall prevail.

Bentham Science Publishers Pte. Ltd.
80 Robinson Road #02-00
Singapore 068898
Singapore
Email: subscriptions@benthamscience.net

BENTHAM SCIENCE

CONTENTS

**CHAPTER 15 THE CO$_2$ MITIGATION AND EXERGO AND ENVIRON-ECONOMICS
ANALYSIS OF BIO-GAS INTEGRATED SEMI-TRANSPARENT PHOTO-VOLTAIC
THERMAL (BI-ISPVT) SYSTEM FOR INDIAN COMPOSITE CLIMATE**.. 363
Gopal Nath Tiwari, Praveen Kumar Srivastava, Akhoury Sudhir Kumar Sinha and *Arvind Tiwari*

PREFACE

Solar energy is the most abundant renewable energy that can meet all the energy requirements without getting depleted and harming the environment. A lot of research is going on to harness solar thermal energy for space heating, water distillation, and electricity generation. So, there is a need for a resource that can provide knowledge regarding solar energy and research on solar thermal systems. This will help develop more efficient solar systems that are more compact, reliable, portable, and able to eliminate the dependency on conventional energy sources for meeting the energy requirement for various daily life and industrial purposes.

This book encapsulates the solar thermal systems available to meet the need of food, fresh water, cooking food, heating water *etc*. The fundamentals of thermodynamics, heat transfer and solar energy are covered in Chapter 1 and Chapter 2. The basics of some solar thermal devices along with their thermal modeling are covered in Chapter 3. The basics of solar still, its thermal modeling, applications, development in past few years and potential of solar distillation system in India is discussed in Chapter 4-6. The design, development and applications of solar cookers along with their thermal modeling are covered in Chapter 7-8. Thermal modeling of semi-transparent PVT systems and its application is discussed in Chapter 9 and Chapter 10 covers the development in solar photovoltaic technology. Chapter 11 and Chapter 12 discusses about thermal modeling of greenhouse solar dryer and case study on hybrid active greenhouse solar dryer. Chapter 13 covers the thermal analysis of photovoltaic thermal (PVT) air heater employing thermoelectric module (TEM). The applications of various solar systems in building sectors and the development in this field are covered in Chapter 14. Chapter 15 deals with exergo and environ- economics analysis of bio-gas integrated semi-transparent photo-voltaic thermal (Bi-iSPVT) system for Indian composite climate.

The book has a broad scope and helps students and researchers in universities, industries, and national and commercial laboratories to learn the fundamentals and in-depth knowledge regarding thermal modeling and developments in solar thermal systems in the past few years. It is a research-oriented book in which different researchers have contributed in the form of different chapters. I hope that the book will provide sufficient knowledge regarding solar systems and will not discourage the readers. This book can be used as a reference tool for teaching the solar energy and thermal modeling of solar thermal systems to the students and research fellows in universities and research organizations

Manoj Kumar Gaur
Madhav Institute of Technology & Science
Gwalior
India

Brian Norton
Tyndall National Institute
University College Cork
Cork
Ireland

&

Gopal Nath Tiwari
Jan Nayak Chandrashekhar University
Ballia
India

ACKNOWLEDGEMENTS

This book has been possible due to the dedicated efforts of many individuals who have contributed to this book: "Solar Thermal Systems: Thermal Analysis and its Applications".

My intellectual debt is to Dr. G.N. Tiwari, my mentor, who has always given his time to discuss detailed and constructive comments to clarify concepts, explore particular facets of insight work, and explain the rationales for specific recommendations.

My sincere thanks and deep gratitude to Prof. Brian Norton, who has been kind enough to accept my request to be the co-editor of this book. Without his guiding force and constant support, this book would not have been possible.

I would also like to thank the authors from various parts of India and the world who contributed their time and valuable experiences and, more importantly, shared their learning, which will lead to valuable discussions and engagement among the scientific community. I also thank the reviewers who contributed their time and valuable insights in reviewing the chapters.

I want to acknowledge Bentham Publication for presenting me with the kind opportunity to bring this book to realization and for their counsel during the final stages of completing the book. I want to thank Humaira Hashmi for assisting with administrative issues.

I would also like to thank my colleagues Dr. C.S. Malvi, Dr. S.K. Dubey, and Dr. J.L. Bhagoriya. I would like to show my appreciation to my Ph.D. students, Vikas Kumar Thakur and Rishika Shah, for their assistance. I am particularly grateful to Pushpendra Singh for his excellent editorial support and coordination.

I would like to express my sincere thanks to Dr. R.K. Gupta for his blessings and support throughout my life. I owe an enormous debt of gratitude to Dr. R.K. Pandit, Director, MITS Gwalior, from whom I have always received generous support and encouragement in all my academic endeavors in this esteemed institute.

Finally, I want to express my special thanks to my family, my parents Shri Shukhad Gaur and Smt. Archana Devi Gaur and my in-laws Shri P.K. Saxena and

Smt. Seema Saxena, whose blessings have proved to be a guiding light, my wife, Dr. Parul Saxena, for inspiring me to fulfill my aspirations and my children, Abhishek and Aarnav, for keeping me motivated in my journey. I would also like to thank all the faculty and staff members of the Mechanical Department of MITS, Gwalior, for their support.

List of Contributors

A.K. Dhamneya Department of Mechanical Engineering, Madhav Institute of Technology & Science, Gwalior-474005, India

Abhishek Saxena Department of Mechanical Engineering, Moradabad Institute of Technology, Moradabad 244001, India

Akhoury Sudhir Kumar Sinha Rajiv Gandhi Institute of Petroleum Technology, Jais, Amethi (UP), India

Anil Kumar Tiwari Department of Mechanical Engineering, National Institute of Technology, Raipur, India

Amit Shrivastava Department of Mechanical Engineering, Madhav Institute of Technology and Science, Gwalior, India

Anil Kumar Depratment of Mechanical Engineering, Delhi Technological University, Delhi 110042, India
Centre for Energy and Environment, Delhi Technological University, Delhi 110042, India

Arvind Tiwari Madhav Institute of Technology & Science, Gwalior, Madhya Pradesh, India

Brian Norton Technological University Dublin, Ireland

C.S. Malvi Madhav Institute of Technology & Science, Gwalior, Madhya Pradesh, India

Deepak Madhav Institute of Technology & Science, Gwalior, Madhya Pradesh, India

Desh Bandhu Singh Department of Mechanical Engineering, Graphic Era Deemed to be University, Bell Road, Clement Town, Dehradun, Uttarakhand, 248002, India

Erdem Cuce Low/Zero Carbon Energy Technologies Laboratory, Faculty of Engineering, Recep Tayyip Erdogan University, Zihni Derin Campus, 53100 Rize, Turkey
Department of Mechanical Engineering, Faculty of Engineering, Recep Tayyip Erdogan University, Zihni Derin Campus, 53100 Rize, Turkey

G.N. Tiwari Bag Energy Research Society, Jawahar Nager (Margupur), Chilkhar-221701, Ballia UP, India

Gaurav Saxena Department of Mechanical Engineering, Madhav Institute of Technology and Science, Gwalior, India

Gopal Nath Tiwari Department of Physical Sciences and Humanities, Rajiv Gandhi Institute of Petroleum Technology, Jais (UP), India

Kuber Nath Mishra Department of Mechanical Engineering, National Institute of Technology, Raipur, India

M.K. Gaur Department of Mechanical Engineering, Madhav Institute of Technology and Science, Gwalior, India

M.K. Sagar Department of Mechanical Engineering, Madhav Institute of Technology and Science, Gwalior-474005, India

Neha Dimri Laboratorie LOCIE, Université Savoie Mont Blanc (USMB), France

Pinar Mert Cuce   Department of Energy Systems Engineering, Faculty of Engineering, Recep Tayyip Erdogan University, Zihni Derin Campus, 53100 Rize, India
Low/Zero Carbon Energy Technologies Laboratory, Faculty of Engineering, Recep Tayyip Erdogan University, Zihni Derin Campus, 53100 Rize, Turkey

Prashant V. Baredar   Energy Centre, Maulana Azad National Institute of Technology, Bhopal - 462003, M.P., India

Praveen Kumar Srivastava   Institute of Bio-sciences and Technology, Shri Ramswaroop Memorial University, Barabanki, (UP), India

Pushpendra Singh   Department of Mechanical Engineering, Madhav Institute of Technology and Science, Gwalior, India

R.K. Pandit   Department of Architecture, Madhav Institute of Technology and Science, Gwalior, India

Rajkumar Malviya   Energy Centre, Maulana Azad National Institute of Technology, Bhopal - 462003, M.P., India

Rishika Shah   Department of Architecture, Madhav Institute of Technology and Science, Gwalior, India

Sampurna Panda   Energy Poornima University, Jaipur, Rajasthan, India

Shubham Kashyap   Department of Mechanical Engineering, National Institute of Technology, Raipur, India

Shubham Srivastava   Madhav Institute of Technology & Science, Gwalior, Madhya Pradesh, India

Taranjeet Sachdev   Department of Mechanical Engineering, National Institute of Technology, Raipur, India

Veeresh Vishwakarma   Energy Centre, Maulana Azad National Institute of Technology, Bhopal - 462003, M.P., India

Vikas Kumar Thakur   Department of Mechanical Engineering, Madhav Institute of Technology and Science, Gwalior-474005, India

CHAPTER 1

Introduction

Pushpendra Singh[1]*, M.K. Gaur[1]

[1]*Department of Mechanical Engineering, Madhav Institute of Technology and Science, Gwalior, India*

Abstract: This chapter covers the basics of thermodynamics and heat transfer, which helps in understanding the heat transfer mechanism and thermal modeling of various solar thermal systems covered in further chapters. The laws of thermodynamics and heat transfer are also covered in the chapter. The general terms related to thermodynamics and heat transfer are also defined in this chapter, as these terms will be used frequently in upcoming chapters.

Keywords: Basics of heat transfer, Basic terminology, Laws of heat transfer, Thermodynamics.

1. THERMODYNAMICS AND HEAT TRANSFER

Thermodynamics is a branch of science that deals with heat and work interaction and its effect on the system and surroundings. The amount of heat transferred from the system or heat transferred to the system and the amount of work done on the system or work done by the system are studied in thermodynamics. However, in thermodynamics, we cannot determine the rate of heat transfer from the system or to the system. Heat transfer is interrelated to thermodynamics as it is based on the second law of thermodynamics only.

The heat supplied to various solar systems by various modes is determined using the concept of heat transfer like heat transfer by solar radiation to inside air of dryer, heating of basin of solar still by absorbing the solar radiation, heat loss from solar systems to surrounding, heat carried away by hot water supplied through solar water heater, *etc.* [1,2].

---

*Corresponding author **Pushpendra Singh:** Department of Mechanical Engineering, Madhav Institute of Technology and Science, Gwalior, India; E-mail: pushpendra852@gmail.com

Manoj Kumar Gaur, Brian Norton & Gopal Tiwari (Eds.)

2. GENERAL TERMS RELATED TO THERMODYNAMICS

The various terms related to thermodynamics that are frequently used are defined below [3]:

2.1. System

The system is simply the quantity of matter upon which the study is focused. In the context of solar systems, it can be a crop dried inside the dryer, water flowing inside the heat pipe, vapor inside the solar still, air or food inside the solar cooker, *etc.* The hypothetical system having an imaginary boundary is called control volume and the quantity of matter inside the control volume is called control mass.

2.2. Surrounding

Everything external to the system is called a surrounding. In other words, the point up to which the effect of the system is observed is called as surrounding, and the remaining is the environment.

2.3. Boundary

The boundary separates the system from the surroundings and it can be rigid or flexible depending upon the system. For example, the transparent cover of the dryer acts as a boundary that separates the inside air from the surrounding air. The water mass inside the heat pipe is an example of a control volume having an imaginary flexible boundary. The boundary is also classified as an adiabatic boundary and diathermic boundary. There is no transfer of heat through an adiabatic boundary, while a diathermic boundary allows the transfer of heat from the system to the surroundings or *vice versa*.

2.4. Types of System

The system is classified into three types: open, closed, and adiabatic system. In an open system, both heat and mass transfer takes place, while in a closed system, there is a transfer of heat only. In an adiabatic system, neither the heat nor the mass is transferred from or to the system.

2.5. Heat

The form of energy that transfers due to temperature differences between the two bodies is termed as heat or thermal energy. In terms of thermodynamics, it is defined as the energy in transition across the boundary between the thermodynamic system and its surroundings without any transfer of mass. Heat always flows from high temperature body to lower temperature body. It is generally measured in calories, Joules, or btu.

2.6. Temperature

Temperature is a measure of the degree of hotness or coldness of anybody. The temperature is also defined as the average kinetic energy of molecules in a body. In simple words, the temperature of any substance denotes the amount of thermal energy inside that substance. The most common instrument used to measure temperature is a thermometer. The temperature is generally measured in degree Celsius (°C), Kelvin (K), and Fahrenheit (F).

2.7. Entropy

Entropy simply denotes the molecular disorder or randomness in the system. It is a property of the system, so it depends on the state, not on the path followed to reach that particular state.

2.8. Enthalpy

Enthalpy is also a property of a thermodynamic system and it is basically the summation of the internal energy *(U)* of the system and product of pressure *(p)* and volume *(V)*. It is expressed as,

$$H = U + pV \tag{1}$$

Change in enthalpy of the system can also be defined as the total heat supplied to the system at constant pressure and temperature.

2.9. Latent Heat

At constant temperature, the amount of heat absorbed or released by the substance during a change of its phase is called latent heat. The heat required for changing the solid to liquid phase or *vice versa* is called the latent heat of fusion. When liquid changes its phase to gas or *vice versa* is called the latent heat of vaporization. Various phase change materials are used in solar systems to store the excess solar thermal energy during the sunshine period and supply that stored energy during the off sunshine period. In this way, the solar systems can be used in the off sunshine period also. The SI unit of latent heat is kJ/kg.

2.10. Sensible Heat

The heat absorbed or released by the substance due to which there is no change in its phase and only the temperature of the substance changes is called sensible heat. The sensible heat of any system can be simply calculated using the expression as follows:

$$Q = m_s C \Delta T \tag{2}$$

Where m_s is the mass of substance in kg, C is the specific heat of that substance in kJ/kgK, and ΔT is the variation in temperature of the substance due to the addition or release of heat in K.

2.11. Specific Heat

The amount of heat required by 1 kg of substance (unit mass) to increase its temperature by 1°C (unit temperature) is called the specific heat of that substance. It is usually expressed in kJ/kgK. It is an intensive property, and for gases, it is of two types, specific heat at constant pressure (C_p) and specific heat at constant volume (C_v). Generally, at NTP, the specific heat of water is 4.18 kJ/kgK, while for air C_p is 1.005 kJ/kgK, and C_v is 0.718 kJ/kgK. Specific heat of any substance is dependent on temperature, so in the study of solar systems, the correlations in terms of temperature are used to calculate the specific heat.

2.12. Specific Humidity

The moisture existing in the unit mass of air is called specific humidity or humidity ratio or absolute humidity. It is generally denoted by ω and expressed in kg/kg of dry air. Mathematically it is calculated using the following expression:

$$\omega = \frac{m_m}{m_a} = 0.622 \times \left(\frac{P_m}{P - P_m}\right) \tag{3}$$

Where m_m is the mass of moisture in the air in kg, m_a is the mass of dry air in kg, P_m is the partial pressure of moisture present in the air in the bar, P is the total pressure of air in bar.

2.13. Relative Humidity

The relative humidity is the ratio of the mass of water vapour in the air to the mass of water vapour in saturated air at the same temperature and pressure. It is denoted by ϕ and expressed in percentage as it is a dimensionless quantity. It is calculated using the relation,

$$\varphi = \frac{m_m}{m_{m,s}} = \frac{P_m}{P_{ms}} = \frac{\mu}{1 - (1 - \mu)\frac{P_{ms}}{P}} \tag{4}$$

Here, P_{ms} is the partial vapour pressure at saturation in the bar, $m_{m,s}$ is the mass of water vapour in saturated air in kg, and μ is the degree of saturation.

Simply the relative humidity of air shows the extent to which it is saturated. For example, if the relative humidity of air is 80%, this means that air gets 80% saturated and is able to carry further 20% water vapour in it. The relative humidity is an ambient parameter that strongly depends on the air temperature as the higher is the air temperature, lower is the relative humidity and *vice versa*. The relative humidity is an important parameter that is measured during the heat transfer analysis of all solar systems.

2.14. Boiling

Boiling is the process of conversion of liquid into vapour due to the addition of latent heat to it. The liquid on any surface starts boiling only when it reaches its saturation temperature and then further addition of latent heat of vaporization changes the phase of liquid to vapour. The boiling process is also of two types namely pool boiling and forced boiling.

In pool boiling, the hot surface is placed inside the stable liquid. In pool boiling, there is no bulk movement of fluid; only the liquid near the heated surface moves due to free convection in the form of bubbles. While in forced boiling, there is bulk movement of liquid over the heated surface.

2.15. Condensation

Condensation is just the reverse process of boiling. When the vapour contacts the surface at a temperature lower than its saturation temperature, then it changes its phase from vapour to liquid by losing its latent heat of vaporization and this process is called condensation. The condensation is also of two types namely drop-wise condensation and film-wise condensation [4].

In drop-wise condensation, the liquid condenses in the form of a droplet and flows down under gravity without wetting the surface. While in film-wise condensation, the liquid wets the surface and the condensate spreads over the surface in the form of thin-film and flows down under gravity. The rate of heat transfer in drop-wise condensation is always higher than the film-wise condensation.

2.16. Zeroth Law of Thermodynamics

Zeroth law of thermodynamics states that "if two individual bodies are in thermal equilibrium with the third body, they are also in thermal equilibrium with each other." This law becomes the basis of the measurement of temperature.

2.17. First law of Thermodynamics

First law of thermodynamics is also called as law of conservation of energy. It can be stated as, "If the system undergoes the thermodynamic cycle, the net heat supplied to the system from its surrounding is equal to the net work done by the system on its surrounding." While for a thermodynamic process, the net heat supplied to the system is equal to the summation of change in its internal energy and work done by the system on its surrounding.

For cycle, $$\oint dQ = \oint dW \tag{5}$$

For a Process, $$Q = \Delta U + W \tag{6}$$

Thus in a process, there must be some loss in heat supplied to the system during conversion of heat to useful work.

The First law of thermodynamics is also stated as "energy can neither be created nor be destroyed, but it can be converted from one form to other."

The thermal modeling of solar systems is carried out by application of the First law of thermodynamics only. The energy balance equation for the particular solar system is developed using the concept that the amount of energy supplied to the solar system must be equal to the amount of energy going out of the solar system either in terms of useful energy or in terms of heat loss to the surrounding. This helps in determining the heat losses occurring in the solar system.

2.18. Second Law of Thermodynamics

The Ist law of thermodynamics explains the conversion of heat into work and *vice versa*, but it does not explain whether the process is possible or not and also it does not tells that in which direction the heat transfer takes place. All these limitations are overcome in the IInd law of thermodynamics. There are two statements of the second law of thermodynamics which are stated below:

2.19. Kelvin-Planck Statement

It states that, "It is impossible to construct an engine, which while operating in a cycle produces no effect other than the transfer of heat from a single reservoir and produces work."

This law tells that no system can be 100% efficient. The heat supplied to any system cannot be utilized 100%; there must be some loss of heat.

2.20. Clausius Statement

It states that "It is impossible to construct a device which operates on a cycle and produces no other effect than the transfer of heat from a cooler body to a hotter body."

This law tells that without the aid of external work, it is not possible to transfer the heat from a colder body to a hotter body.

2.21. Third Law of Thermodynamics

According to IIIrd law of thermodynamics, "the temperature of any system cannot be reduced to absolute zero in a finite number of processes."

2.22. Modes of Heat Transfer

Heat transfer is based on the second law of thermodynamics, *i.e.* heat always flows from a body at a higher temperature to a body at a lower temperature. Heat transfer occurs in three modes, namely conduction, convection and radiation. The similarity in all three modes is that there must be a temperature differential to transfer the heat in the direction of decreasing temperature. In solar systems also, heat transfer can take place by any mode or combination of different modes.

3. CONDUCTION

Conduction is the mode of heat transfer that takes place within a medium or between different mediums in direct physical contact with others. Conduction heat transfer takes place in two ways, firstly the heat transfer by means of free electrons within the medium and secondly, by lattice vibration. The common examples of heat transfer through conduction in solar systems are heat transfer through the wall of solar stills and solar cookers, heat transferred through wall of glass cover in solar stills and solar dryers *etc.*

4. CONVECTION

In convection, the heat from the hot plate to cold fluid or *vice versa* takes place due to bulk displacement of fluid flowing over the wetted surface. Heat transfer through convection occurs in fluids only. Convection is of two types namely free or natural convection and forced convection. In free convection, the fluid movement occurs due to density difference or buoyancy effect. The hot fluid rises upward and cold fluid comes downward to replace the void created by hot fluid. In forced convection, the fluid is moved with the aid of some external devices like a fan, blower, pump *etc.* to force the fluid to move over the hot surface.

In solar systems, some common examples of free convection are flow of air in passive solar dryers, movement of water in water in tube type evacuated solar collector *etc.*, and examples of forced convection are the movement of air in active greenhouse solar dryer by using fans, circulation of hot water of flat plate solar collector to solar still by using pump *etc.*

5. RADIATION

Radiation is a mode of heat transfer in which the transmission of heat takes place in the form of electromagnetic waves. Like other modes of heat transfer, the medium is not a necessary requirement for the transfer of heat through radiation as radiation heat transfer occurs most effectively in a vacuum. All bodies above the absolute zero temperature are capable of emitting radiant energy. The best example of radiant heat transfer is the solar radiation that is emitted by the sun and, after traveling through space reaches the Earth's surface. Solar radiation acts as a source of input thermal energy to all solar systems.

6. GENERAL TERMS RELATED TO HEAT TRANSFER

The common terms relating to heat transfer that will be used in further chapters are defined below so as to make a better understanding of these commonly used terms.

7. THERMAL CONDUCTIVITY

Thermal conductivity is the property of a material, and it shows the rate at which the conductive heat energy is transferred through the material. Based on Fourier's law of conduction, thermal conductivity is also defined as the rate of heat conducted through the unit area of material maintained at unit temperature difference across the unit thickness of material. It is denoted by k and generally expressed in W/mK or W/m°C. The thermal conductivity of some of the commonly used materials in solar systems is shown in Table **1**.

Table 1. Thermal conductivity at normal temperature and pressure [5].

Material	Thermal Conductivity (W/mK)
Copper	385
Aluminium	225
Brass	107
Cast Iron	55 - 65
Steel	20 - 45
Concrete	1.20
Glass	0.75
Water	0.55 – 0.7
Air	0.024

8. CONVECTIVE HEAT TRANSFER COEFFICIENT

Convective heat transfer coefficient can be defined as the amount of heat transfer through convection per unit time, per unit area and per unit temperature difference between solid surface and fluid. Unlike thermal conductivity, it is not the property of fluid as it depends on various thermophysical properties of fluid like density, viscosity, thermal conductivity, specific heat and coefficient of expansion. Nature of fluid flow *i.e.* laminar or turbulent, roughness, geometry and orientation of solid surface also affects the convective heat transfer coefficient of fluid. It is denoted by h and expressed in W/m^{2}K or W/m^2°C.

9. OVERALL HEAT TRANSFER COEFFICIENT

In the problems in which heat transfer takes place through the metallic surface between the two fluids, it is convenient to adopt an overall heat transfer coefficient (U) which gives the heat transmitted per unit area per unit time per degree temperature difference between the bulk fluids on each side of the metal. Overall heat transfer is usually considered when heat transfer is taking place through more than one mode. Let fluid A and B are flowing on both the sides of a flat metallic surface; the overall heat transfer coefficient is calculated as:

$$\frac{1}{U} = \frac{1}{h_A} + \frac{\delta}{k} + \frac{1}{h_B} \tag{7}$$

Here, δ is the thickness of the solid surface, k is the thermal conductivity of the solid surface, h_A and h_B is the convective heat transfer coefficient of fluid A and B respectively.

10. EMISSIVITY

Emissivity of a material is the ratio of thermal energy radiated from the material to the thermal energy radiated from the perfect emitter (black body) at the same temperature and wavelength and under the same viewing conditions. The emissivity of a material represents the effectiveness of the material to emit thermal radiation. It usually varies from 0 for the perfect reflector to 1 for the perfect emitter.

The emissivity of the material is dependent on the type of material, type of surface and also on the temperature of surface and wavelength. The clean and polished surface materials have a low emissivity, while roughened and oxidized surfaces will have high emissivity.

11. IRRADIATION

The total amount of radiation striking the surface per unit area is called irradiation. It is denoted by G and expressed in W/m^2.

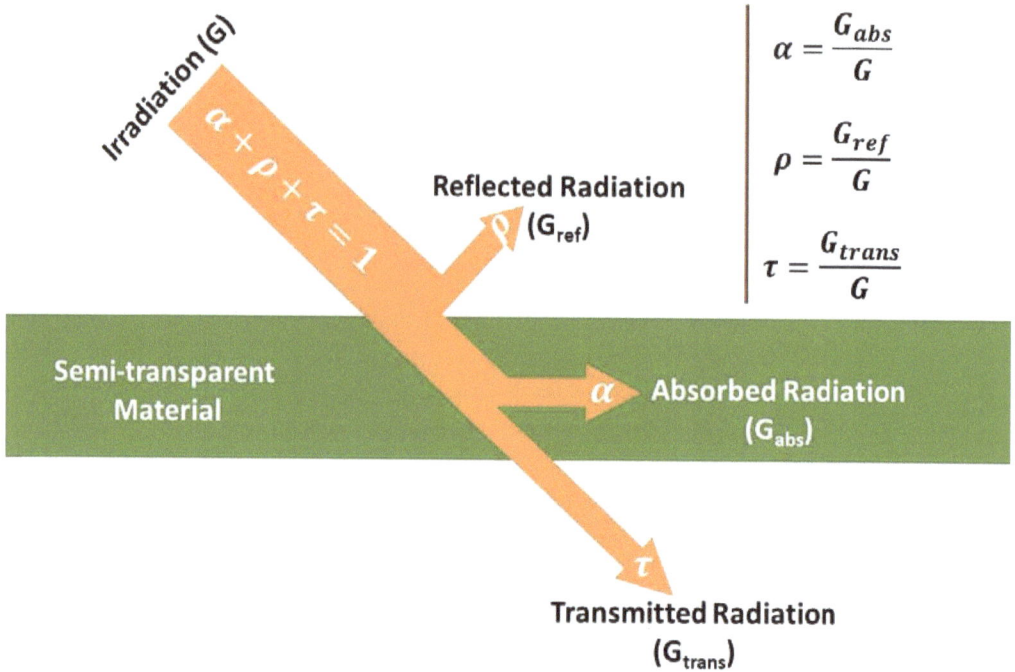

Fig. (1). Absorptivity, reflectivity and transmissivity.

12. ABSORPTIVITY, REFLECTIVITY AND TRANSMISSIVITY

The portion of irradiation absorbed by the surface is termed as absorptivity (α), and a portion of irradiation reflected and transmitted by the surface is termed as reflectivity (ρ) and transmissivity (τ). The value of α, ρ and τ for anybody lies between 0 and 1. Also, as shown in Fig. **(1)**,

$$\alpha + \rho + \tau = 1 \tag{8}$$

For black body, $\rho = 0 \;\; and \;\; \tau = 0$ $\therefore$ $\alpha = 1$

For opaque body, $\tau = 0$ $\therefore$ $\alpha + \rho = 1$

For white body, $\alpha = 0 \;\; and \;\; \tau = 0$ $\therefore$ $\rho = 1$

13. LAWS OF HEAT TRANSFER

There are different laws associated with the different modes of heat transfer. The laws related to conduction, convection and radiation heat transfer are explained below [6,7]:

14. FOURIER LAW OF CONDUCTION

According to this law, *"the rate of heat conducted through a body is directly proportional to the area normal to the direction of heat flow and the temperature gradient across the walls of that body."* Mathematically it is represented as,

$$\dot{Q}_{cond} = -kA\frac{dT}{dx} \tag{9}$$

Here dT/dx is the temperature gradient and x is the thickness of the material.

As heat flows in the direction of decreasing temperature gradient, therefore the value of temperature gradient is negative. Hence to counteract this, a negative sign is provided in the equation so that value of heat transfer becomes positive.

15. NEWTON'S LAW OF COOLING FOR CONVECTION

According to this law, *"the rate of heat transfer through convection is directly proportional to the area of solid surface wetted by the fluid flowing over it and temperature difference between the solid surface and the moving fluid."*

Mathematically it is expressed as,

$$\dot{Q}_{conv} = hA\left(T_s - T_f\right) \tag{10}$$

Here, h is the convective heat transfer coefficient, A is the area of solid surface, T_s and T_f are the surface and fluid temperature respectively.

16. STEFAN BOLTZMANN LAW OF RADIATION

According to this law, *"the amount of radiant energy emitted by the body per unit area is directly proportional to the fourth power of the absolute temperature of that body."* Mathematically it is expressed as,

$$\dot{Q}_{rad} = \sigma A T^4 \tag{11}$$

Here, σ is the Stefan-Boltzmann constant whose value is 5.67×10^{-8} W/m^{2}K^4 and T is the absolute temperature of the radiating surface in K.

This equation is valid for the ideal surface and used for calculating the maximum energy that can be emitted from the surface. For real surface, the above equation is written as,

$$\dot{Q}_{rad} = \varepsilon \sigma A T^4 \tag{12}$$

Where ϵ is the emissivity of the real surface.

17. SUMMARY

In this chapter, the basics of thermodynamics and heat transfer are introduced. Thermodynamics deals with the property in an equilibrium state, while the heat transfer helps in determining the rate at which the heat is transferring from one point to another. The general terms of thermodynamics and heat transfer like specific heat, thermal conductivity, heat transfer coefficient, entropy, enthalpy *etc.* are defined in this chapter so as to make a better understanding of further chapters. The Ist law of thermodynamics shows the principle of energy conservation that is used during the energy balance of solar systems. The second law of thermodynamics deals with the possibility of the energy transfer and shows that no system can be 100% efficient as there must be some heat loss associated with it. The third law of thermodynamics shows that it is not possible to reduce the system temperature to absolute zero as the entropy becomes constant, which violates the Clausius Inequality statement that the entropy of the universe always increases. The different laws for the three modes of heat transfer are also covered in this chapter.

CONSENT FOR PUBLICATION

Not applicable.

CONFLICT OF INTEREST

The authors declare no conflict of interest, financial or otherwise.

ACKNOWLEDGEMENTS

Declared none.

REFERENCES

[1] B. Norton, *Harnessing Solar Heat,* 1st ed Springer International Publishing, 2019.

[2] G.N. Tiwari, P. Barnwal, S.C. Solanki, and M.K. Gaur, *Solar Energy, Problems, Solutions & Experiments, First.* Anamaya Publishers: New Delhi.

[3] Y.A. Cengel, and M.A. Boles, *Thermodynamics, An Engineering Approach.* McGraw Hill Education.

[4] R.K. Rajput, *Heat and Mass Transfer*S. Chand and Company Ltd.: New Delhi.

[5] D.S. Kumar, *Heat and Mass Transfer.* S.K. Kataria & Sons: New Delhi, 2012.

[6] Y.A. Cengel, and A.J. Ghajar, *Heat and Mass Transfer,* McGraw Hill Education.

[7] P.K. Nag, *Heat and Mass Transfer.* McGraw Hill Education.

CHAPTER 2

Basics of Solar Energy and Various Sun-Earth Angles

Brian Norton[1*]

[1] *Technological University Dublin, Ireland*

Abstract: This chapter discusses the causes and effects of solar energy incidents on Earth's spectral, skyward angular, intensity, and luminous characteristics. A particular perspective is how these characteristics influence and determine how particular practical devices, systems, and applications harness available solar energy. The uses of solar energy to produce heat, generate electricity *via* photovoltaics, and provide daylight are considered. Moreover, diverse factors limiting the applicability of specific solar energy conversion devices or approaches are also discussed.

Keywords: Radiation, Solar energy, Solar instruments, Sun-Earth angle.

1. MATCHING SOLAR COLLECTORS TO PREVAILING SOLAR ENERGY CONDITIONS

Fundamentally, solar energy is the origin of nearly all energy sources used by people; it fuels life through photosynthesis, provides warmth, and heats the land and sea, creating atmospheric pressure differentials that generate wind and rain. It does the latter at various temporal scales affecting both climates and weather. The Sun is thus not only directly used by photovoltaic and solar thermal devices but is also the source for all wind, hydroelectric, and bioenergy renewable energy resources. Solar energy directly provides [1]:

(a) Heat gain and daylight directly through the windows of buildings.

(b) Heating of heat transfer fluids. This can be water for sanitation, air for crop drying, or direct use in distillation or cooking. High-temperature heat transfer fluids can drive power conversion systems to produce electricity. Thermal energy storage provides heat to solar devices during inadequate sunshine.

*Corresponding author Brian Norton: Technological University Dublin, Ireland; E-mail: brian.norton@ TUDublin.ie

Manoj Kumar Gaur, Brian Norton & Gopal Tiwari (Eds.)

(c) Electricity *via* photovoltaic cells.

A solar energy collector intercepts and converts incident solar radiation into a usable form of electricity, heat, or light. This energy can:

(a) directly reduce specific concurrent energy demand; for example, daylight displacing electric lighting.
(b) directly meet a specific concurrent energy demand; for example, photovoltaic electricity, solar heat, or solar thermal electricity production.
(c) be stored for subsequent discharge to meet a later energy demand; for example, as sensible, latent, or thermochemical heat stores and/or as electricity in batteries.
(d) be directly used in an end application; for example, solar drying, solar cooking, passive solar distillation, solar detoxification, and/or
(e) be further converted; for example, *via* photovoltaic powered electrolysis to produce hydrogen.

In flat-plate and evacuated-tube thermal solar collectors, a selective solar radiation absorber material absorbs incident solar energy, converting it to thermal energy and heating that absorber plate. Tubes or ducting in the absorber plate constitute a specialized heat exchanger that removes the heat from the plate in the form of a liquid or gas. The latter is conveyed to heat storage or directly to the load. A transparent aperture in front of a flat absorber plate and opaque insulation on the back of the absorber allows solar radiation to be collected while reducing heat losses from a collector. A contiguous glass tube enclosing an evacuated space surrounding the absorber [2] performs these functions.

Photovoltaic modules, flat plate collectors, and evacuated tube collectors absorb direct sunlight, diffuse sunlight from the sky, and reflect sunlight from the ground. They seldom track the Sun's daily path across the sky. In fixed mounting, the tilt is usually provided toward the south at an angle equal to the latitude to minimize the average angle between the Sun's rays and the surface for optimal solar energy interception.

A total of 120,000 terawatts of solar energy strike the Earth's surface. However, this incident solar energy has a low power density with annual daily averages between $400 Wm^2$ and $600 Wm^2$, so its use to meet larger energy loads thus requires the use of available suitable unshaded outdoor space on buildings and specifically devoted areas of land. For smaller loads, additional purposing to incorporate solar energy harnessing devices enables autonomous powering of outdoor systems, such as streetlights. When maximizing the use of an area to harness solar energy, shading can arise from:

- Thermal collectors or photovoltaic modules shading each other at lower sun angles in large arrays of successive rows.
- Buildings features, such as chimneys, shade roof-mounted solar energy collection systems.
- Surrounding trees or other buildings for both roof-mounted and ground-located solar energy collection systems.
  Concentrating solar collectors are developed for high-temperature thermal applications. All higher-concentration optical concentrating collectors can only concentrate incident beam normal solar radiation. Concentrating solar energy systems must track the Sun's azimuthal motion across the sky to align at normal incidence with the changing incident angle of beam solar radiation. Concentrators generally used for higher temperature applications are [3]:

- Parabolic troughs that concentrate incident solar radiation onto the axis of the trough. A tubular receiver carrying heat transfer fluid placed along this axis absorbs the concentrated solar radiation and heats the fluid. A parabolic trough tracks the sun about one, usually east-to-west aligned, axis.
- Parabolic dishes track the azimuthal motion of the Sun about two axes to concentrate incident solar radiation to a point. An insulated cavity at that point absorbs the concentrated radiation.
- Central receiver systems consist of an extensive array of independent heliostat flat mirrors. Each heliostat separately moves differently about two axes throughout the day to reflect incident solar energy to maintain a high-intensity cumulative reflection from all the heliostats on a receiver located at the top of a "solar tower."

Full two-axis tracking collects the most direct solar radiation. However, it can be expensive to operate and maintain two-axis tracking systems. This may not be viable, particularly in climates that have a short duration of periods with high direct components of solar radiation for significant parts of the year. Full two-axis tracking requires a greater distance between each solar energy collection device than would be the case for single-axis tracking solar collectors. Shows the amount of incident solar energy received by different tracking strategies than two-axis tracking.

In large-scale solar energy installations, inter-row shading at low sun angles enables maximum utilization of the available land area. At latitudes outside the tropics, the greater electricity production at high Sun angles from a maximized photovoltaic array more than offsets the smaller loss of generation in the early mornings and late afternoons, as the solar radiation intensity is also low for the lower sun angles experienced at such times. Such installations also experience partial shade from passing clouds. Photovoltaics integrated to provide power to mobile devices and vehicles are susceptible to shading; for such systems, the duration of the solar shaded time determines battery size and weight.

Solar thermal power generation concentrating collectors can only harness direct solar radiation. Thus, they deploy successfully only where direct normal incident solar radiation intensity is sufficient to produce the desired instantaneous temperatures to provide reliable power generation capacity factors [4].

The latter often means producing sufficient surplus energy during solar energy collection periods that can be stored either as sensible or latent heat or in batteries as electricity to maintain power production when low solar radiation conditions prevail and at night. Appropriate locations for high concentration solar energy collection must have sufficient duration and intensity of direct normal solar radiation. These are broadly delineated.

Photovoltaic arrays are southerly-facing assemblies of modules that each contain individual photovoltaic cells electrically connected in parallel and series. Physically compressed within a sandwich structure, each cell has a front glass or plastic surface. Solar energy incident on that front surface passes through the cover to reach a photovoltaic semiconductor material. Those parts of the incident solar spectrum with photon energies that correspond to particular band gaps in the semiconductor produce a potential difference that, in a completed circuit, provides a DC electrical energy to a load [5]. A silicon PV cell typically demonstrates an efficiency decline of 0.5% for every 1°C cell rise in cell temperature. To realize their best achievable operational efficiency, photovoltaic modules thus need to either avoid absorbing solar heat or reject as much absorbed solar heat as practicable to the local environment or a heat sink. High PV cell temperatures cause lower efficiency of conversion from solar energy to electricity, while around twenty percent of incident solar energy gets converted into electricity, and more than 80% of incident solar energy heats a PV module; it produces temperatures that reduce the semiconductor band gap, adversely affecting the open-circuit voltage. Convective heat transfer to air, water flow in photovoltaic thermal solar collectors, various heat sinks, and enhanced heat transfer techniques have been used to cool photovoltaics. When cooling PV modules using a phase-change-material heat sink, solidification of the melted phase change material to regenerate it for use on the next day may be difficult to achieve in climates with warm night time, ambient temperatures, and cloud cover, inhibiting radiative cooling to the night sky.

In a photovoltaic module, a shaded photovoltaic cell becomes reverse-biased when the short-circuit current produced by that cell becomes less than the module string current. A cell-bypass diode is included in the module electrical circuit to protect shaded cells within a substring from potentially experiencing failure. A cell-bypass diode redirects current generated by the other, un-shaded cells to flow around the shaded substring, limiting the overall impact on module performance of reverse biasing on the shaded, thus current-mismatched cells. A cell-bypass diode is connected in parallel with the substring of cells, so the cell-bypass diode forward characteristic becomes the reverse characteristic of that particular cell substring. To protect a substring of cells, the breakdown voltage of the reverse-biased cell must be greater than the sum of the voltages of those cells in the substring plus the bypass diode voltage [6]. The bypass diode limits the amount of power dissipated in a single cell. The most vulnerable connection in any particular substring is the cell that exhibits the highest breakdown voltage.

2. SOLAR RADIATION AT THE EXTREMITY OF THE EARTH'S ATMOSPHERE

Sitting at the center of the Solar System as it rotates in an arm of the Milky Way galaxy, the Sun radiates electromagnetic energy across a spectrum corresponding to a surface temperature of about 6,000 K. Fifty-four percent of the emitted solar spectrum is in the infrared region. Of the remainder, forty-five percent lies in the visible spectral range between 0.3 micrometers and 0.7 micrometers. Only 1% of the Sun's spectral energy lies at wavelengths shorter than the visible range. The intensity of solar radiation at the Sun's surface is approximately 6.33×10^7 W/m². The solar radiation intensity diminishes geometrically by the square of the distance traveled. Therefore, when solar radiation arrives at the outer extremity of the Earth's atmosphere, the extra-terrestrial solar radiant energy incident on a one square meter surface area has an intensity of only 1367 W, referred to as the "Solar Constant."

The solar radiation at the outer extremity of the Earth's atmosphere varies by ±3.4 percent because of the Earth's elliptical orbit around the Sun. The solar radiation reaches its maximum at the perihelion around the 4th of January each year when the Earth's orbit passes closer to the Sun. The minimum intensity of solar radiation occurs at the aphelion on the 5th of July each year. Between these limits, the annual variation of extraterrestrial solar radiation intensity at the extremity of the Earth's atmosphere is given by;

$$I_o = I_{sc}\left[1 + 0.034cos\left(\frac{360N}{365.25}\right)\right] \ , (W/m^2)$$
(1)

Where I_o is solar radiation intensity at the outer extremity of the Earth's atmosphere, and N is the sequential day number starting on 1st January each year.

For a plane horizontal surface at the outer extremity of the Earth's atmosphere, the solar radiation intensity is given by the simple geometric relationship;

$$I_{o,h} = I_o cos\theta_z \ \ (W/m^2)$$
(2)

Where θ_z is the angle between a horizontal surface at the outer extremity of the Earth's atmosphere and the projection of the solar zenith angle.

The intensity of solar radiation received on a horizontal surface outside the Earth's atmosphere is an upper limit of radiation on any horizontal surface below the Earth's atmosphere. It is thus a common denominator used in many approaches to normalizing measured or predicted solar energy system performance outputs to enable their objective comparison.

The cumulative intensity of solar energy (Ho,h) is calculated at a horizontal surface outside the Earth's atmosphere by simply integrating the relevant sequence of instantaneous solar radiation intensities received over a given elapsed period. So for a full day;

$$H_{o,h} = \int_{sunrise}^{sunset} I_{o,h} dt \ \ \ \ \ (J/m^2)$$
(3)

While the Daily cumulative solar radiation intensity is thus given by;

$$H_{o,h} = \frac{86,400 I_o}{\pi}(\omega_s sin\varphi sin\delta + cos\delta cos\varphi sin\omega_s) , \ \ \ \ \ (J/m^2)$$
(4)

Where ϕ and δ are the applicable latitude and declination angles, respectively, and ω_s is the sunset hour angle.

3. GROUND-LEVEL SOLAR RADIATION

On Earth's surface, solar radiation has lower intensities, different colors, and a different spectrum than above the atmosphere. This is because at the surface of the Earth:

- A small fraction of solar energy leaving the Sun's surface reaches Earth's surface, becoming less intense.

- Solar energy intensity varies cyclically as the Earth rotates daily about its polar axis.

- Absorption and reflection of radiation in the Earth's atmosphere reduce the solar energy arriving at the Earth's surface.

- Scattering and absorption of solar radiation as it has passed through the Earth's atmosphere has reduced the intensity of particular absorbed wavelengths and diffused beam radiation giving anisotropic skyward angular spectral variations.

- Prevailing weather conditions limit solar radiation, particularly the amount of direct radiation in many climates in winter, for many days in a row.

Consequently, the annual spatial distribution of solar energy across the Earth's surface on a horizontal plane displays complex, stochastic behavior. An instantaneous illustration of solar energy's spatial complexity is shown in Fig. (**1**). An example for Cairo, Egypt, of a complete set of hourly global horizontal insolation data over a day is shown in Fig. (**2**).

On the clearest days, attenuation in the atmosphere reduces the daily cumulative solar radiation intensity incident on the Earth's surface by about thirty percent. However, this decrease can reach ninety percent on the cloudiest days and approach similar values on sunny days in locations with high local air pollution. Fig. (**3**) shows absorption and scattering phenomena that affect solar radiation intensity as solar energy passes through the Earth's atmosphere.

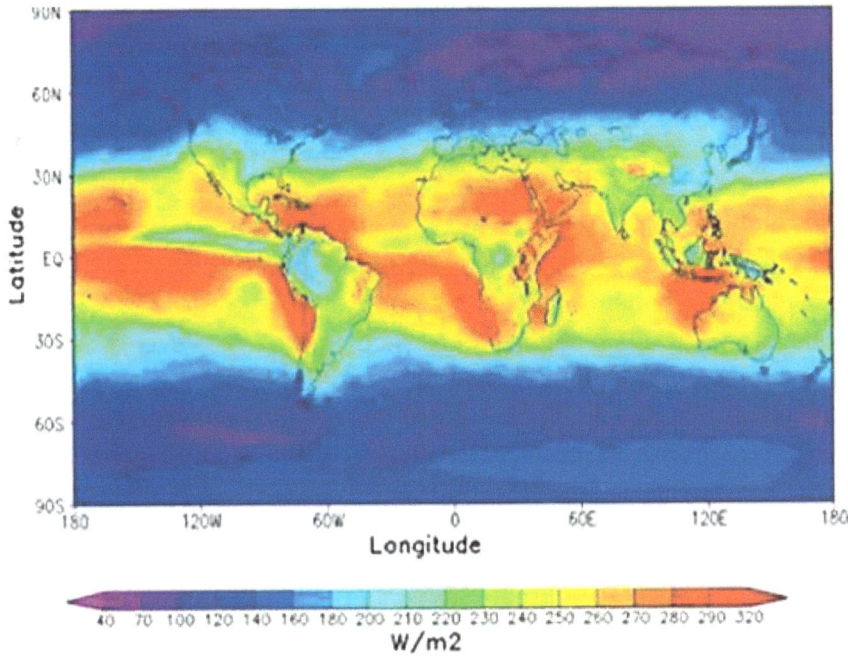

Fig. (1). Annual global mean downward solar radiation at the Earth's surface.

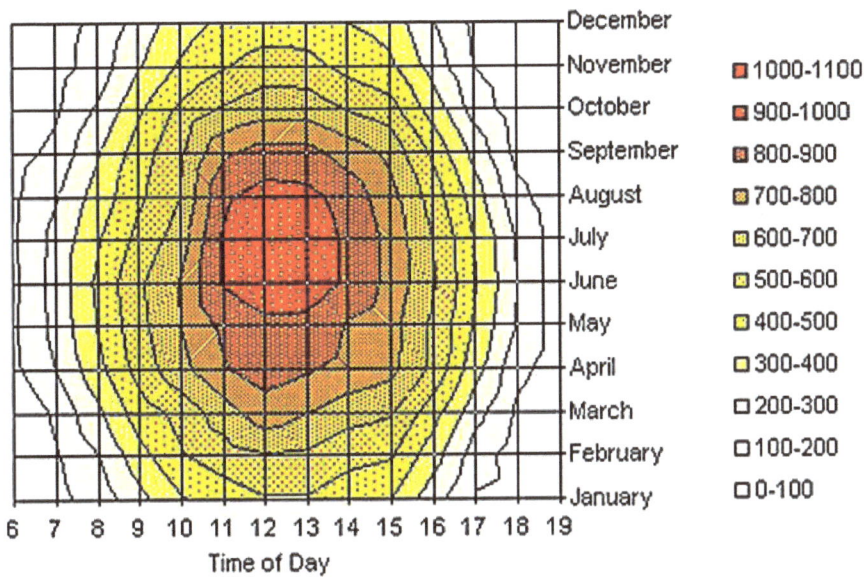

Fig. (2). Global horizontal insolation for Cairo, Egypt in Wm-2.

On the clearest days, direct solar radiation represents about 80% of the total hemispherical solar radiation received on the Earth's horizontal surface. Diffuse radiation exists due to the scattering of solar radiation by nitrogen, oxygen, water molecules, water droplets, and dust particles in the Earth's atmosphere [8]. The extent and effect of such scattering varies with the:

- Time and location depend on the amount of water and dust in the atmosphere,
- Thickness of the atmosphere traversed by the solar radiation (that depends on latitude and time of day)
- Altitude of the surface receiving solar energy above sea level.

The diffuse component invariably constitutes about twenty percent of total incident solar radiation, as it is an inevitable consequence of atmospheric scattering of solar radiation. Indeed, when solar radiation has no diffuse component, skies appear black during the daytime, with stars continuously visible. This phenomenon is seen on the moon, where there is no atmosphere to scatter solar radiation.

There is often no discernible direct component to the incident solar radiation on cloudy, overcast and foggy days. Between these extremes, the clearness index, K_T has been used to provide a normalized comparison of sky conditions. A clearness index is the ratio of global horizontal solar radiation at a particular location to the solar radiation intensity received on a horizontal surface outside the Earth's atmosphere above that location,

$$K_T = \frac{H_{t,h}}{H_{o,h}} \tag{5}$$

Measurement of diffuse solar irradiance can be expensive due to the instruments required and the need to adjust shadow rings. However, even when such measurements are made, they remain location-specific. In many climates, particularly at times of changeable weather conditions, it is not reliable to extrapolate measurements from a single location to a larger surrounding region, as can be seen from past measurements. Incident solar radiation is subject to weather and atmospheric influences with a broad statistical range of interacting varying parameters. As the time increment sought for solar radiation prediction becomes smaller, precise prediction of intensity becomes less likely. Algorithms can thus only predict the geographical distribution of hour-by-hour solar radiation intensities and corresponding diffuse fractions within broad statistical deviations. The use of cloud transmission and clearness indices from satellite images of the Earth enables more accurate incident direct radiation intensity estimation.

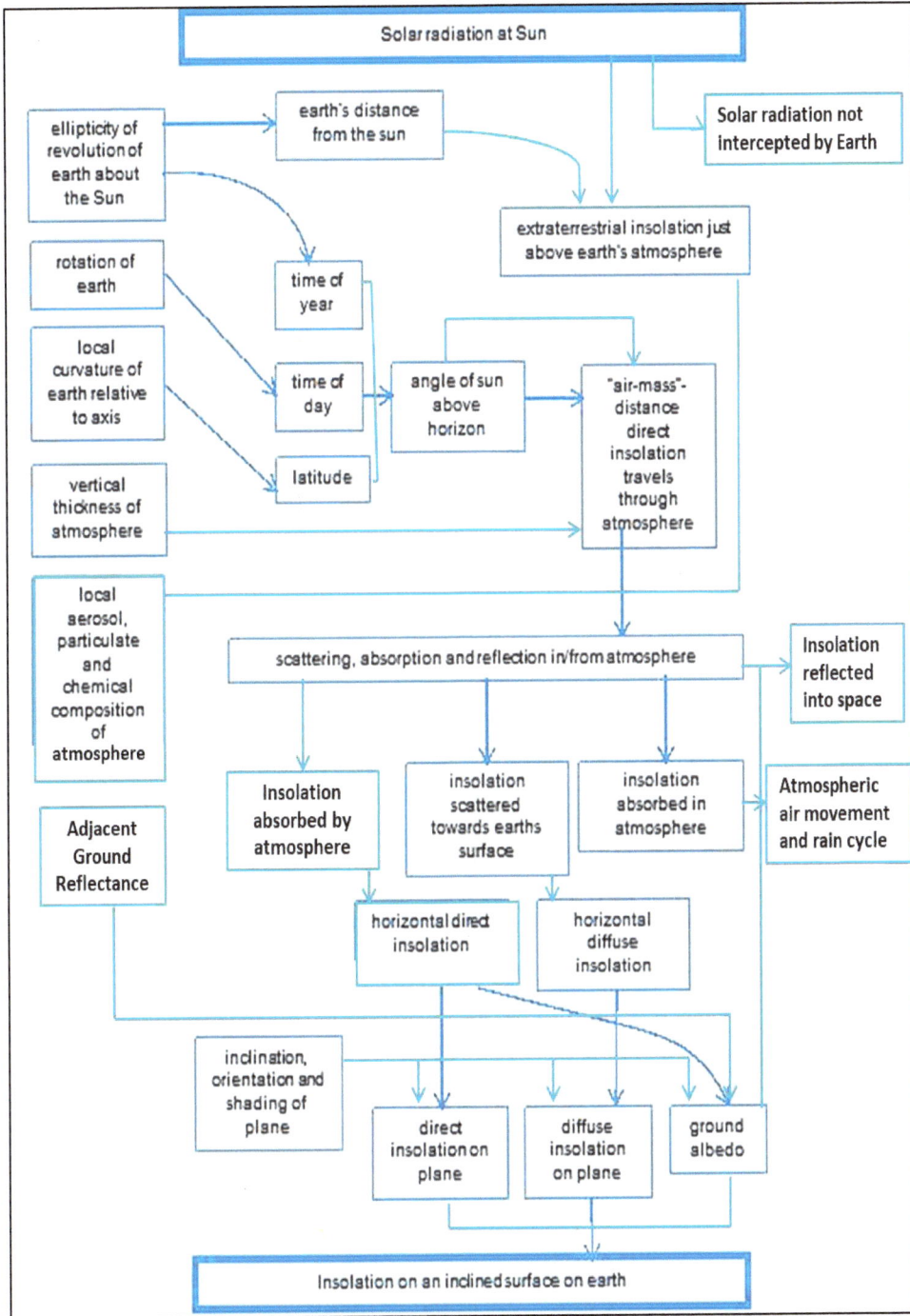

Fig. (3). Absorption and scattering of solar energy in Earth's atmosphere [7].

Diffuse radiation comprises a circumsolar region, the bright area around the Sun, a horizontal band at the horizon due to brightening from solar radiation incidents by scattering over the horizon, and an anisotropic skyward hemisphere. A tilted shaded surface will receive reflected radiation and isotropic and horizon brightening diffuse elements. Areas shaded by trees can receive between 30-70% of the total available radiation, regardless of the magnitude of the diffuse fraction. Advanced photovoltaic array shade modeling software often accommodates the permanent leaf cover associated with evergreen and conifer trees and deciduous trees shedding their leaves in winter. Entering a tree canopy transmittance for each month of the year in such software enables the seasonal variability of shading from trees from leaf growth and defoliation to be considered in shading calculation.

Diffuse solar radiation intensity measurements on a horizontal surface plane do not include ground-reflected albedo radiation. The surrounding ground surface would fall outside the field of view of a horizontally-mounted solar radiation pyranometer. For solar radiation pyranometers mounted on a tilted surface plane, total and diffuse radiation intensity measurements will include skyward diffuse radiation and reflected ground albedo diffuse radiation components. If measurements of diffuse radiation intensity on a tilted plane are not available, the diffuse-tilted portion and reflected-tilted components can be calculated using contemporaneous horizontal-plane global and diffuse radiation intensity data. Normally, it is assumed that the diffuse radiation component's intensity at a given place and time is isotropic, that it has the same intensity irrespective of its incident direction from the hemisphere above the horizontal plane. Without including reflected ground radiation intensity, the intensity of such an isotropic diffuse solar radiation component on a tilted surface plane G_{dt}, is given by:

$$G_{dt} = 1 - cos\beta \cdot 2G_d \qquad\qquad (6)$$

Where β is the tilt angle of a solar energy receiving surface plane counted from $\beta = 0°$ for a surface plane facing the zenith, and G_d is the intensity of the diffuse component of solar radiation on a horizontal plane.

The intensity of ground reflected radiation is dependent on global solar radiation intensity, the reflectance (R) of the local ground surface and the inclination angle of the radiation receiving surface. Ground reflectance depends on the properties of surface materials and, in some climates, can vary considerably and often quite abruptly over a year. For example, in climates with snow precipitation, ground reflectance is much higher in winter with fresh snow, as snow has a 0.8 reflectance. In summer, the same location with grass will receive less the half the winter ground reflected radiation, as the grass has a reflectance of 0.25. Assuming that the reflected radiation intensity at a given place and time is isotropic, that it is equal in intensity from all directions in the hemisphere above the horizontal plane, then the intensity of ground reflected radiation on a tilted plane G_{rt}, is given by,

$$G_{rt} = 1 - cos\beta \cdot 2RG \qquad (7)$$

Where G is the intensity of horizontal global solar radiation.

The spectrum of solar radiation arriving at the Earth's surface changes due to Rayleigh, scattering by particulate (water and dust) and absorption by ozone, water vapor, and carbon dioxide. These processes depend on the atmosphere:

- Specific constituents
- Thermo-physical conditions
- Thickness that the solar radiation has passed through.

Atmospheric thickness is measured by "air mass." Air mass is the ratio of the distance that the direct component of solar radiation travels through the Earth's atmosphere to the distance the direct component of solar radiation would have traveled if the Sun were directly overhead. Solar radiation incident directly overhead is defined to traverse through an air mass of unity at sea level; at altitudes above sea level, the air mass is reduced by the ratio of the local atmospheric pressure to atmospheric pressure at sea level. An air mass of unity only occurs at tropical latitudes as solar geometry permits the Sun to be directly overhead at solar noon. Solar radiation incident from a zenith angle of 60 degrees traverses an air mass of two, as it passes through twice the thickness of the atmosphere compared with the Sun being directly overhead. Fig. (4) shows the effect of the Earth's atmosphere on the solar radiation spectrum for different air masses. The air mass at any zenith angle θ_z can be calculated by;

$$air\ mass = \frac{1}{cos\theta_z + 0.50572(96.07995 - \theta_z)^{-1.6364}} \qquad (8)$$

Where θz is the solar zenith angle (in degrees).

As it passes through the atmosphere, solar radiation experiences [9];

- Water vapor and carbon dioxide absorption bands in the infra-red region at a wavelength greater than 0.7 micrometers,

- Reduction in blue and violet light at wavelengths from 0.3 micrometers to 0.4 micrometers arising from particulate and Rayleigh scattering and,
- Reductions in solar ultraviolet radiation at wavelengths less than 0.3 micrometers due mostly caused by absorption by ozone in the upper atmosphere.

At sunrise and sunset, when the zenith angle is 90°, solar radiation passes through an air mass of 38; therefore, at sunrises and sunsets, parts of the sky towards the east or west horizons respectively thus appear to be red.

Photovoltaic materials absorb distinct wavelengths of the solar radiation spectrum, as shown in Fig. (**5**). As a solar spectrum changes over a day, a particular photovoltaic material's absorption of solar energy will also change. In addition, the output increases with intensity but with a diminishing rate as the higher intensity solar radiation gives higher photovoltaic temperatures that lower the module efficiency typically by 0.5% for each 1^0C rise in temperature. Convective heat transfer to air, water flow in photovoltaic thermal solar collectors, and various heat sink and enhanced heat transfer techniques are used to cool photovoltaic modules.

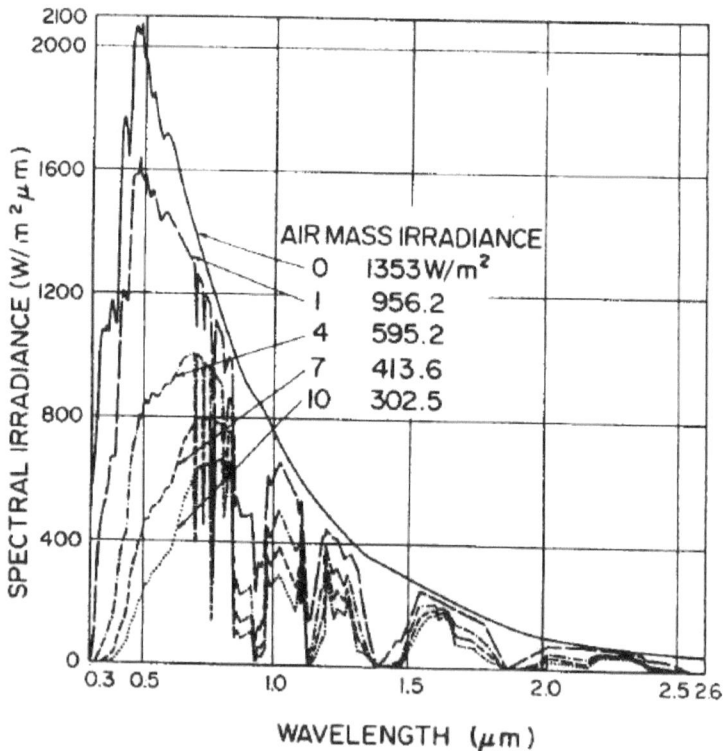

Fig. (4). Solar spectral irradiance for different air masses.

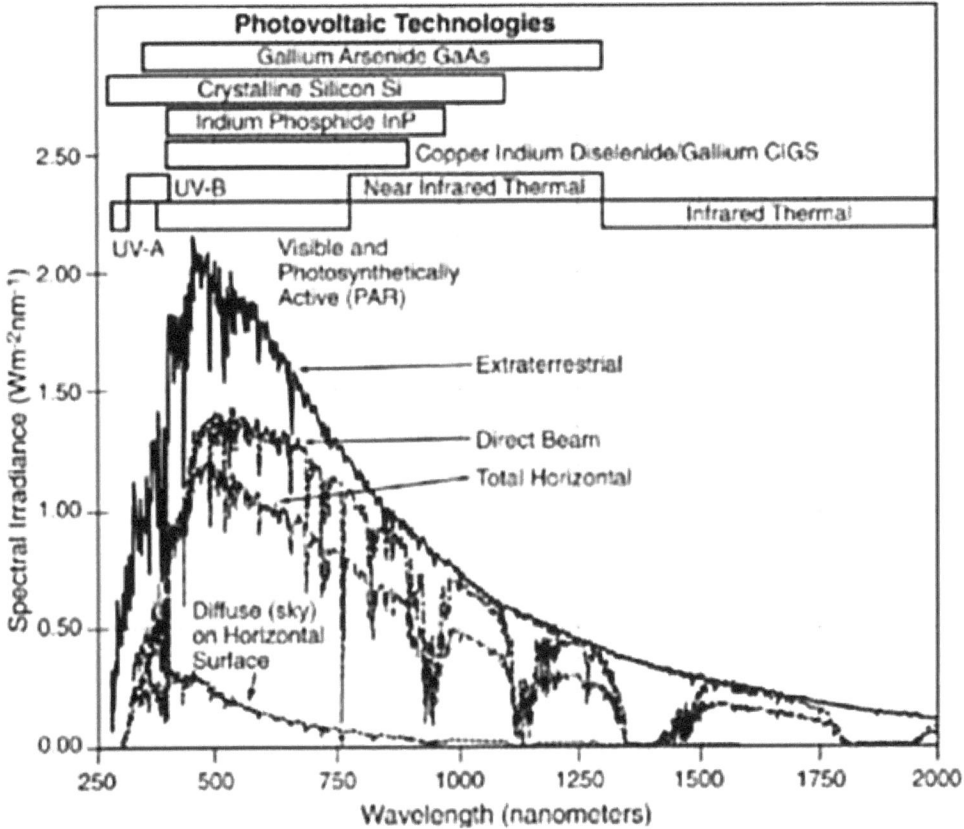

Fig. (5). Solar energy spectral absorption for a range of photovoltaic materials.

4. DAYLIGHT

The intensity of any light source, including daylight provided by solar energy, is given by the Poynting vector, which defines the power output per unit area (i.e., in Wm^2) of any electromagnetic wave [10]. Light in the visible solar spectrum, as shown in Fig. (5), provides daylight; the extent of the interior lighting provided by daylight depends on the brightness, or perceived power, of the prevailing daylight, projected into the room. The luminous flux measured in lumens (lm) gives the light's perceived power. Human eyes are most sensitive to light with a wavelength of 555 nanometres; this corresponds to the color green. One watt carried by the light of 555 nanometres wavelength has a luminous flux of 683 lm. The luminous flux per solid angle or luminous intensity (measured in Candela) represents the directionality of the energy radiated by the light. Illuminance (lnm^{-2}) is the luminous flux per unit of light-receiving area.

Luminous efficacy relates daylight luminous flux to solar radiation intensity. Thus, luminous efficacy facilitates the calculation of daylight illuminances from solar radiation data. Actual luminous efficacies are different at specific locations and for a particular location range considerably over a day and across the year. This is due to the incident solar spectrum altering as the solar radiation encounters;

- Different air masses,
- Changes in atmospheric concentrations of aerosol and water vapor and/or
- Variations in the prevailing type, extent and height of clouds.

The luminous efficacy of direct components of incident radiation increases typically from about 70 lm/W to about 105 lm/W with increasing solar altitude. The average skyward luminous efficacy of diffuse components of incident solar radiation varies from approximately 130 lm/W for a clear blue sky to approximately 110 lm/W for an overcast cloudy sky.

In hot climates, as shown in Fig. (**6**), to avoid overheating but gain daylight, a smaller spectral range of incident solar energy would normally be admitted into a building through glazing than the case during heating seasons in colder climates.

Daylight is the most influential cue to the night-and-day cycles of the human circadian pacemaker. This is key to the healthy regulation of hormonal rhythms that ensure effective cognitive performance. Experiencing sufficient daylight has many beneficial qualities that include (i) supporting vision, (ii) making vitamin D, (iii) making the temperament-enhancing neurotransmitter serotonin, (iv) supporting proper eye growth, and (v) helping to maintain hygiene, as the ultraviolet spectral part of daylight can destroy bacteria and, to some extent, damage viruses.

There is less need for artificial lighting when people do visual tasks close to windows. Deeper daylight projection into buildings combined with controls that turn off unnecessary artificial lighting reduces electricity use. As illustrated in Fig. (**6**), the use and modification of daylight by individuals, buildings, and wider surroundings has multifaceted, intertwined, and often unconscious, biological, mental and social consequences.

Even under overcast skies, daylight projecting into buildings can provide sufficient illumination for many activities for most of the day. Indeed, adequate consideration of daylight has been crucial to the successful design of many buildings. At higher latitudes, daylighting design strategies for buildings include:

- lengthening shallow-plan buildings along an east-west axis,
- letting daylight to come in at height into rooms,
- locating windows on multiple sides of rooms to alleviate discomfort from glare when the direct radiation incident at low sun angles,
- painting interior surfaces in light colors and/or
- Installing light-reflecting blinds, light-deflecting glazing or light-transmitting light pipes.

An implementation of daylight deflecting blind is illustrated in Fig. (**7**).

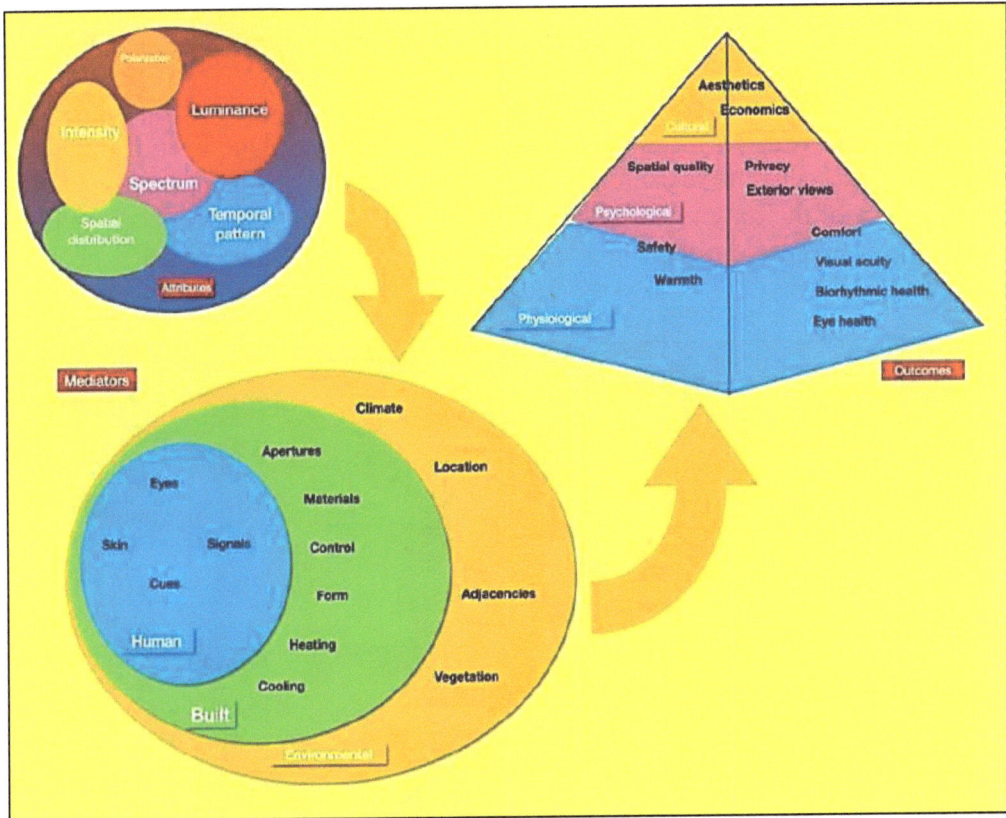

Fig. (6). Aspects of people, buildings and the wider environment that mediate daylight's key attributes give a range of physiological, psychological, and cultural outcomes.

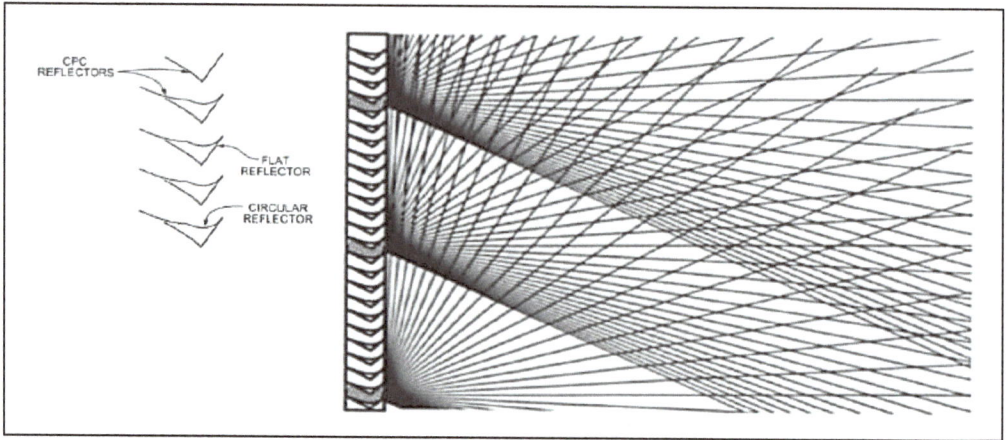

Fig. (7). Passage of the reflected direct component of solar radiation from daylight deflecting blind.

With prismatic light-deflecting glazing, the daylight spectrum projected in the building varies with the distance of the daylight-receiving plane from the window. In the particular illustrative examples of prismatic daylight projection shown in Fig. (**8**), the color of the red motor vehicle outside is projected at locations close to the window. However, from the same window, deeper in the same room, a neutrally colored projection of the clear sky is experienced.

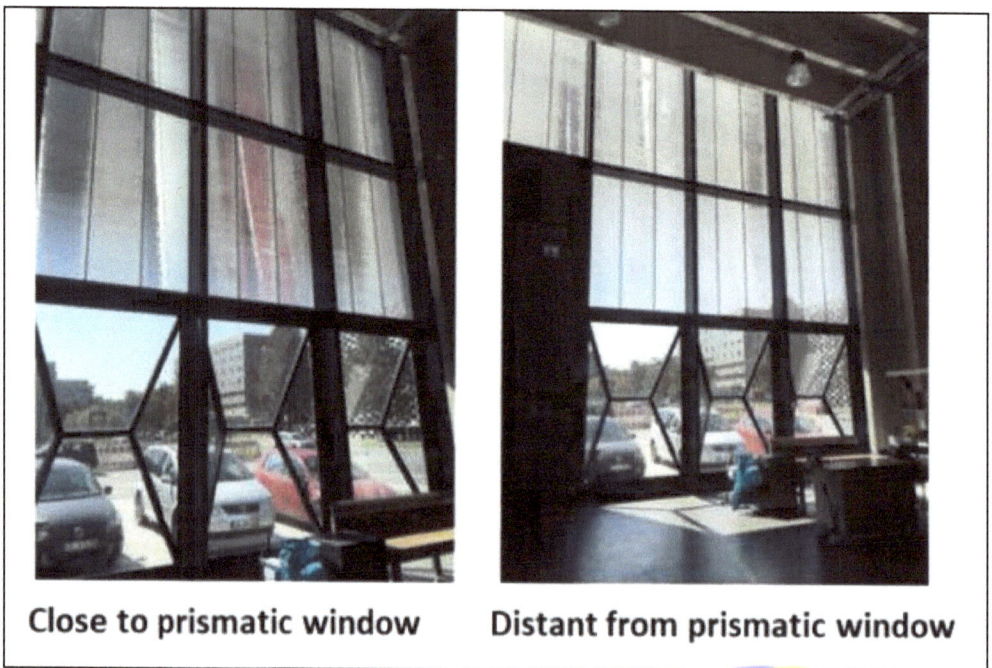

Close to prismatic window   **Distant from prismatic window**

Fig. (8). An example of prismatic glazing.

5. APPARENT MOTION OF THE SUN ACROSS THE SKY

Every 365.25 days, as depicted in Fig. (**9**), the Earth elliptically orbits around the Sun at an average distance of 1.496 x 10^{11} m. The Earth's maximum distance from the Sun, or aphelion, of 1.52×10^{11} m (94.4×10^6 miles) is on the third day of July. The Earth's minimum distance from the Sun, the perihelion, happens on the second day of January when the Earth is 1.47×10^{11} m (91.3×10^6 miles) from the Sun. The Earth rotates in day-long cycles with sidereal solar noon defined as half the elapsed time for one complete rotation about its axis. Increasing by 15 degrees every hour, the solar hour angle *(ω),* shown in Fig. (**10**), is the angle between the meridian of a solar energy receiving surface plane and the meridian whose plane contains the Sun. The solar hour angle is zero at sidereal solar noon, at which the Sun's position is highest in the sky.

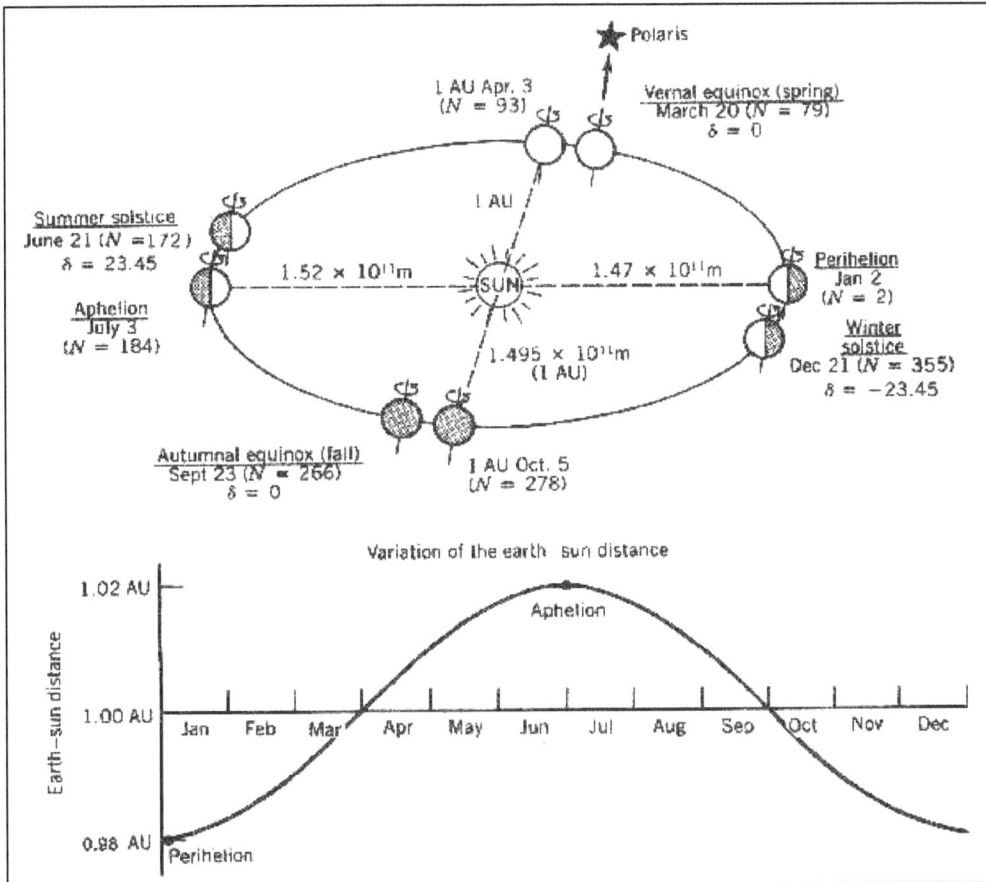

Fig. (9). Earth-sun distances, equinoxes and solstices [11].

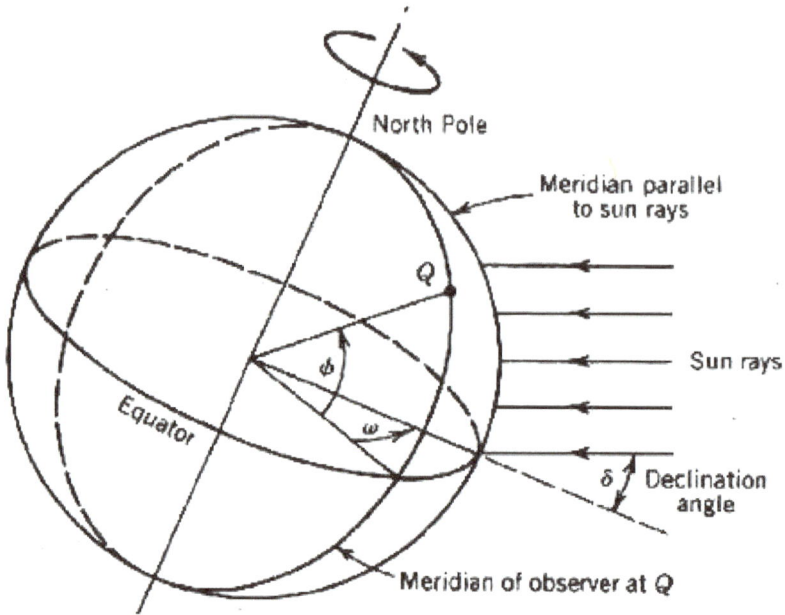

Fig. (10). The hour angle *(ω)* in the context of other solar angles [11].

Local clock time is defined by statutory time zones. "Solar time" is defined by at 12:00noon the Sun's position being precisely due south. Solar time is thus longitude dependent. Therefore, unless the solar energy collecting surface plane is located at the defining meridian of a time zone, solar time always differs from local clock time. This difference is particularly marked for time zones with a large longitudinal extent, for example, in China. The difference between solar and local time also changes abruptly by one hour twice a year in time zones that alter clocks for "summertime" and "winter time" by one hour.

Hour angles and solar time represent the same progression of the Sun through the day, either as an angle or time, respectively. They are interrelated simply by the expression:

$$\omega = 15(t_s - 12) \ , \qquad degrees \tag{9}$$

Where t_s is sidereal solar time (in hours)

Days have variable lengths due to the Earth's;

- Orbit around the Sun being elliptical,

- Axis being tilted to the plane of its orbit around the Sun.
- Rotation at an irregular rate around its axis of rotation.
- Perturbation on its axis-of-rotation.

Therefore, a "mean solar day" has a different period from a "sidereal day" because, when the Earth makes one axial revolution, the Earth has also concurrently continued to progress in its orbit about the Sun. The deviation, illustrated in Fig. (**11**), between mean solar time and true sidereal solar time on a given date is the "equation of time."

It is important to take an accurate goniometrical account of this deviation:

- In calculations of solar energy incidence angles in solar energy applications where the solar radiation transmittance through an aperture or surface absorption of solar radiation varies significantly with incident angle.
- Particularly for solar energy collection devices locations with high direct components of incident solar radiation.
-

An angle between a projected straight line between the centers of the Earth and the Sun and Earth's equatorial plane is termed a solar declination angle (δ). As portrayed in Fig. (**12**), when the northern half of the Earth is inclined towards the Sun, the Earth's equatorial plane is inclined 23.45° to a line extending between the centers of the Earth and the Sun. The tilt of the polar axis to the ecliptic plane by 23.45°gives rise to the seasons as the Earth revolves about the Sun. At the summer solstice at noon on June 21, the declination angle is 23.45° when the Sun is at its highest point. On 23rd September, the autumn equinox and on the 22nd March, vernal equinox, a straight line extending from the Earth to the Sun lies on the equatorial plane, so the δ is zero.

For a particular selected day in a year, the solar declination angle is calculated from;

$$sin\delta = 0.39795cos[0.98563(N - 173)] \tag{10}$$

The "Tropic of Cancer" at 23.45 degrees latitude north and the "Tropic of Capricorn" at 23.45 degrees latitude south signify the maximum tilts of the north and south poles toward the Sun. Only in tropical locations between these latitudes can the Sun be found directly overhead in the sky at solar noon. The "Arctic Circle" at 66.55° north and Antarctic Circle at 66.55° south is the lowest latitudes where there are 24 hours of daylight or 24 hours of darkness at the summer or winter solstices, respectively.

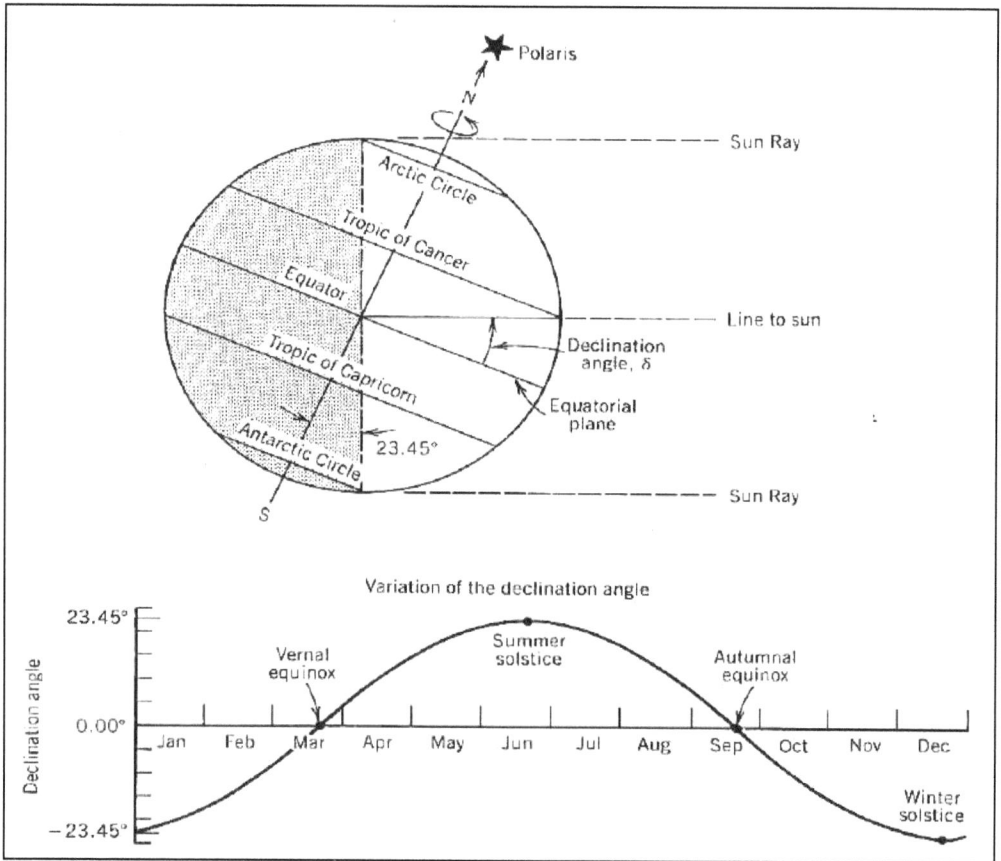

Fig. (11). Annual variation of declination angle [11].

At latitudes above the tropics, in summer, sunrise and sunset happen behind a south-facing solar energy-receiving surface, as shown in the Sun path diagram in Fig. (**12**) for locations at the latitude of 40 degrees north.

In summer months, the vertical facades of buildings located at higher latitudes can experience a combination of the Sun rising from towards the north-east then setting towards the north-west respectively, the Sun being 20 degrees above the horizon when it is due east and low diffuse fractions. This leads to early-morning and late-evening solar heat gains through windows that can beneficially contribute to heating a building if ambient temperatures are below indoor comfort temperature. A disadvantage is that if the direct component of the incident solar radiation is high, glare from the low angle of the Sun can cause discomfort to the building's occupants near east and west-facing windows in the mornings and afternoons, respectively.

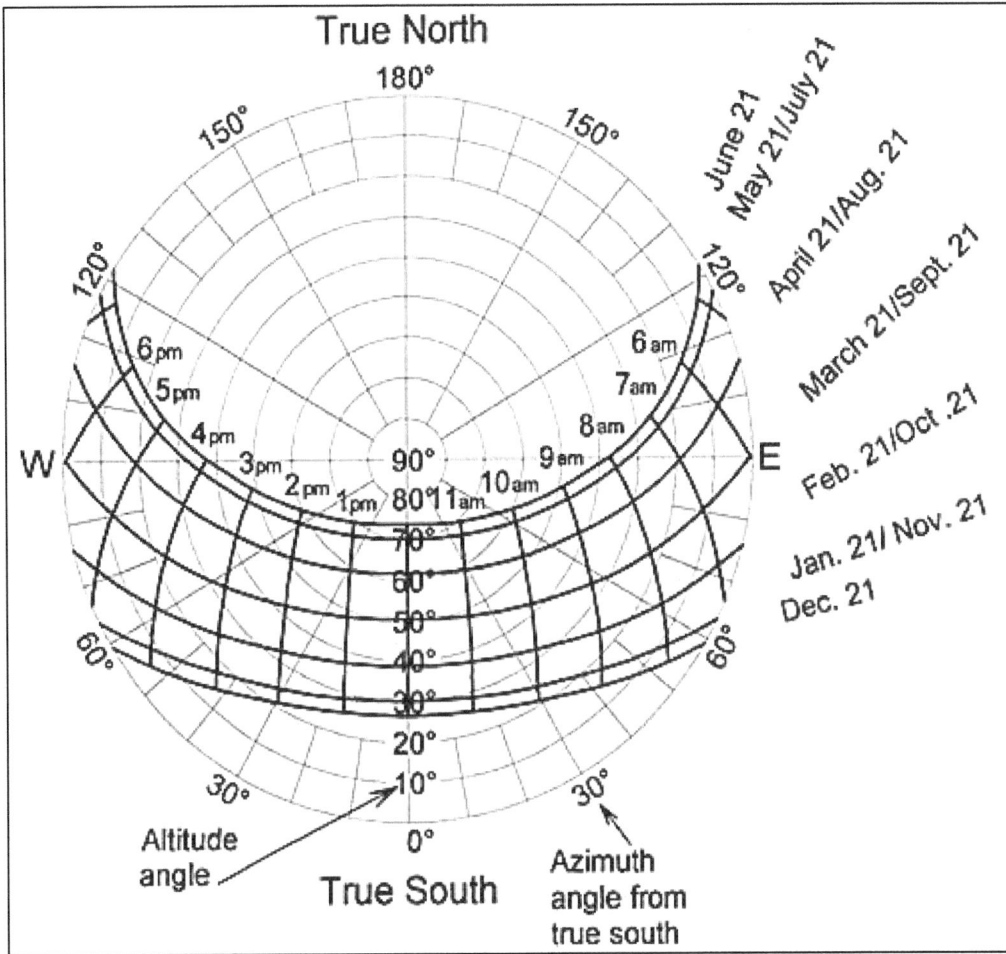

Fig. (12). Sun path diagram for 40 degrees North latitude.

The "solar altitude angle" (α) and the "solar zenith angle" (θz) (which is simply the complement of the solar altitude angle) are merely alternative ways to represent the Sun's position. The solar altitude angle (α) is defined as the angle between the direct solar component and a horizontal solar energy receiving surface plane, as shown in Fig. (**13**). A solar azimuth angle (A) is conventionally calculated in a clockwise cycle on the horizontal plane, from the north-pointing axial line to a projection of the incident angle of the direct solar radiation component.

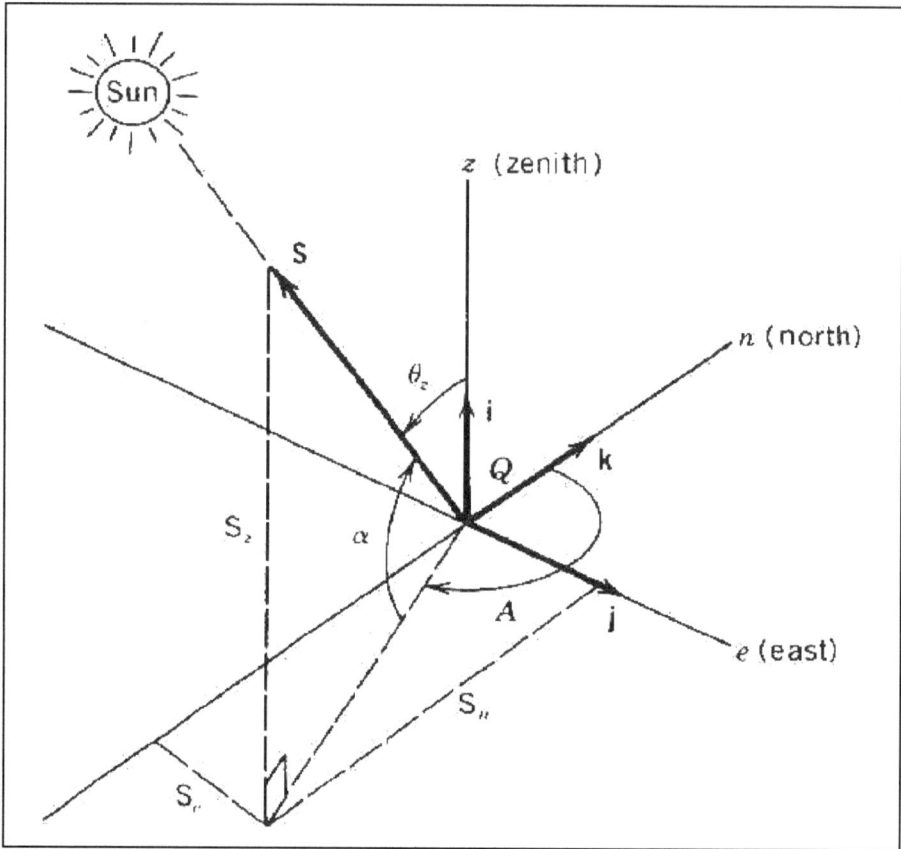

Fig. (13). Solar azimuth angle *A*, solar altitude angle *(α)* and solar zenith angle *(θ_z)* for a direct solar radiation component S incident at a point Q on a horizontal plane. Unit vectors *i, j, and k* are indicated on their respective ordinates [11].

Solar altitude and azimuth angles are geometrically related to the latitude *(ϕ)*, hour angle *(ω)* and declination *(δ)* by,

$$sin\alpha = sin\delta sin\varphi + cos\delta cos\omega cos\varphi \tag{11}$$

$$cos\alpha sinA = -cos\delta sin\omega \tag{12}$$

$$cos\alpha cosA = sin\delta cos\varphi - cos\delta cos\omega sin\varphi \tag{13}$$

The angle of incidence between the incident angle of the direct component of solar radiation and an angle normal to a plane solar energy receiving surface plane is given by,

$$cos\theta_i = cos\beta(sin\delta sin\varphi + cos\delta cos\varphi cos\omega) - cos\delta sin\omega sin\beta sin\gamma$$
$$+ sin\beta cos\gamma(sin\delta cos\varphi - cos\delta cos\omega sin\varphi) \qquad (14)$$

Tilt angle *(β)* is zero for a plane horizontal solar energy receiving surface plane, and

$$cos\theta_i = sin\alpha \qquad (15)$$

For a plane vertical solar energy receiving surface plane,

$$cos\theta_i = cos\alpha cos(\gamma - A) \qquad (16)$$

For a tilted south-facing solar energy receiving surface,

$$cos\theta_i = sin\alpha \, cos\beta - cos\alpha \, sin\beta \, cosA \qquad (17)$$

6. MEASUREMENT OF INCIDENT SOLAR RADIATION

Pyranometer is mostly used to measure total global solar radiation intensity from all skyward directions in the hemisphere above the horizontal plane. The most common pyranometer uses a thermopile comprising multiple thermocouples connected in series attached to a thin blackened absorbing surface shielded with a glass dome to inhibit convective loss and insulated against conductive heat losses [12]. When a pyranometer absorbs solar energy, the thermopile absorber surface reaches a temperature proportional to the incident radiant energy. A calibration converts the measured temperature to global solar irradiance. Pyranometers are also used to measure ground-reflected irradiance received by inclined surfaces.

A normal incidence pyrheliometer measures the direct normal incident solar radiation solely. A thermopile pyranometer is located at the end of a cylindrical tube in this instrument. This tube tracks the Sun to remain aligned normal to the solar position. A two-axis tracking mechanism maintains the Sun's disc within this acceptance. The aspect ratio of a pyrheliometer tube accepts radiation from a cone of about 5 degrees. As direct solar radiation from the Sun's disc occupies approximately 1/2 degree of this cone, the measurement provided by a normal incidence pyrheliometer also includes most circumsolar solar radiation.

Pyranometer can measure the diffuse component of the global horizontal radiation by using a shadowing device to block direct irradiance. A shadow band is often used to avoid moving a shadowing disc throughout the day. Adjustment of this band maintains its position in the ecliptic plane of the direct incident solar radiation. Corrections are required to prevent the shadow band from blocking the sky part. Fig. (**14**) shows the top and side views of a typical pyranometer. Fig. (**15**) shows two different approaches to measuring the direct and diffuse solar radiation intensity separately.

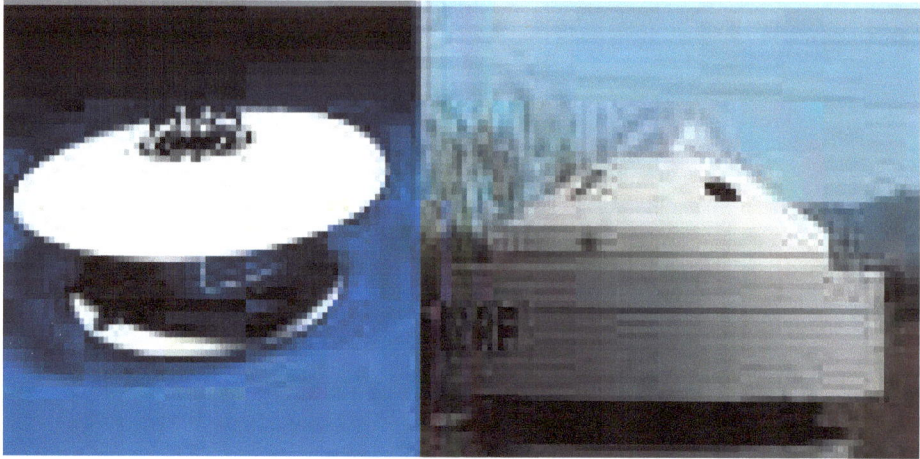

Fig. (14). Top and side views of a pyranometer for measuring total hemispherical solar radiation.

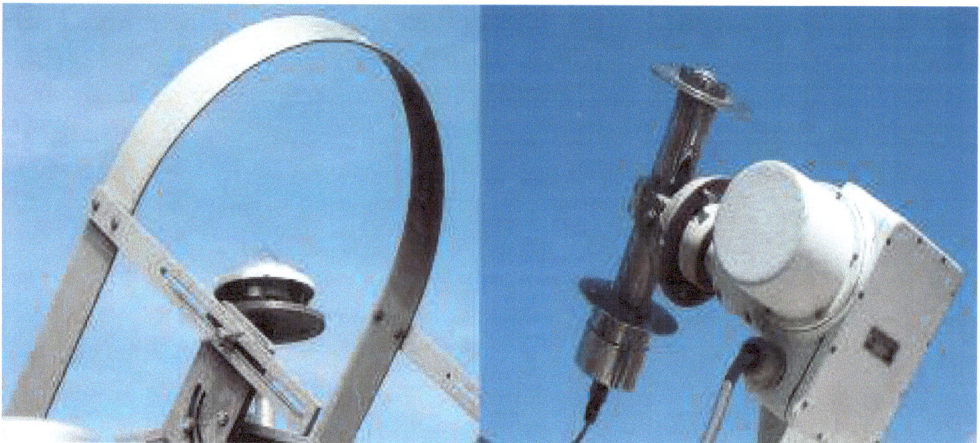

Fig. (15). Right photograph of a pyranometer with a shadow band for measuring the intensity of the diffuse solar radiation. Left, a photograph of a normal incident pyrheliometer for measuring the intensity of the direct solar radiation.

Increasing data-sampling frequency inherently adds resolution to solar radiation intensity measurements as it removes errors associated with extrapolating between less frequent solar radiation intensity measurements. Similarly, hourly solar radiation intensity data points that are the average of frequent measurements made in that hour represent the solar energy received in each hour than a single measurement made on the hour. Monitoring the inclined plane incident solar radiation directly removes errors in the translation of horizontal measurements to an inclined solar energy collection plane. This is particularly the case where the ground reflectance changes over the year and for vertical surfaces, for which ground reflected radiation albedo can constitute a significant part of the total incident radiation. This obviates calculating the diffuse and reflected tilted-plane solar radiation from horizontal measurements, which can also incur errors.

CONCLUSION

Understanding how incident solar radiation changes throughout the year in a particular location is a prerequisite to the successful design of solar energy applications. For engineered systems, such as solar thermal collectors and photovoltaic arrays, design specifications are provided by relevant international standards, manufacturers' specifications, and relevant technical literature. This information can usually be applied in a wide range of climatic conditions. In contrast, designing exemplar buildings is usually climate-specific when using solar energy to heat, cool, and provide daylight in buildings. However, they are fixed in particular cultural, social and building contexts. Additional care is recommended in interpreting and inferring the applicability of particular low-energy building solutions outside the often tacit, original assumptions associated with social, economic, and climatic contexts. This chapter provides some of the insights necessary to successfully implement such care.

CONSENT FOR PUBLICATION

Not applicable.

CONFLICT OF INTEREST

The authors declare no conflict of interest, financial or otherwise.

ACKNOWLEDGEMENTS

The author would like to acknowledge the Tyndall National Institute and Technological University Dublin, supported by MaREI, the SFI Research Centre for Energy, Climate and Marine [Grant No. 12/RC/2302_P2].

REFERENCES

[1]   S.A. Kalogirou, *Solar energy engineering: processes and systems.* Academic Press, 2013.

[2]   J.F. Kreider, and F. Kreith, *Solar Heating and Cooling,* 2nd ed McGraw-Hill: New York, 1982.

[3]   J.A. Duffie, W.A. Beckman, and N. Blair, *Solar engineering of thermal processes, photovoltaics and wind.* John Wiley & Sons, 2020.
       http://dx.doi.org/10.1002/9781119540328

[4]   A. Rabl, *Active solar collectors and their applications.* Oxford University Press, 1985.

[5]   A. Reinders, P. Verlinden, W. Van Sark, and A. Freundlich, *Photovoltaic solar energy: from fundamentals to applications.* John Wiley & Sons, 2017.

[6]   T.M. Letcher, V.M. Fthenakis, Eds., *A Comprehensive Guide to Solar Energy Systems: With Special Focus on Photovoltaic Systems..* Academic Press, 2018.

[7]   B. Norton, *Harnessing solar heat.* Springer Science & Business Media, 2013.

[8]   P.J. Lunde, *Solar Thermal Engineering.* John Wiley & Sons: New York, 1980.

[9]   M. Iqbal, *An introduction to solar radiation.* Elsevier, 2012.

[10] R. Kittler, M. Kocifaj, and S. Darula, *Daylight science and day lighting technology.* Springer Science & Business Media, 2011.

[11] W.B. Stine, and R.W. Harrigan, *Solar Energy Fundamentals and Design..* Office of Science and Technology Information, US Department of Energy, 1985.

[12] D.Y. Goswami, F. Kreith, and J.F. Kreider, *Principles of solar engineering.* CRC Press, 2000.

CHAPTER 3

Internal Characteristics of Double-base Array

Shubham Kashyap[1], Kuber Nath Mishra[1], Taranjeet Sachdev[1] and **Anil Kumar Tiwari[1,*]**

[1]*Department of Mechanical Engineering, National Institute of Technology Raipur, India*

Abstract: The thermal applications of solar energy have gained momentum after the revolution in the field of PV technology in recent decades. The worldwide quest to harness the thermal component of solar energy, which constitutes the major part of the incident, solar radiation incident globally, has led to the development of numerous thermal devices and applications that harness and store or utilize the same with never before seen efficiency. A few applications and devices are discussed here in this chapter. The first part of the chapter presents a brief discussion of the solar pond and its features along with thermal modeling of the system, followed by the thermal modeling and discussions about solar cooling systems. The later part of the chapter describes solar refrigerators and solar concentrators. The application and devices discussed here are of prime importance in developing basic and advanced solar thermal devices to harness solar thermal energy efficiently for human needs.

Keywords: Application, Solar cooling, Solar devices, Solar thermal.

1. GENERAL INTRODUCTION TO THE SOLAR POND

The concept of natural heating of water reservoirs existed before humans came into existence and can be regarded as one of the key factors responsible for the creation of life on the planet we live on. An optimum water temperature across the oceans is not only essential for the sustenance of marine flora and fauna, but as per the recent century's turn of events, our very existence may also depend upon the temperature of the oceans. The temperature rise of the oceans may be worrisome, but there are numerous applications of a water reservoir with hot water. In its simplest form, a water reservoir that receives heat from the sun and stores the same for later use may be regarded as a solar hot water pond or simply a solar pond. It

Corresponding author Anil Kumar Tiwari: Department of Mechanical Engineering, National Institute of Technology Raipur, India; E-mail: anil.kr.tiwari@gmail.com

Manoj Kumar Gaur, Brian Norton & Gopal Tiwari (Eds.)

may be any water reservoir with a black bottom in order to achieve heating capability. However, the heat received in the above case will be lost to the ambient due to convective and buoyancy effects that cause the hot water to rise to the surface. To avoid convective loss, a layer of still water can be maintained above the hot water using a salt concentration gradient along with the depth, a physical partition between the convective and non-convective layers or using a viscosity gradient along with the depth. Under the current discussion, the three systems are discussed briefly, and the more widely accepted salt gradient stabilization method is discussed in detail across this chapter.

1.1. Partitioned Salt-stabilized Pond

The simplest yet most effective way to counter the heat loss due to all convective zone type ponds is to inhibit the mixing of the convective and non-convective zones using a physical barrier or partition. Here, the partition used may be a transparent cellophane sheet that separates the convective zone from a non-convective zone. This kind of arrangement eliminates the need for heat exchangers for thermal extraction and eliminates the need for brine solution as a thermal stabilizer. The large surface area, maintaining transparency by avoiding the fouling of the transparent sheet are the few critical challenges that limit the practical implementation of this type of pond.

1.2. Viscosity Stabilized Pond

This system is similar to the salt concentration stabilized system discussed in the next subsection. The key difference is using gels and thickeners instead of salts to separate convective and non-convective zones. Polymers, detergent oil, water gels, *etc.*, are the required thickeners.

1.3. Stabilization using a Salt Concentration Gradient

The convective effect obtained in the pond water due to the temperature rise near the bottom of the pond is the main cause for concern here since the hot water tends to rise to the top, hence losing the heat gained to the ambient. This is a cyclic process, and hence at the end of the day, the water can lose the heat gained during the day. This also calls for measures to impede the convective heat loss through vertical currents along the depth to maximize heat storage. One of the most effective means of achieving this is to provide a density gradient between the top and bottom layers of the pond so that the warm water may never attain a density lower than that of the top layer and hence does not tend to rise to the surface. This can be done using a salt concentration gradient between the layers of water. As shown in Fig. (1), the layer of water closest to the bottom has the highest salt concentration, while the layer at the top is freshwater.

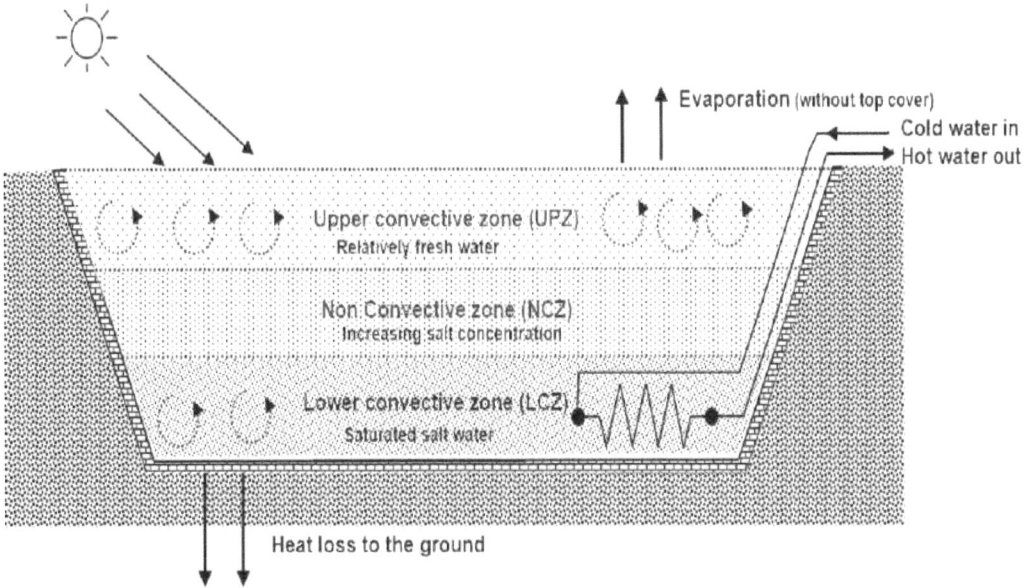

Fig. (1). Solar pond with a salt concentration gradient.

The salt concentration gradient can be achieved using pumping brine with varying salt concentrations through floating diffusers. The heat extraction from such a pond may be done using a heat exchanger in a closed or open loop. In the case of a closed-loop heat exchanger, the brine may be circulated across two heat exchangers where one of them is placed at the bottom of the pond, and the other is at the place of heat requirement. The open-loop system, on the other hand, withdraws the brine from the convective layer and pumps it back to the pond bottom after passing it through the heat exchanger.

Fig. (2) can be used to predict thermal energy efficiency of an ideal pond (solar). Instantaneous performance characteristics like thermal efficiency cannot be determined for such a pond since the mass of water, and resulting storage capacity are huge. The figure thus can be utilized for an average annual insolation level of

$200\mathrm{W/m^2}$ for varying extraction depths. It can also be inferred from the figure that for flux intensities other than 200 $\mathrm{W/m^2}$, the temperature rise of the pond is directly proportional to the insolation.

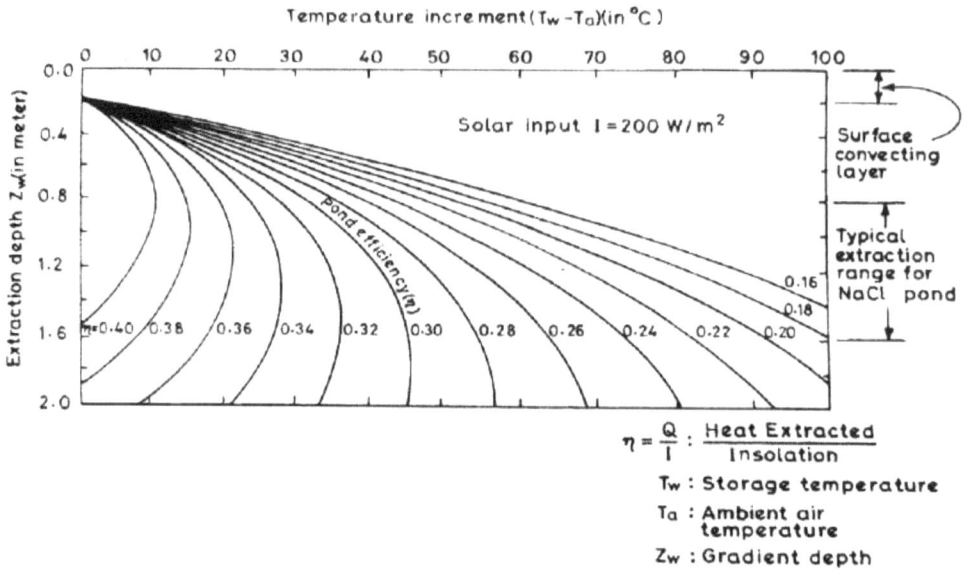

Fig. (2). Extraction depth *vs.* temperature increment for different pond efficiency [26].

1.4. Thermal Stability Conditions

Salt constitutes the major and most expensive component of a solar pond; thus, the knowledge of the salt gradient requirement and the technology to establish and maintain it is required for efficient pond operation. In the solar pond, the density (ρ) of the fluid is a function of salt concentration, S, and temperature T. In the case of pond stability against vertical convection, the magnitude of salt density gradient ($\partial\rho/\partial S$) due to salt concentration gradient ($\partial S/\partial x$) must be greater than the negative density gradient ($\partial\rho/\partial T$) produced by temperature gradient ($\partial T/\partial x$).

Considering the X-axis along the vertical direction and measuring positive downloads, the equilibrium can be expressed as,

$$\frac{\partial\rho}{\partial S}\cdot\frac{\partial S}{\partial x} \geq -\frac{\partial\rho}{\partial T}\cdot\frac{\partial T}{\partial x} \tag{1a}$$

$$\frac{\partial S}{\partial x} \geq \frac{\alpha_c}{\beta}\cdot\frac{\partial T}{\partial x} \tag{1b}$$

Here $\alpha_c = -(1/\rho)(\partial\rho/\partial T)$ is being treated as a coefficient of thermal expansion and $\beta = (1/\rho)(\partial\rho/\partial S)$ as the coefficient of expansion for salt. The equation for the static stability criterion:

$$(\Delta S)_{min} = \frac{\alpha_c\Delta T}{\beta} \tag{1c}$$

This denotes the minimum concentration difference for the fluid to be stable against vertical convection occurring due to a temperature difference ΔT between these points.

2. CALCULATIONS FOR SOLAR RADIATION

Calculation of a declination angle (d) [1],

$$d = 23.45 \times sin\left((284 + N)\frac{360}{365}\right) \tag{2}$$

Here, d represents declination angle, and N depicts the day of the year

Hour angle at sunrise and sunset (h_s) can be determined by,

$$h_s = \pm[cos^{-1}(-tan\,l\,tan\,d)] \tag{3}$$

h_s is the hour angle at sunrise or sunset, and l is the latitude angle of the place.

Hour angle h can be determined by apparent solar time t_s in an hour [2],

$$h = 15 \times (t_s - 12) \tag{4}$$

h is found to be positive for the morning and negative for the afternoon.

Calculating the angle of incidence for a horizontal plane (θ_i)

$$\cos \theta_i = \cos l \cos d \cos h + \sin l \sin d \tag{5}$$

θ_i is the incidence angle for a horizontal plane, l is the latitude angle corresponding to the place, d is the declination angle, and h is the hour angle.

The angle of refraction (θ_r) can be found by following Snell's law,

$$n_i \sin(\theta_i) = n_r \sin(\theta_r) \tag{6}$$

Hence θ_r will become,

$$\theta_r = \sin^{-1}\left(\frac{\sin(\theta_i)}{n_r}\right) \tag{7}$$

$n_i = 1$ and $n_r = 1.33$ [3].

The ratio of the parameters, *i.e.*, first hourly total insolation to second total daily insolation (r_t), can be evaluated by,

$$r_t = \frac{\frac{\pi}{24}(a + b\cos h)(\cos h - \cos h_s)}{\sin h_s - \frac{\pi h_s}{180}\cos h_s} \tag{8}$$

Here, *a* and *b* are coefficients, and the value of *h* is considered at the midpoint of the time (hour) for which determination is to be made.

The value of coefficients a and b can be calculated as follows: -

$$a = 0.4090 + 0.5016 \times \sin(h_s - 60) \tag{9}$$

$$b = 0.6609 + 0.4767 \times \sin(h_s - 60) \tag{10}$$

3. ENERGY BALANCE IN SOLAR POND UCZ

Applying energy balance for UCZ [4],

$$\rho_{UCZ} C_{p,UCZ} t_{UCZ} A_{UCZ}\left(\frac{dT_{UCZ}}{d\tau}\right) = Q_{s,UCZ} + Q_{cond} - Q_{losses,UCZ} \tag{11}$$

In Eq. 11, the left-hand side shows the heat stored in the zone. $Q_{s,UCZ}$ represents the magnitude of solar insolation absorbed in UCZ. Q_{cond} represents thermal energy transfer by conduction. $Q_{losses,\ UCZ}$ represents thermal energy losses from UCZ at time τ.

Total solar insolation captured in UCZ can be calculated by the following,

$$Q_{s,UCZ} = Q_{s,UCZ,in} - Q_{s,UCZ,out} \tag{12}$$

$Q_{s,UCZ}$ represents solar radiation absorbed in UCZ. $Q_{s,UCZ,in}$ represents solar radiation entering the UCZ. $Q_{s,UCZ,out}$ depicts the solar insolation leaving the UCZ.

Calculating the intensity of solar insolation entering the solar pond (I_0) [5]

$$I_0 = (1 - R) \times I \tag{13}$$

I_0 represents the intensity of solar radiation entering the solar pond, R represents the coefficient of reflection, and I represent the intensity of solar insolation striking the UCZ plane.

The coefficient of reflection (R) can be calculated by following [1,6]

$$R = 0.5 \times \left[\frac{sin^2(\theta_i - \theta_r)}{sin^2(\theta_i + \theta_r)} + \frac{tan^2(\theta_i - \theta_r)}{tan^2(\theta_i + \theta_r)} \right] \tag{14}$$

The magnitude of solar radiation entering the UCZ,

$$Q_{s,UCZ,in} = I_0 A_{UCZ} = (1 - R)I A_{UCZ} \tag{15}$$

The fraction of solar radiation reaching any surface h_x [7],

$$h_x = 0.36 - 0.08\ ln\left(\frac{x}{cos\ \theta_r}\right) \tag{16}$$

The above Eq. 16 is valid, corresponding to the depth of water in the range of 1 cm to 10 m, where depth x is expressed in the meter.

The magnitude of solar radiation leaving the UCZ,

$$Q_{s,UCZ,out} = I_0 h_{x,UCZ,out} A_{UCZ} = (1 - R)I h_{x,UCZ,out} A_{UCZ} \tag{17}$$

Solar radiation absorbed in UCZ will be obtained when the value of $Q_{s,UCZ,in}$ from Eq. 15 and the value of $Q_{s,UCZ,out}$ from Eq. 17 is substituted in Eq. 12.

$$Q_{s,UCZ} = (1 - R)I A_{UCZ} \left[1 - h_{x,UCZ,out} \right] \tag{18}$$

The conduction thermal energy transfer from LCZ to UCZ (Q_{cond}) [4,8],

$$Q_{cond} = U_t A(T_{LCZ} - T_{UCZ}) \tag{19}$$

U_t represents the overall thermal energy transfer coefficient.

Calculating the value of the overall thermal energy transfer coefficient (U_t),

$$U_t = \frac{1}{\left(\frac{1}{h_1}\right) + \left(\frac{t_{NCZ}}{k_w}\right) + \left(\frac{1}{h_2}\right)} \tag{20}$$

U_t is the overall thermal energy transfer coefficient, $h_1 = 56.58$ W/m^2 °C, $h_2 = 48.279$ W/m^2 °C and $k_w = 0.596$ W/m °C. t $_{NCZ}$ represents the thickness of NCZ.

Thermal energy losses from the plane of UCZ (Q $_{losses, UCZ}$),

$$Q_{losses,UCZ} = A_{UCZ} \left(Q_{conv} + Q_{evap} + Q_{rad} \right) \tag{21}$$

$Q_{losses, UCZ}$ represents thermal energy losses from the plane of UCZ. Q_{conv} represents convection heat losses from UCZ. Q_{evap} represents evaporation heat losses from UCZ, and Q_{rad} represents radiation heat losses from UCZ.

Convective heat losses from UCZ (Q_{conv}),

$$Q_{conv} = h_c(T_{UCZ} - T_a) \tag{22}$$

Here, h_c represents the convective thermal energy transfer coefficient between UCZ and air.

Convective heat transfer coefficient between UCZ and air (h_c) [9],

$$h_c = 5.7 + 3.8 \times V_{air} \tag{23}$$

V_{air} represents the average daily air velocity.

Thermal energy losses from a surface due to evaporation (Q_{evap}) [4,9,10],

$$Q_{evap} = \frac{h_c \lambda (P_{UCZ} - P_{vap})}{1.6 \times C_a \times P_{atm}} \tag{24}$$

Where h_c represents the thermal energy transfer coefficient (convection) among UCZ and air, λ represents latent heat of water vaporization equal to 2250 kJ/kg. P_{atm} represents the atmospheric/ambient pressure in mm of Hg. C_a represents the humid air-specific heat equal to 1.0216 kJ/kg °C. P_{UCZ} is the vapour pressure of water at UCZ temperature in mm of Hg. P_{vap} represents the partial pressure of water vapour at ambient outdoor air temperature in mm of Hg.

The vapour pressure of water at UCZ temperature and expressed in mm of Hg (P_{UCZ}),

$$P_{UCZ} = exp\left(18.403 - \frac{3885}{T_{UCZ} + 230}\right) \tag{25}$$

The partial pressure of water vapour at atmospheric temperature and expressed in mm of Hg (P_{vap}),

$$P_{vap} = \phi_h exp\left(18.403 - \frac{3885}{T_a + 230}\right) \tag{26}$$

Thermal energy losses by evaporation can also be calculated in an alternative way which is given below [11],

$$Q_{evap} = h_e[(T_{UCZ} - \phi_h T_a)F_1 - F_2] \tag{27}$$

Here, h_e represents the evaporative heat transfer coefficient. $F_1 = 2.933$, and $F_2 = 39.11505\ (1 - \phi_h)$.

Heat transfer coefficient by evaporation (h_e),

$$h_e = 8.88 + 7.82 \times V_{air} \tag{28}$$

Calculating thermal energy loss by radiation from UCZ (Q_{rad}),

$$Q_{rad} = \varepsilon_w \sigma \left((T_{UCZ} + 273.15)^4 - \left(T_{sky} \right)^4 \right) \tag{29}$$

Where ε_w represents the emissivity of water equal to 0.83 [4,10], σ represents the stiffen Boltzman constant equal to 5.67×10^{-8} W/m^{2}K^4. T_{sky} represents the sky temperature in kelvin.

Calculating sky temperature in kelvin (T_{sky}),

$$T_{sky} = 0.0552 \, (T_a + 273.15)^{1.5} \tag{30}$$

4. APPLICATIONS OF THERMAL ENERGY BALANCE FOR NON-CONVECTIVE ZONE (NCZ)

Applying energy balance for each layer of NCZ,

$$Q_{cond} = k_{NCZ} A_{NCZ} \left(\frac{dT_{NCZ,m}}{dt} \right) + Q_{s,NCZ,m} - Q_{losses,NCZ,m} \tag{31}$$

Here, k_{NCZ} represents the thermal conductivity of water solution. $dT_{NCZ,m}$ represents the temperature difference among layers. dT represents the thickness of each layer.

The technique to calculate the magnitude of solar insolation absorbed in each layer of NCZ is similar to the methodology adopted in the UCZ.

$$Q_{s,NCZ,m} = Q_{s,NCZ,m,in} - Q_{s,NCZ,m,out} \tag{32}$$

$$Q_{s,NCZ,m,in} = (1 - R) I h_{x,NCZ,m,in} A_{NCZ} \tag{33}$$

$$Q_{s,NCZ,m,out} = (1 - R) I h_{x,NCZ,m,out} A_{NCZ} \tag{34}$$

In the Eq. 32-34, $h_{x,NCZ,m,in}$ and $h_{x,NCZ,m,out}$ represents the fraction of solar insolation arriving at the top and bottom of the plane, m, respectively.

Calculating solar insolation absorbed in each plane of NCZ by putting a value of $h_{x,NCZ,m,in}$ from Eq. 33 and $h_{x,NCZ,m,out}$ from Eq. 34 in Eq. 32.

$$Q_{s,NCZ,m} = (1 - R)IA_{NCZ}[h_{x,NCZ,m,in} - h_{x,NCZ,m,out}] \tag{35}$$

5. APPLICATIONS OF THERMAL ENERGY BALANCE FOR LOWER CONVECTIVE ZONE (LCZ)

Applying energy balance on LCZ, the following Eq. has been obtained: -

$$\rho_{LCZ}C_{p,LCZ}t_{LCZ}A_{LCZ}\left(\frac{dT_{LCZ}}{d\tau}\right) = Q_{s,LCZ} - Q_{cond} - Q_{load} - Q_{losses,LCZ} \tag{36}$$

Calculating the magnitude of solar energy stored in LCZ ($Q_{s, LCZ}$),

$$Q_{s,LCZ} = Q_{s,LCZ,in} - Q_{s,LCZ,out} \tag{37}$$

The value of $Q_{s,LCZ,out} = 0$ because it is considered that solar insolation is completely absorbed in this region.

The magnitude of solar thermal energy stored in LCZ ($Q_{s,LCZ}$),

$$Q_{s,LCZ} = Q_{s,LCZ,in} = (1 - R)IA_{LCZ}h_{x,LCZ,in} \tag{38}$$

Here, $h_{x,LCZ,in}$ depicts the fraction of solar insolation reaching the top.

The ground thermal energy losses (Q_g),

$$Q_g = U_g A(T_{LCZ} - T_g) \tag{39}$$

U_g represents the overall thermal energy loss coefficient.

Overall thermal energy loss coefficient (U_g) [4,10],

$$U_g = \frac{1}{\left(\frac{1}{h_3}\right) + \left(\frac{x_g}{k_g}\right) + \left(\frac{1}{h_4}\right)} \tag{40}$$

The value of h_3 and h_4 is considered to be 78.12 W/m^2 $^{\circ}$C and 185.8 W/m^2 $^{\circ}$C.

6. APPLICATIONS OF NON-CONVECTIVE SOLAR POND

6.1. Space Heating

Solar ponds can be effectively implemented for space heating applications, as shown in Fig. (**1**). The hot water obtained from the pond may be directly sent to the AHU unit, as in the case of other HVAC systems. The renewable source of heating and simple construction make this an attractive-to-use application.

6.2. Greenhouse Solar Pond Heating System

A greenhouse space may be heated by adding a shell and tube heat exchanger and a heat pump system along with the pond heating system. The hot brine from the pond is circulated across the shell and tube heat exchanger for heat extraction while the heat pump system assists in maintaining the pond above operating temperature. Direct brine pumping is preferred here since the heat exchanger area required at the pond bottom may be impractically large due to the smaller temperature gradient between the space and pond brine.

6.3. Electricity Generation

The solar pond coupled with a closed Rankine cycle is utilized for electrical power generation in its more recent form. Apart from the above applications, the solar pond may be applied to the following:

- Salt and mineral extraction,
- Solar engines based on the Rankine cycle,
- Solar refrigeration
- Space and swimming pool heating,
- Agricultural crop and horticulture product drying,
- Direct hot water production,
- Water distillation,
- Large scale laundry and textile manufacturing and
- Food processing *etc.*

7. VARIOUS SPACE COOLING CONCEPTS

Thermal cooling through passive means for reducing heat load in a building utilizes numerous natural heat dissipating techniques, *viz.* ventilation, evaporative cooling, radiations in an infrared band to the ambient (sky), and earth coupled cooling. Some researchers [12–27] have performed research work in the field of ventilation by solar energy and earth air tunnel coupling.

The first prime step toward passive heat dissipation is the decrease in unnecessary (supplementary) thermal loads, *viz.*, the exterior (outside) loads due to ambient conditions and interior loads due to the presence of occupants in the buildings. The climate (ambient) dependent loads comprise conduction of thermal energy *via* the building envelope, infiltration of outdoor air, and entry of small wavelength insolation directly. The nature of the cooling concept may be direct and/or indirect.

The direct cooling techniques consist of (i) ventilation/infiltration, (ii) air vents\wind tower as well as (iii) Earth shelter, whereas the indirect cooling techniques consist of (i) direct/indirect evaporative type cooling/roof still (pond) (ii) exterior and interior shading/position changeable thermal/solar insulation and also (iii) earth air tunnel heat exchanger. All these techniques have been elaborated on in the upcoming sections.

7.1. Cooling by Evaporation

In the direct type of evaporative (passive cooling) systems, the representative room (interior) air directly touches the water envelope. The evaporation of water in the air raises the humidity level. In such situations, a mini-size building can be cooled by positioning wetted pads/cloths in the exposed windows in the wind direction.

Water in the form of a thin film (water sprayed intermittently) or flowing water on the outer surface of the roof is also one of the techniques of indirect evaporative cooling. When the water captures the thermal energy from the exterior roof surface, it causes cooling of the bottom ceiling surface, which behaves like a radiative cooling plate for the interior space. The indoor space temperature is reduced without increasing the humidity level. The indirect evaporative cooling technique, which consists of a roof pond, is highly efficient if the roof is covered with greenery and position changeable thermal insulation.

In the case of heating, q is enhanced while it is decreased for cooling. For the condition of evaporative cooling of walls/roof, the rate of heat flux reduced can be estimated by utilizing the below-mentioned expression

$$\dot{q} = -K \frac{\partial T}{\partial x}\bigg|_{x=0} = h_1(T_{sa} - T|_{x=0}) \tag{41}$$

$$\dot{q} = \frac{K}{L}(T|_{x=0} - T|_{x=L}) \tag{42}$$

$$\dot{q} = h_{si}(T|_{x=L} - T_b) \tag{43}$$

Above Eq. 41-43 can be obtained and written in the form

$$\dot{q} = U_w(T_{sa} - T_b) \tag{44}$$

Where

$$U_w = \left[\frac{1}{h_1} + \frac{1}{K} + \frac{1}{h_{si}}\right]^{-1} \tag{45}$$

7.2. Ventilation/Infiltration

Infiltration refers to entering outside (exterior) air through the door and/or window opening and cracks, and spaces near the doors and windows, into the living space.

The infiltration can be because of (i) the pressure variation created by the temperature and humidity difference between outside air and that inside the building, (ii) the movement of occupants, and (iii) wind pressure.

Ventilation causes convective heat losses, which significantly affect the air exchange rate, the difference in temperature between the interior and exterior sides of the representative room, and the specific heat capacity (sensible) of air.

The ventilation losses can be given by,

$$Q_v = 0.33NV(T_i - T_a) \tag{46}$$

In Eq. 46, N represents the frequency of air exchanges per hour, V represents the cubic capacity of the building in m^3, and T_i represents the temperature of air present on the interior side of the building in °C.

Wind conditions are significantly affected by surface texture and local topography; if the space of the building is kept six times the height (dimension), a gridiron pattern yields inappropriate air stream movement with a non-variable flow and elimination of stagnant zones.

Windows dominantly influence ventilation. The rate of ventilation is significantly influenced by climate wind direction, the dimension of inlet and outlet apertures, the cubic capacity of the room, shading devices, and interior separations.

7.2.1. Solar Chimney

The solar chimney is a device for creating ventilation. Fig. (**3**) depicts the diagram (thermal) of a typical solar chimney. Ventilation is made in the solar chimney with the help of thermal energy obtained from capturing the solar radiation in the blackened absorber surface. Due to capturing solar radiation by the absorber, its temperature rises, and this absorber transfers its heat to the air. Ventilation occurs when the temperature of chimney air becomes higher than the ambient air temperature.

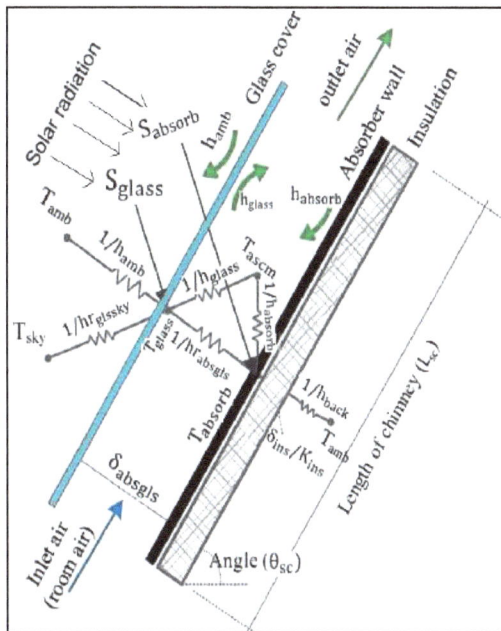

Fig. (3). Thermal diagram of solar chimney.

Considering energy balance on the glazing cover of the solar chimney,

$$S_{glass}A_{glass} + hr_{absgls}A_{absorb}(T_{absorb} - T_{glass}) = h_{glass}A_{glass}(T_{glass} - T_{ascm}) + U_{glassamb}A_{glass}(T_{glass} - T_{amb}) \tag{47}$$

Considering energy balance on the air available inside the solar chimney,

$$h_{absorb}A_{absorb}(T_{absorb} - T_{ascm}) + h_{glass}A_{glass}(T_{glass} - T_{ascm}) = m_{asc}C_{asc}(T_{ascout} - T_{ascin}) \tag{48}$$

Considering energy balance on the absorber surface of a solar chimney,

$$S_{absorb}A_{absorb} = h_{absorb}A_{absorb}(T_{absorb} - T_{ascm}) + hr_{absgls}A_{absorb}(T_{absorb} - T_{glass}) + U_{absorbamb}A_{absorb}(T_{absorb} - T_{amb}) \tag{49}$$

From Eq. 47-49, the below mentioned simultaneous Eq. 50-52 are obtained,

$$a_1T_{glass} + b_1T_{ascm} + c_1T_{absorb} = R_1 \tag{50}$$

$$a_2T_{glass} + b_2T_{ascm} + c_2T_{absorb} = R_2 \tag{51}$$

$$a_3T_{glass} + b_3T_{ascm} + c_3T_{absorb} = R_3 \tag{52}$$

7.2.2. Wind Tower

Cooling of air, circulation of air in the building, and harnessing wind energy can be simultaneously performed in wind towers. The availability of wind and the time of day substantially determine the operation of a wind tower. At the time of the day, warm outdoor air enters the wind tower passing through the opening positioned on the sides, and air temperature is decreased as processed through the cooled wind tower. The cool air is denser than the hot air and moves down through the tower, generating a downdraft. When the wind is present, the downdraft is faster. The operation of a wind tower is replicated like a chimney at night time. The cooled night air is heated with the help of thermal energy stored in the tower during daytime.

A multi-storied apartment is not suitable for wind tower application, and the mechanism of the wind tower is well applicable in the individual units.

7.2.3. Earth Air Tunnel

A few meters beneath the ground surface, a constant temperature of the earth is observed. Earth – air tunnel utilizes this constant temperature for processing the air. This temperature is constant with respect to time for a whole year. When air passes through, a buried pipe becomes cooled in summer when the earth's temperature is lower than the air, and in winter, the air is heated by taking heat from the earth/ground. Heat transfer between the air flowing inside a tube or tunnel and the surrounding earth is significantly affected by some influencing parameters like area of the pipe, length (dimension) of tunnel and depth of the tunnel below ground surface, dampness property of the earth, inlet air humidity and its speed affect the transfer of heat among air and the surrounding soil. Table **1** shows the ground temperature for different surface conditions.

Table 1. Ground temperatures corresponding to different surface conditions at a buried depth of 4 meters.

Surface Type	Temperature of Ground
Wet Shaded	21.0 °C
Wet Sunlit	21.5 °C
Dry Sunlit	27.5 °C

Fig. (**4**) represents the sectional (cross) representation of an innovative earth tunnel beneath the ground at 4 meters buried depth, and air harnessed from the atmosphere is permitted to circulate through it. The geometry/shape of the air tunnel is considered to be cylindrical of radius r and corresponding length (L). When the air flows *via* the air tunnel, a heat exchange from an interior exposed surface of the air tunnel to a passing air by convection (forced) is observed. According to the condition (temperature) of air, the air temperature is either increased or decreased. If the interior surface of the tunnel is colder than the air flowing through it, then thermal energy is exchanged from the interior surface to air for increasing temperature. This happens in the colder season. For the summer season, it is *vice versa*.

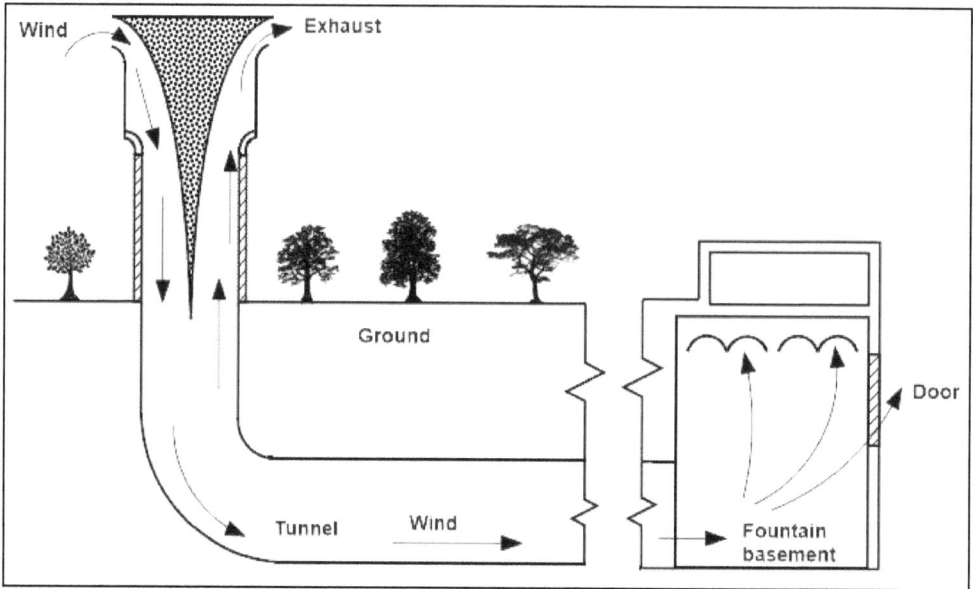

Fig. (4). Coupled arrangement of wind tower with innovative earth tunnel heat exchanger

In Fig. (**5**) elemental segment of the tunnel is depicted. Referring to Fig. (**5**), for an elemental segment, the dx energy balance can be applied as

$$\dot{m}_a C_a \frac{dT(x)}{dx} dx = 2\pi r h_c (T_0 - T(x)) dx \qquad (53)$$

Where $h_c = 2.8 + 3.0$ V, V is the speed of air passing *via* the air tunnel, $\dot{m}_a = \pi r^2 \rho V$ and C_a are the rate of mass transfer and heat capacity of air, and $T(x)$ represents the temperature of the air as a function of x.

Eq. 53 can be evaluated corresponding to $T(x{=}0) = T_{fi}$ and can be presented as,

$$T(x) = T_0 \left(1 - e^{-\frac{2\pi r h_c}{m_a C_a} x} \right) + T_{fi} e^{-\frac{2\pi r h_c}{m_a C_a} x} \qquad (54)$$

Now, $\quad T_{fo} = T(x = L) \qquad (55)$

Thermal energy taken away in unit time by the passing air is,

$$\dot{Q}_U = \dot{m}_a C_a (T_{f0} - T_{fi}) \tag{56}$$

Or,
$$\dot{Q}_u = \dot{m}_a C_a (T_0 - T_{fi}) \left[1 - e^{-\frac{2\pi r h_c L}{m_a C_a}} \right] \tag{57}$$

The energy available in one hour $= Q_U \times 3600$ J (58)

The volume (magnitude) of hot air in one hour $=$ (59)
$\pi r^2 V \times 3600$

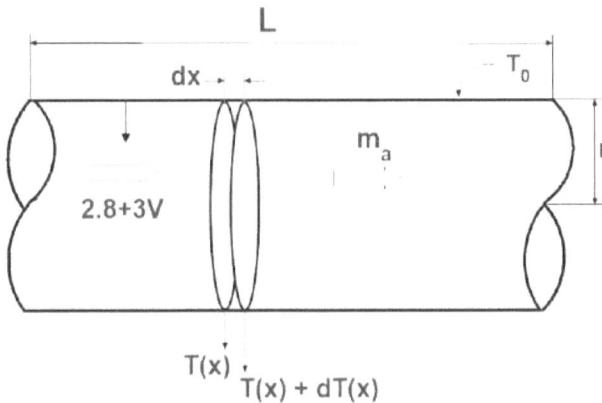

Fig. (5). Mathematical model of earth air tunnel.

If V_0 is the cubic capacity of a representative room whose temperature has to rise, then the frequency of air change can be given by,

Number of air changes per hour,
$$N = \frac{\pi r^2 V \times 3600}{V_0} \tag{60}$$

For $L \to \infty$ means very large tunnel length, then

$$T_{fo} = T_0 \quad \text{and} \quad Q_U = m_a C_a (T_0 - T_{fi}) \tag{61}$$

This shows the withdrawal of maximum heat to increase the temperature of the living space.

For L → 0 means small tunnel length, then

$$T_{fo} = T_{fi} \tag{62}$$

$$Q_U = 0 \tag{63}$$

This shows no removal of thermal energy from the air tunnel, and therefore, it is required to optimize the velocity of air, radius, and length for increasing/decreasing the temperature of an occupied zone. The deviation of Q_U corresponding to L for winter and the summer season for a standard set of various parameters has been represented in Fig. (6).

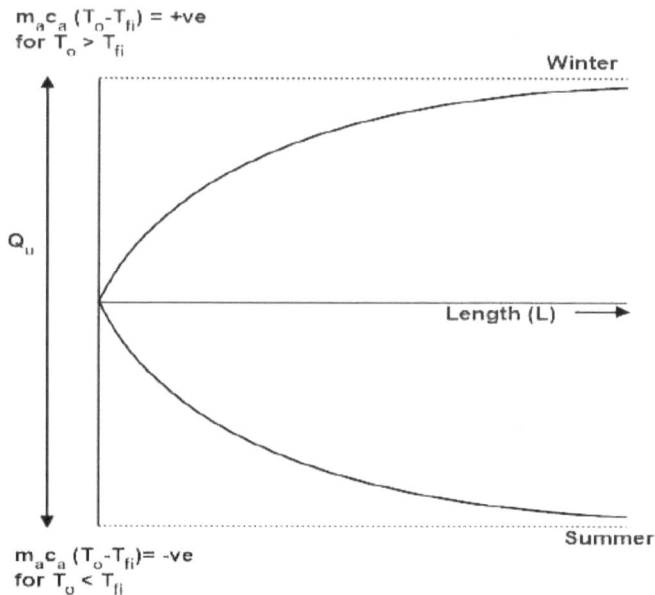

Fig. (6). Variation of Q_u with L.

Fig. (6) shows that the value of Q_U will become positive for the winter (colder) season due to the small value of outdoor air temperature ($T_a = T_{fi}$) than underground subsoil temperature (T_0). Further, the value of Q_U will become negative because of a higher value of outdoor air temperature. It is also important that the length (dimension) of the air tunnel must be optimized for a suitable volume of an occupant zone.

A coupled arrangement of a wind tower with an earth tunnel is efficient in raising the draft of air inside the earth tunnel for enhancing the heat-dissipating rate of air present in the building.

7.2.4. Air Vent

In some places, dusty air creates an unpleasant atmosphere making working on a wind tower impossible. They are suitable for individual units and work in a good manner in hot and dry and hot and humid climates. A vent consists of a hole in the apex of a cylindrical roof or domed consisting of a securing cap over it. Openings in the securing cap on the vent send the wind around it. As the air transfers on the surface having a curvature, its speed raises, thus reducing its pressure at the apex of the roof having a curvature; this causes inducing the warm air beneath the roof to pass out through the vent. Fig. (**7**) shows an operational diagram of the air vent. In this way, the air is passing *via* the representative room. Vents for air are generally kept over the occupant zone to reduce the temperature of air passing *via* the room.

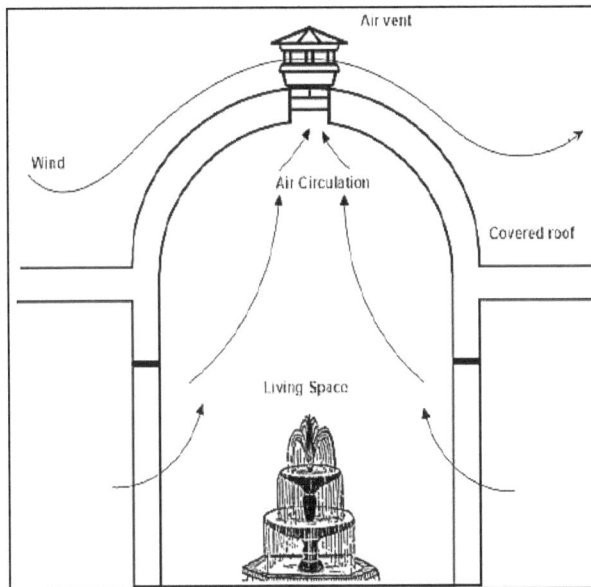

Fig. (7). Operational diagram of air vent.

7.2.5. Shading

The window is another crucial building component that greatly affects the thermal condition. The design of window openings has to be considered concerning sunlight, ventilation, and air motion. Designing windows concerning sunshine can be performed in two ways; first, by developing devices for shading to capture radiation from admitting, and the second thing is the proper designing of openings to allow sufficient sun lighting of the interior space. Numerous options are available for windows consisting:

(i) Self-inflating curtain: This type of curtain contains several layers of thin, flexible material with low emissivity and high reflectivity. Radiation increases the temperature of the air between the laminas, which causes a rise in pressure and a reduction in the density in the top portion of the system. The pressure forces the layers away and produces an entry of fresh air from the lower portion. Thus the arrangement consisting of good reflecting layers separated by air gaps offers suitable insulation, and whenever there is no need for the insulation, then the air is removed from the sides. The magnitude of energy supply rate *via* a self-inflating curtain is calculated as,

$$q = (1 - f)\tau \cdot I(t) - U(T_b - T_a) \tag{64}$$

In Eq. 64, f represents the shading factor with a value of 1 for complete (perfect) shading and a value lower than unity for a partial type of shading.

(ii) Window quilt shade: The quilt contains fine layers sandwiched and assembled together using a welder (ultrasonic fiber type). The quilt is confined to a decorative kind of fiber (polyester).

(iii) Venetian blind in between the glazing: This is an efficient arrangement to decrease the thermal energy loss *via* a glazed window (double type). In this condition, the characteristic measurement (size) of the unit is low; therefore, the convective thermal energy exchange is terminated. In this condition, q is equal to that in Eq. with a smaller value of τ as well as U.

(iv) Transparent heat mirrors: A technique to decrease the loss of energy (thermal) from the glazed surface is to cover the glass with a thin film that substantially reflects the radiation in an infra-red zone from the glass surface. However, this thin film coating also decreases the transmissivity of the glazing for solar insolation, and a matching compromise has to be made. The thin-film coating may contain a single or several layers of various materials, deposited by vacuum type evaporation method or spray method. The thermal mirror yields a much lesser loss of thermal energy and more transmission than the multipane system.

(v) Heat trap: A suitable thickness of thermal insulating substance with good transmissivity can decrease the thermal energy transfer rate.

(vi) Optical shutter: An optical shutter contains one layer of cloud gel and three layers of transparent sheets. It is non-transparent at elevated temperatures and highlights intensities. It can be utilized for decreasing air conditioning demand and preventing/reducing excessive heating in greenhouses, as well as solar collector arrangements.

7.2.6. Rock Bed Regenerative Cooler

In Fig. (**8**), a cross-sectional view of a typical regenerative cooler is represented. It utilizes two beds made of rock set side by side as depicted in Fig. (**8**).

It also functions as a heat exchanger separated by an air gap. A damper has been utilized between 2 rock beds to provide direction of the entering air from the house towards temperature reducing rock bed by water (liquid) spray. The temperature of the rock beds is decreased alternatively. During cooling, it also absorbs H_2O in liquid form. The temperature of the air moving through a dry rock bed (which is already cooled in an initial cycle) is reduced by transferring its thermal energy to the rock bed, and the cold air is permitted to enter the room.

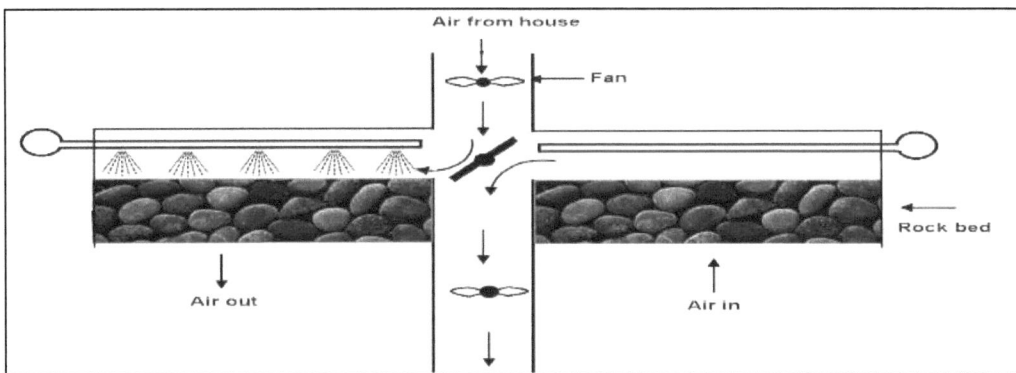

Fig. (8). A regenerative cooler.

The moist air generated during the evaporation process from the rock bed is ventilated to the outdoor. After the rock bed becomes hot, the damper orientation is reversed from further cooling. Simultaneously, another cold rock bed is utilized for reducing the temperature similarly.

7.2.7. Radiative Cooling

It has been observed that the sky temperature is always less than the outdoor air temperature by 12°C with a cloud-free night sky. In the northern hemisphere, the sky is cool enough most of the time, even during the sunshine hour. It behaves like a lower temperature heat sink corresponding to outdoor air in the day/night time. A flat horizontal plane is the most efficient radiative heat transfer arrangement.

Exposed flat horizontal planes liberate thermal energy to outdoor air by convection and radiation till the temperature of the horizontal surface approaches equal to the dry-bulb temperature of the air. Also, there are thermal energy losses from outdoor air/surface to the clear sky due to large-wavelength radiation transfer. When the heat transfer between the surface and clear sky decreases the temperature of the roof plane to the nearby air WBT, moisture from the air starts condensing on the exterior roof surface. This condensation of moisture will further reduce the temperature of the roof due to quick thermal energy loss from the room to the roof plane. In Fig. (**9**), open-loop radiative cooling is depicted.

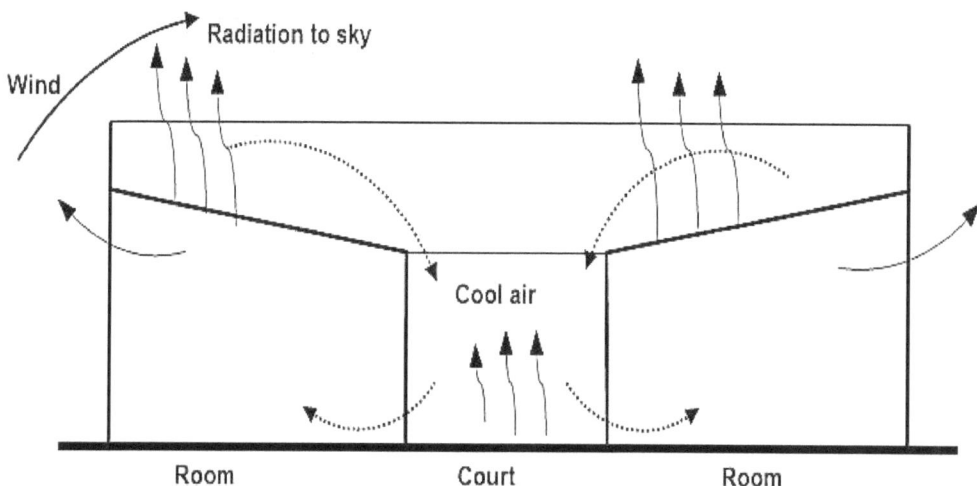

Fig. (9). Open-loop radiative cooling.

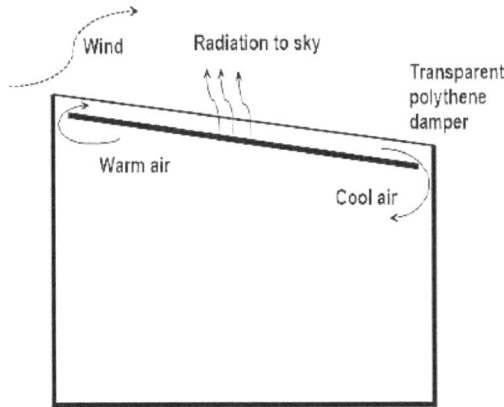

Fig. (10). Closed-loop radiative cooling.

If the surface makes different angles with horizontals, as depicted in Fig. (**9**), the cold air over the inclined plane will move towards an internal courtyard due to a higher density. Then the moving cold air passes into the room *via* the opening at the bottom level, as depicted in Fig. (**9**). However, this concept does not effectively work in the case of wind because the winds will become the transporter of cold air. The roof surface must be covered with a polythene sheet (transparent) to prevent this. It has been found that the polythene sheet also has a very limited life span. This problematic condition can be eliminated by enveloping the roof surface with a corrugated thin metal film with the opening at the bottom and top-end from the interior side of the roof, which is depicted in Fig. (**10**).

8. SOLAR REFRIGERATION

The solar cooling device is basically a flat plate collector working at lower temperatures with the help of a working fluid refrigerant. The working principle of vapor absorption and vapor compression refrigeration is used to obtain household refrigeration, air conditioning, *etc*. In warm climates, solar cooling plays a vital role to achieve thermal comfort in buildings and can be accomplished by following cycles:

(a) Solar absorption process

(b) Solar desiccant cooling

(c) Solar-mechanical processes

8.1. Solar-Absorption Process

The basic working principle of the solar absorption process is presented with the help of Fig. **(11)**. Vessel A contains vapor refrigerant bound with the absorbent, whereas vessel B has refrigerant only. The valve provided in the flow line remains closed when the temperature is equal in containers; however, the vapor pressure in vessel B will be more than vessel B. As the valve is opened, the vapor refrigerant starts to flow from vessel B towards vessel A because of the pressure difference. The vapor pressure in vessel A becomes more than the equilibrium vapor pressure, and therefore condensation starts in it. This condensation causes an increase in the temperature of the refrigerant in vessel A due to the release of latent heat when the vapor refrigerant condenses. The opposite process takes place in vessel B and thus decreases the temperature.

Due to the transfer of mass that takes place towards vessel A from vessel B, vapor pressure tends to be equal in both vessels. It causes a difference in the temperature of refrigerants contained in vessels, A and B. At this stage, heat is provided in vessel B from the surroundings, and due to this, pressure difference also takes place. In vessel B, vapor pressure increases beyond the vapor pressure in vessel A, which causes the transfer of mass towards vessel A, and with this mass transfer, the flow of heat also takes place.

The absorption process can be used for cooling or heating by using a concerned vessel.

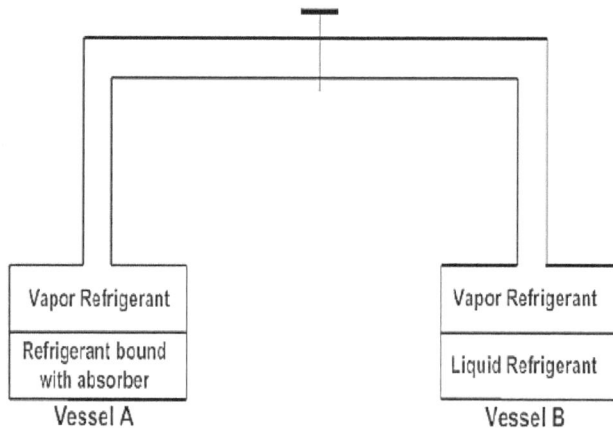

Fig. (11). Working principle of the absorption process.

To restore the heat pumping capacity, the system needs to be regenerated. The regeneration is achieved by inverting the mass and thermal energy exchange from vessel A to vessel B. Cooling and heating with energy storage form three key characteristics of the absorption process, and its success depends on the right selection of refrigerant and absorbent, and this selection is made based on chemical and thermo-physical properties of the fluid. The combinations of lithium–bromide–water (LiBr–H_2O) and ammonia–water (NH_3–H_2O) are generally used due to their favorable properties.

A solar absorption cooling system can be operated in two ways, which are:

(i) Continuous coolers that have construction and work similar to conventional units;

(ii) Intermittent coolers.

Lithium bromide water (LiBr–H_2O) based systems require water to cool the absorber and condenser; therefore, the need for a cooling tower also exists. The drawback involved with the use of water as a refrigerant is the need for a water (liquid) cooled condenser to achieve the necessary temperature for the air conditioning. Additionally, this combination is corrosive in nature and thus harmful to construction materials. Therefore, the Ammonia–water system is widely used for industrial purposes, absorption-based refrigeration, and air conditioning.

8.2. Solar-Desiccant Cooling

Solar-desiccant cooling is a system based on an open cycle arrangement that utilizes water as a refrigerant, and it remains directly in contact with air in contrast to the solar absorption unit. Evaporative cooling, combined with dehumidification of air, utilizes a desiccant (liquid or solid) and creates the targeted cooling. In this system, refrigerant is not allowed to recycle after participating in the refrigeration effect, and thus an open cycle is followed.

The two types of solar desiccant cooling are discussed below:

8.2.1. Solid-desiccant Cooling

The main constituents of the solid desiccant cooling system are shown in Fig. (**12**). In the working of the system, warm and moist air passes over the slowly revolving desiccant wheel, and dehumidification of air takes place because of the adsorption of water. After dehumidification, hot air passes through the heat-recovery wheel, thus pre-cooling the air.

The pre-cooled air is now supplied to the humidification tower, where it is additionally cooled with humidification and thus attains the required value of moisture content and temperature. Hereafter, the air is supplied to the building. In the building, warm and humid air coming out from the building is then fed to a humidifier, and humidification takes place up to a saturation state for maximum cooling capability. Now, it is provided for heat recovery, and after that, the air is allowed to flow through the dehumidifier before exiting. In order to obtain continuous cooling, the sorption system is heated in the temperature range of 50°C - 75°C for regeneration.

In case of high humidity weather, as in coastal areas, an improved design of the desiccant system is needed to bring down the humidity to a favorable level related to direct type evaporative cooling. For this condition, additional heat recovery media like a wheel or extra coolers for air can be used to achieve the desired state of the air. The innovative approach for improving the heat transfer efficiency of the cooling system is dehumidification by absorptive coating on the walls of the heat-exchanger and performing air cooling with the help of an air-to-air heat exchanger.

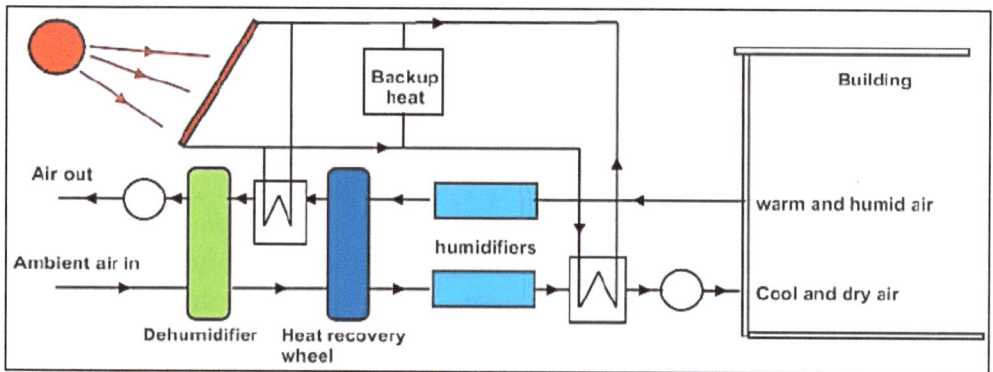

Fig. (12). Solar desiccant cooling system.

8.2.2. Liquid-desiccant Cooling

Liquid desiccant cooling is an innovative technique that uses liquid water–lithium chloride (H_2O–Li–Cl) as the sorption material. These systems have the following advantages:

(i) It provides higher dehumidification of air than solid desiccant cooling units when operated over the same temperature variation.
(ii) It shows better storage of energy in comparison to concentrated solutions.

8.3. Solar Mechanical Cooling

The solar mechanical cooler comprises a solar-assisted Rankine cycle heat engine and a conventional air-conditioning system. The main limitations of conventional solar air-conditioning systems are dependency on solar energy to fulfill mechanical energy requirements and the need for improvement in the design of conventional air conditioning systems to adapt to variable load operation. The schematic of the solar cooling system working with the Rankine cycle is shown in Fig. (**13**).

In this system, the heat engine develops mechanical work by using heat energy stored in an insulated storage water tank with the help of a heat exchanger. The efficiency of solar thermal systems decline with an increase in temperature; however, it is the opposite for the heat engine, where higher temperatures provide higher efficiency.

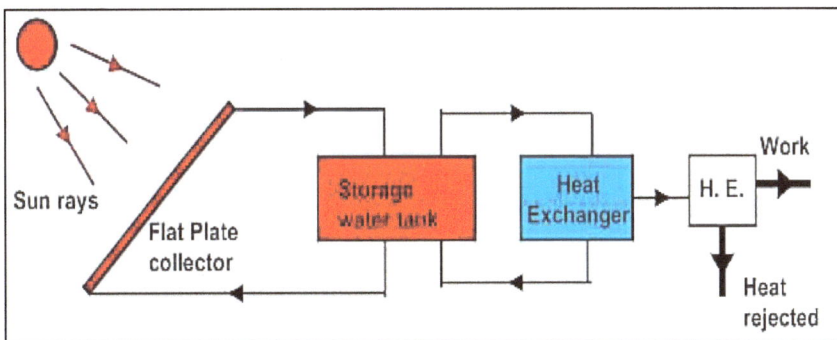

Fig. (13). Solar based Rankine cycle cooler.

8.4. Comparison of Solar Cooling Technologies

Following solar cooling technologies have been compared by [28] for the performance:

(a) Selective solar flat plate collector
(b) Parabolic-trough collector
(c) Evacuated tube solar collector
(d) Photovoltaic collector

The performance of the above-mentioned technologies has been compared to the constant collection temperature defined by the following conditions:

(a) Adsorption at a temperature of 70 °C.
(b) Single effect absorption at a temperature of 90 °C.
(c) Double effect absorption at a temperature of 160 °C.
(d) GAX ammonia absorption is air-cooled at a temperature of 160 °C.
The comparison for daily useful thermal energy in kWh per m^2 per day available at different working temperatures and coefficient of performance are tabulated in Tables **2** and **3**, respectively.

Table 2. Comparison of daily useful heat energy (kWh/m²/day).

Temperature (in °C)	Selective FPC	Parabolic Trough Concentrator	Evacuated Tubular Collector
70	3.63	4.50	3.34
90	2.98	4.29	3.23
60	0.89	3.54	2.87

Table 3. Comparison of coefficient of performance.

Technology	COP
Single effect adsorption (Li–Br) using heat energy	0.8
Double effect absorption (Li–Br) using heat energy	1.2

(Table 3) cont.....

GAX NH$_3$ absorption air-cooled using heat energy	0.6
Adsorption, electricity	0.4

9. SOLAR CONCENTRATOR

A solar concentrator plays its role in focusing the solar radiation on a small plane with the help of suitable reflecting or refracting means. For effective concentration of solar energy, a sun-tracking arrangement and focusing device are also needed. These mentioned components of solar concentrators enable the device to reach a temperature up to 3000 °C, thus making it suitable for thermal as well as photovoltaic applications of solar energy. The main advantages of solar concentrators are –

1. Increased intensity of solar energy
2. Reduced heat loss due to smaller area used for concentration of solar energy
3. Very high delivery temperature
4. Reduced cost as it can replace the large receiver

The above-mentioned advantages make solar concentrators very suitable for thermal utilization of solar energy; however, optical losses, dependability on beam radiations only, and the complexity of design and proper maintenance are the challenges associated with solar concentrators.

9.1. Important Parameters

The important parameters that describe the characteristics of solar concentrators are shown in Fig. (14) and are described below-

(a) **Aperture area:** The area located at the front portion of a solar concentrator that receives the beam radiation.
(b) **Acceptance angle:** The maximum angle up to which the path of incident beam insolation may diverge from normal to the plane of aperture and finally reach the absorber or receiver. The solar concentrators with larger acceptance angles are required to move seasonally, whereas, for smaller acceptance angles, continuous movement of the concentrator is required to track the sun.

(c) **Absorber or receiver area:** The total area absorbing the beam radiation and thus provides necessary heat energy to the concentrator.

(d) **Geometric concentration ratio:** It is defined as the ratio of two things, the first aperture area to the second absorber area, and its value varies from unity (for flat plate collector) to several thousand (for a parabolic dish)

(e) **Local concentration ratio:** The ratio of beam radiation at any point on the absorber to the beam radiation at the entrance of the concentrator when the whole system is not uniformly illuminated.

(f) **Intercept factor:** It is the fraction of the total focused beam intercepted by the absorber of a given size, and its value depends on the dimensions of the absorber. For a typical concentrator structure, it is more than 0.9.

(g) **Optical efficiency:** The ratio of the net thermal energy available for the absorber to the available beam radiation on the absorber after absorption and transmission.

(h) **Instantaneous thermal efficiency:** The ratio between useful energy on the absorber to the beam energy available on the aperture.

(i) **Concentration ratio:** The ratio of aperture area to the absorber area.

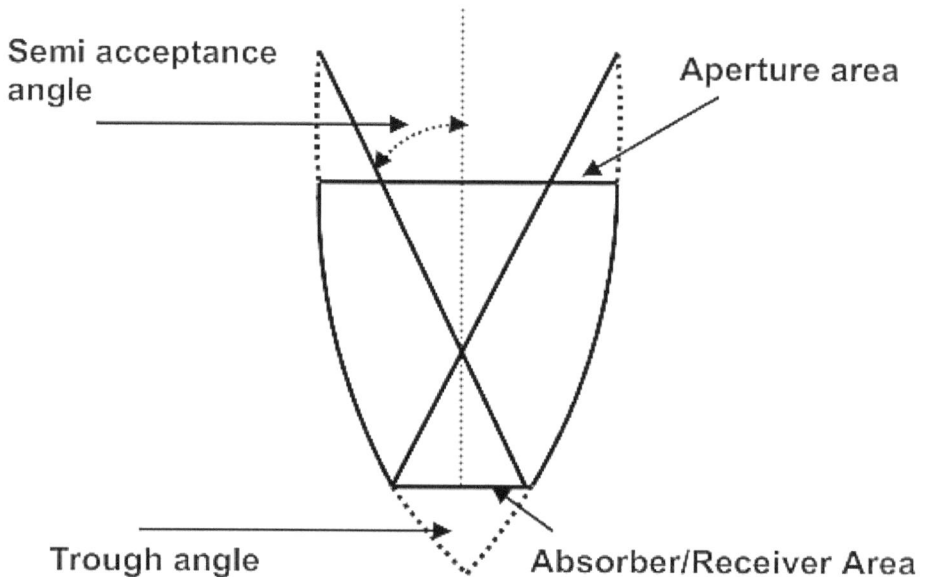

Fig. (14). Characteristic parameters of solar concentrator.

9.2. Solar Concentrators Classifications

Broadly, solar concentrators can be categorized as:

(a) Tracking type - which can be further categorized depending on continuous or intermittent tracking used and single-axis or two-axis tracking applied.
(b) Non-tracking type - The axis of a concentrator is fixed in this type, and there is no moving component.
Additionally, solar concentrators may be categorized depending on optical constituents as:

(i) Reflecting/refracting type – The concerned surfaces related to reflection or refraction can be in the form of a single piece or the form of a composite surface.
(ii) Imaging or non-imaging type
(iii) Line-focusing or point-focusing type

9.3. Types of Solar Concentrator

There are several means to enhance the beam radiation on absorbers, and the following section deals with different concentrators to briefly discuss them-

9.3.1. One/Single-Axis Tracking Type Solar Concentrators

One-axis tracking type solar concentrators generally utilized to attain moderate concentration are:

9.4. Fixed Mirror Type Solar Concentrator

This concentrator has a fixed mirror with long and thin flat strips organized on a reference cylinder and a receiver capable of tracking, as shown in Fig (15). Each strip is adjusted to maintain focal distance in the array to be two times the radius (r) of the cylinder on which strips are placed. This arrangement provides a fine focal line alongside the same circular path during day time motion of the sun. The concentration ratio achieved with this arrangement is approximately equal to the number of mirror elements. These concentrators have provided an overall thermal efficiency between 40 and 50%.

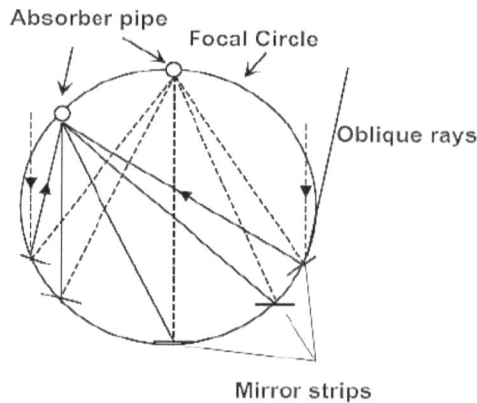

Fig (15). Fixed-mirror solar concentrator.

9.5. Cylindrical Parabolic Solar Concentrator

The main constituents of this type of solar concentrator are a cylindrical parabolic reflecting surface and metal made tubular-shaped receiver placed on the focal plane, as presented in Fig (**16**). The receiver is black coated and enclosed by the concentrating surface, and rotation about a single axis enables it to follow the sun. During flow through the absorber tube, the temperature of heat-transfer fluid increases. The main parameters which may be used to define concentration are the diameter of the aperture, rim angle, and size of the receiver. A cylindrical parabolic type solar concentrator has achieved a solar concentration ratio in the range of 5-30.

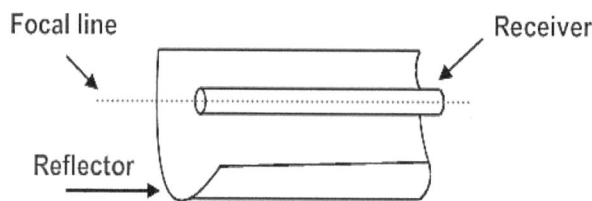

Fig. (16). Cylindrical parabolic solar concentrator.

These concentrators may be oriented east/west, north/south, or in a polar direction. The mentioned two directions result in more incidence angle cosine losses, whereas the third direction can provide more beam radiation/unit area and thus ensure better performance.

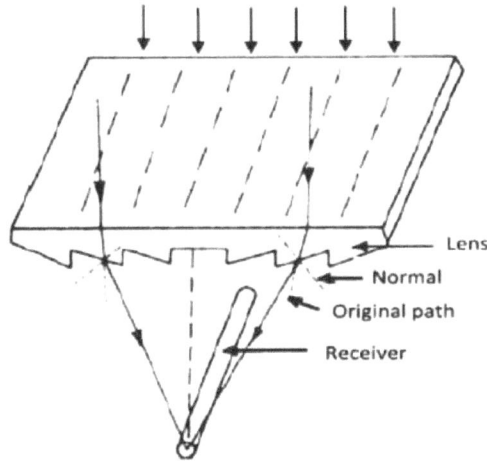

Fig. (17). Linear Fresnel lens solar concentrator [29].

9.6. Linear Fresnel Lens Solar Concentrator

Linear Fresnel lens solar concentrator is equipped with linear grooves provided on one side of the refracting material, as shown in Fig (**17**). The groove angles are selected so that the lens converges the incident rays, which are available normally. Materials like glass or plastic are commonly used as refraction materials in the fabrication of Fresnel lenses. However, plastic is preferred due to its ease of manufacturing, and also it is economical. The Fresnel lenses may be installed with sun-facing grooves or downward-facing grooves. In the first case, grooves prevent some part of the incident beam from transmitting to the focus. In the second case, solar concentrators have more surface reflection losses which causes low thermal efficiency.

9.6.1. Two-Axes Tracking Concentrators

In solar concentrators, double curvature is used to achieve higher concentrations that require two axes tracking the sun. The important two axes tracking concentrator are briefly discussed here:

9.7. Paraboloidal Dish Solar Concentrator

In this type of concentrator, a parabola rotates about the optical axis to achieve a higher concentration ratio. In working of paraboloidal dish solar concentrator, beam flux is focused on a point in paraboloid because of compound curvature, but the sun's finite angular substance finally creates the image. Additionally, if the surface is not to be completely parabolic, that causes a comparatively larger image due to variation in the direction of the beam rays. The thermal losses involved in this type of concentrator are primarily due to radiation which can be reduced by reducing the aperture area. The optimum intercept factor for such concentrators is approximately 0.95–0.98. The main advantages of paraboloidal dish types of solar concentrators are higher collection heat transfer efficiency and enhanced heat energy, making them suitable for high-temperature requirements.

9.8. Central Tower Receiver

The central tower receivers consist of a centrally placed stationary receiver and heliostat, which consists of several mirrors fixed on a supporting frame, as shown in Fig. (**18**). Heliostat reflects the beam radiation toward the receiver. The concentration ratio having a value of more than 3000 has been achieved by this concentrator. The absorbed thermal energy is extracted from a centrally placed receiver and supplied at a temperature and pressure condition required for power generation by running the turbines. In this system, heat losses are reduced significantly due to the elimination of the flow of working fluid over larger distances; however, the need for a large area due to a large number of heliostats is a major drawback.

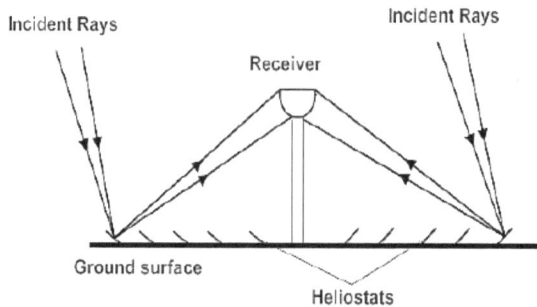

Fig. (18). Central tower receiver.

9.9. Circular Fresnel Lens

These lenses are utilized in solar-based utilities when a higher temperature is required, like a solar furnace. A Fresnel lens is similar to a narrow lens approximation and consists of several zones spread over a few tenths of a millimeter. Within an individual zone, the slope of the lens plane is changed to resemble the usual sphere-shaped lens having an equal focal length. The Fresnel lens (circular in shape) gives a much higher concentration ratio as high as 2000.

9.10. Hemispherical Bowl Mirror

This type of concentrator has a stationary hemispherical mirror, moveable tracking absorber, and a supporting frame, as shown in Fig. (**19**). The hemisphere mirror results in an extremely irregular optical image, but rays that enter the hemispherical surface also cross the paraxial line in a location between the focus and the surface of the mirror due to the symmetry of incident rays. Thus, the absorber works to pivot about the center of the curvature and, therefore, intercept all reflected rays. Two axes tracking is required here to move the absorber with an axis aligned with sun rays passing through the center of the sphere. However, this motion can be achieved another way by using an equatorial mount, and the absorber is driven at an angular velocity of 15°/h around a polar axis. This concentrator provides lowered concentration compared to paraboloids due to the spherical deviation.

9.10.1. Non-tracking Solar Concentrators

These concentrators are installed without tracking for medium temperature requirements; therefore, they are not as expensive as tracking type concentrators. Some of these categories of concentrators are discussed in brief here:

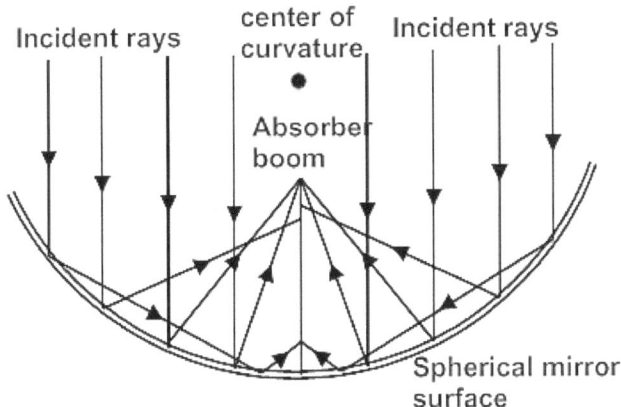

Fig. (19). Hemispherical bowl mirror.

9.11. Flat Receiver Along with Booster Mirror

Fig. (20) depicts a flat plate receiver equipped with flat reflectors placed on its sides to reflect the incident rays towards the absorber. These flat mirrors provide additional radiation and are hence called booster mirrors. The concentration ratio of this concentrator is lower (maximum value <4) but still more than that of the simple flat plate collector. Booster mirrors may be provided on all four sides of the collector, but they also cast a shadow on the absorbing surface when the sun angle has value more than the semi-angle of a flat mirror. When multiple flat plate collectors are used, these mirrors are provided along the two sides only. The benefit of this concentrating device is that it can use diffuse radiation and beam radiation. It also provides more temperature and thermal efficiency than a flat plate collector having the same collection area.

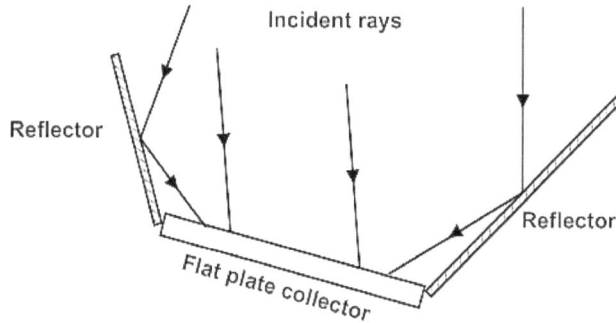

Fig. (20). Flat receiver with booster mirror.

9.12. Compound Parabolic Solar Concentrator

The concentrating device has the maximum possible concentration for a given acceptance angle. Additionally, it has a higher acceptance angle, and there is a need to be turned toward the sun. The basic design consists of two parabolic-shaped portions and placed to maintain the focus of the first portion at the lower point of the second portion and *vice versa*, as depicted in Fig. (21).

The axes of the parabolic portions create an angle with the axis of the concentrator that is equal to the acceptance angle. The receiver is a flat surface situated parallel to the aperture joining foci of the reflecting surfaces. The reflections of incident beam radiations depend on incident angle, collector depth, and concentration ratio [30].

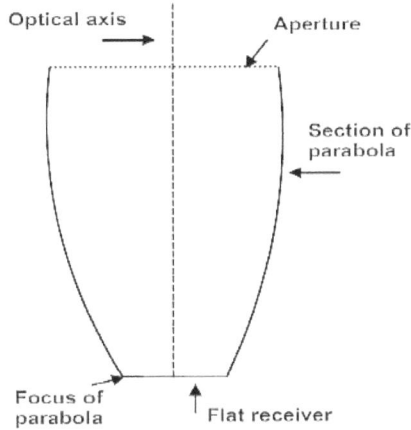

Fig. (21). Compound parabolic solar concentrator.

Over the years, several modified designs of compound parabolic solar concentrators have also been suggested, and the main modifications are listed below:

(i)   Receiver shapes in fins and circular pipes have been suggested to enhance optical and thermal performance.
(ii) Reduction in the height of the concentrator to reduce the size and required cost
(iii) Use of compound parabolic solar concentrator as a second-stage concentrator.

It has been served with a maximum concentration ratio of almost six because a higher value of concentration ratio will also require a larger reflecting surface area, and hence it is not in practice.

9.13. Theoretical Solar Image

In the image formation of the sun on the absorber, even a perfectly designed optical system generates a finite size image because the sun is actually not a point source. The distance between the sun and earth causes the solar disc to subtend an angle of 32' at a point on the surface of the earth. If r represents the distance from the point of incidence to focus, and W represents image size, then the expression for W is [29]:

$$W = \frac{2r\,(tan16')}{cos\emptyset} \tag{65}$$

Here, $\emptyset$ is the angle formed by the optical axis with normal (at the point of incidence).

9.14. Thermal Performance

The following reasons are important considerations that must be included while calculating the heat losses from the receiver or absorber.

(i) The shape of the receiver varies significantly, which causes the availability of non-uniform beam flux at the receiver.
(ii) Due to high temperature, thermal losses from edges and conduction losses need to be included.
The mathematical equation for collection heat transfer efficiency has been derived for two conditions:

(a) Operation in natural mode

(b) Operation in forced mode.

9.15. Natural Mode

The instantaneous thermal efficiency is given by the relation,

$$\eta = \frac{q_u}{I_b} \tag{66}$$

Where I_b is incident beam radiation and q_u is the useful thermal energy rate/unit aperture area and can be calculated by the relation,

$$q_u = \frac{Q_u}{A_a} = F'\left[(\rho\alpha\tau)I_b - U_t\left(\frac{1}{C}\right)(T_f - T_a)\right] \tag{67}$$

9.16. Forced Mode

For the forced mode, the equation for heat flow rate by the fluid for absorber having tubular shape can be given as,

$$Q_u = \frac{Q_u}{A_a} = A_a F_R \left[S - U_t \left(\frac{A_r}{A_a} \right) (T_{fi} - T_a) \right] \tag{68}$$

A similar expression for a flat-plate absorber is,

$$Q_u = A_a F_R \left[S - U_L \left(\frac{A_r}{A_a} \right) (T_{fi} - T_a) \right] \tag{69}$$

9.17. Solar Concentration Ratio

To improve the thermal characteristics of a solar-based concentrator, the thermal energy loss rate must be optimized, and it implies that the absorber area needs to be reduced. However, a small size absorber results in a low value for the intercept factor and thus results in poor performance related to optical considerations. Therefore, a compromise has been made in performance related to optical and thermal considerations for a perfect concentration ratio. The maximum value of the concentration of a concentrating collector is obtained by [30-31] and (Rabl and Winston, 1976) and suggested that the upper limit of concentration is inverse of the shape factor for sun and concentrator. This suggestion is found suitable for acceptance angle in the range 0.5°–180°.

9.18. Substances for Solar Concentrators

The use of appropriate substances ensures proper usage of solar thermal energy therefore, different materials associated with components of solar concentrators have been briefly discussed here.

9.19. Reflecting and Refracting Surfaces

The main requirements for solar reflectors are high reflectivity and effective specular reflectance. Glass with silver coating on a rear plane is generally utilized for mirror materials. Front-surface mirrors and aluminum can be utilized for the same purpose also. Aluminum and silver with reflectivity of about 85–90 and 95 %, respectively, have been found as effective surfaces for reflection needed in solar applications; however, silver has not been found enough for front plane mirrors. Glass (glazing) is durable in nature and can be used as a transmitting material. Polymethyl methacrylate is commonly used due to having weather-resistant characteristics. Acrylic is found suitable for Fresnel lenses due to the ease related to molding or extrusion.

9.20. Receiver Covers and Surface Coatings

Glass and clear plastic sheets are commonly utilized in receivers as glazing substances. Low iron content in glass reduces absorption. The surface is etched to decrease losses due to reflection. Plastic is not used as a covering material because of a lower limit of service temperature and degradation due to ultraviolet rays. Absorber in the form of an evacuated tube is also in use because it can avoid heat losses due to convection. Black chrome has been found as the most suitable selective coating as it is capable of providing higher absorptivity and stability even at higher temperatures. Metal-oxide coatings such as black CuO, black Ni, and selective paints like PbS granules in a silicon binder are also used.

9.21. Working Fluids

For proper flow of heat, the working fluid must show stability even at high temperatures. It also should be non-corrosive and cost-effective. Pressurized H_2O, liquid metals, Mobile therm 603, and therminol 55 are the commonly used working fluids.

9.22. Insulation

Insulation is provided for minimizing the thermal loss from the un-irradiated parts of the receiver. It is applied to the tubes through which working fluid flows during operation. Insulation must be economical and strong to withstand even in high-temperature variations. Fiberglass with or without binder, mineral fiber blankets, and urethane foams are the generally used insulations.

CONCLUSION

Researchers around the globe are trying to increase the utilization of solar energy by various methods for all possible applications. Solar pond is one such method to store the solar energy and then further utilize it for solar thermal applications. This chapter covers the details about the solar pond, solar refrigeration, and solar concentrators. A lot of research is available on solar concentrators, but very few are available on solar refrigeration and solar pond. These are topics more to be explored and require more research.

NOMENCLATURE

A	Pond surface area, m^2	W	Pond width
C_a	Heat capacity (humid specific), J/kg °C	d	Declination angle, degree
C_p	Heat capacity (specific), J/kg °C	l	Latitude of location, degree
h	Hour angle, degree	A_a	Aperture area
h_s	Hour angle at sunrise and sunset, degree	A_r	Receiver area
h_c	Thermal energy transfer coefficient (convection) among UCZ and air, W/m^2°C	h_{absorb}	Thermal energy exchange coefficient (convection), W/m^2 K
h_e	Thermal energy transfer coefficient (evaporation), W/m^2 °C	$h_{rabsgls}$	Thermal energy exchange coefficient (radiation), W/m^2 K
h_x	Part of solar insolation that available to depth x in m of pond	h_{glass}	Thermal energy exchange coefficient (convection), W/m^2 K
h_1	Thermal energy transfer coefficient (convection) among UCZ and NCZ, W/m^2°C	m_{asc}	Mass passing of air in solar chimney in unit time, kg/s
h_2	Thermal energy transfer coefficient (convection) among NCZ and LCZ, W/m^2°C	S_{absorb}	Solar insolation gains by the absorber, W/m^2
h_3	Thermal energy transfer coefficient (convection) among LCZ with pond, W/m^2°C	S_{glass}	Solar insolation gains by a glass cover, W/m^2
h_4	Thermal energy transfer coefficient (convection) at the groundwater level, W/m^2 °C	q_u	Useful heat energy rate/unit aperture surface area
I	Solar insolation intensity striking the plane of UCZ, W/m^2	T_{ascout}	Temperature of air at solar chimney outlet, K
I_0	Solar insolation intensity entering the pond plane, W/m^2	V	Cubic capacity of the representative room, m^3

k_g	Ground conductivity (thermal), W/m °C	I_b	Incident beam radiation
k_w	Water conductivity (thermal), W/m °C	T_a	Ambient temperature, K
L	Pond dimension (length), m	A_{glass}	Glazing cover area, m²
m	Layers quantity (number)	A_{absorb}	Absorber plane area, m²
N	Counting number of day of a year	C_{asc}	Heat capacity (solar chimney air), J/kg K
n_i	Refractive index (air)	C	Concentration ratio
n_r	Refractive index (water)	F'	Collector efficiency factor
P_{vap}	Partial pressure vapour (water) at atmospheric temperature, mm of Hg	U_w	Overall thermal energy exchange coefficient, W/m² K
P_{atm}	Atmospheric pressure, mm of Hg	T_{absorb}	Absorber plane temperature, K
P_{UCZ}	Vapour pressure (water) at a temperature of UCZ, mm of Hg	T_i	Air temperature at interior side of the building, K
Q_{cond}	Thermal energy transfer rate (conduction), W	T_{glass}	Glass plane temperature, K
Q_{conv}	Thermal energy transfer rate (convection), W	T_{room}	Air temperature (room), K
Q_{evap}	Thermal energy loss rate (evaporation) from the plane of a pond, W	T_{ascm}	Air average temperature in a solar chimney, K
Q_{rad}	Thermal energy loss rate (radiation) from the surface of a pond, W	T_{ascin}	Temperature of air at solar chimney inlet, K
Q_g	Ground thermal energy loss rate from LCZ, W	F_R	Mass flow–rate factor
Q_{load}	Thermal energy extraction rate from LCZ, W	T_{amb}	Outdoor temperature of air, K
Q_{losses}	Thermal energy loss rate, W	N	Air change per hour
Q_s	Solar radiation absorption rate in a layer, W	W	Image size
R	Reflection coefficient	q	Thermal energy transfer, W

r_t	Ratio of hourly total insolation to total daily insolation	K	Conductivity (thermal), W/m-K
T_a	Atmospheric temperature, °C	*Greek letters*	
T_g	Ground temperature, °C	θ_i	Incident angle, degree
T_{sky}	Sky temperature, K	θ_r	Refraction angle, degree
T	Node temperature, °C	ρ	Mass density, kg/m³
t	Layer thickness, m	ε_w	Water emissivity
U_g	Overall thermal energy exchange coefficient of the ground, W/m² °C	τ	Present time, s
		$d\tau$	Time increment, s
U_t	Overall thermal energy exchange coefficient of the pond, W/m² °C	λ	Evaporation latent heat, J/kg
		ϕ_h	Air humidity
V_{air}	Velocity wind (m/s)	$\varnothing$	Angle formed by optical axis with normal
r	Distance from point of incidence to focus	η	Instantaneous thermal efficiency
S	Absorbed beam radiation rate per unit surface area of un-shaded receiver	*Subscripts*	
T_a	Ambient Temperature	LCZ	Lower convective zone
Tp	Absorber temperature	NCZ	Non-convective zone
T_f	Working fluid temperature	UCZ	Upper convective zone
U_L	Overall heat transfer coefficient		

CONSENT FOR PUBLICATION

Not applicable.

CONFLICT OF INTEREST

The authors declare no conflict of interest, financial or otherwise.

ACKNOWLEDGEMENTS

Declared none.

REFERENCES

[1] J.A. Duffie, and W.A. Beckman, *Solar Engineering of Thermal Processes,* 4th ed John Wiley Sons, Inc., 2013.
http://dx.doi.org/10.1002/9781118671603

[2] S.A. Kalogirou, "Environmental Characteristics", In: S.A. Kalogirou, Ed., *Solar Energy Engineering.,* 2nd ed Academic Press: Boston, 2014, pp. 51-123.
http://dx.doi.org/10.1016/B978-0-12-397270-5.00002-9

[3] Y. Wang, and A. Akbarzadeh, "A study on the transient behaviour of solar ponds", *Energy,* vol. 7, no. 12, pp. 1005-1017, 1982.
http://dx.doi.org/10.1016/0360-5442(82)90084-6

[4] A.H. Sayer, H. Al-Hussaini, and A.N. Campbell, "New theoretical modelling of heat transfer in solar ponds", *Sol. Energy,* vol. 125, pp. 207-218, 2016.
http://dx.doi.org/10.1016/j.solener.2015.12.015

[5] M. Khalilian, "Experimental and numerical investigations of the thermal behavior of small solar ponds with wall shading effect", *Sol. Energy,* vol. 159, pp. 55-65, 2018.
http://dx.doi.org/10.1016/j.solener.2017.10.065

[6] Y.F. Wang, and A. Akbarzadeh, "A parametric study on solar ponds", *Sol. Energy,* vol. 30, no. 6, pp. 555-562, 1983.
http://dx.doi.org/10.1016/0038-092X(83)90067-1

[7] H.C. Bryant, and I. Colbeck, "A solar pond for London?", *Sol. Energy,* vol. 19, no. 3, pp. 321-322, 1977.
http://dx.doi.org/10.1016/0038-092X(77)90079-2

[8] H.M. Ali, "Mathematical modelling of salt gradient solar pond performance", *Int. J. Energy Res.,* vol. 10, no. 4, pp. 377-384, 1986.
http://dx.doi.org/10.1002/er.4440100408

[9] V.V.N. Kishore, and V. Joshi, "A practical collector efficiency equation for nonconvecting solar ponds", *Sol. Energy,* vol. 33, no. 5, pp. 391-395, 1984.
http://dx.doi.org/10.1016/0038-092X(84)90190-7

[10] A. Kumar, K. Singh, S. Verma, and R. Das, "Inverse prediction and optimization analysis of a solar pond powering a thermoelectric generator", *Sol. Energy,* vol. 169, pp. 658-672, 2018.
http://dx.doi.org/10.1016/j.solener.2018.05.035

[11] M.S. Sodha, N.D. Kaushik, and S.K. Rao, "Thermal analysis of three zone solar pond", *Int. J. Energy Res.,* vol. 5, no. 4, pp. 321-340, 1981.
http://dx.doi.org/10.1002/er.4440050404

[12] M.S. Sodha, A.K. Sharma, S.P. Singh, N.K. Bansal, and A. Kumar, "Evaluation of an earth—air tunnel system for cooling/heating of a hospital complex", *Build. Environ.,* vol. 20, no. 2, pp. 115-122, 1985.
http://dx.doi.org/10.1016/0360-1323(85)90005-8

[13] N.K. Bansal, and M.S. Sodha, "An earth-air tunnel system for cooling buildings", *Tunn. Undergr. Space Technol.,* vol. 1, no. 2, pp. 177-182, 1986.
http://dx.doi.org/10.1016/0886-7798(86)90057-X

[14] K.S. Ong, "A mathematical model of a solar chimney", *Renew. Energy,* vol. 28, no. 7, pp. 1047-1060, 2003.
http://dx.doi.org/10.1016/S0960-1481(02)00057-5

[15] F. Al-Ajmi, D.L. Loveday, and V.I. Hanby, "The cooling potential of earth–air heat exchangers for domestic buildings in a desert climate", *Build. Environ.,* vol. 41, no. 3, pp. 235-244, 2006.
http://dx.doi.org/10.1016/j.buildenv.2005.01.027

[16] J. Mathur, N.K. Bansal, S. Mathur, M. Jain, and Anupma, "Experimental investigations on solar chimney for room ventilation", *Sol. Energy,* vol. 80, no. 8, pp. 927-935, 2006.
http://dx.doi.org/10.1016/j.solener.2005.08.008

[17] R. Bassiouny, and N.S.A. Koura, "An analytical and numerical study of solar chimney use for room natural ventilation", *Energy Build.,* vol. 40, no. 5, pp. 865-873, 2008.
http://dx.doi.org/10.1016/j.enbuild.2007.06.005

[18] R. Bassiouny, and N.S.A. Korah, "Effect of solar chimney inclination angle on space flow pattern and ventilation rate", *Energy Build.,* vol. 41, no. 2, pp. 190-196, 2009.
http://dx.doi.org/10.1016/j.enbuild.2008.08.009

[19] M. Maerefat, and A.P. Haghighi, "Passive cooling of buildings by using integrated earth to air heat exchanger and solar chimney", *Renew. Energy,* vol. 35, no. 10, pp. 2316-2324, 2010.
http://dx.doi.org/10.1016/j.renene.2010.03.003

[20] V. Bansal, R. Mishra, G.D. Agarwal, and J. Mathur, "Performance analysis of integrated earth–air-tunnel-evaporative cooling system in hot and dry climate", *Energy Build.,* vol. 47, pp. 525-532, 2012.
http://dx.doi.org/10.1016/j.enbuild.2011.12.024

[21] A.A. Saleem, M. Bady, S. Ookawara, and A.K. Abdel-Rahman, "Achieving standard natural ventilation rate of dwellings in a hot-arid climate using solar chimney", *Energy Build.,* vol. 133, pp. 360-370, 2016.
http://dx.doi.org/10.1016/j.enbuild.2016.10.001

[22] M.A. Hosien, and S.M. Selim, "Effects of the geometrical and operational parameters and alternative outer cover materials on the performance of solar chimney used for natural ventilation", *Energy Build.,* vol. 138, pp. 355-367, 2017.
http://dx.doi.org/10.1016/j.enbuild.2016.12.041

[23] A.K. Tiwari, and R. Sharma, "Techno – economic analysis of ventilation driven through solar metallic wall for a hostel building of National Institute of Technology Raipur India", *Energy Procedia,* vol. 141, pp. 39-44, 2017.
http://dx.doi.org/10.1016/j.egypro.2017.11.008

[24] S. Kashyap, V.K. Gaba, and A.K. Tiwari, "Enhancing passive cooling and natural ventilation for houses of various climatic zones of India", *IEEE 2nd International Conference on Power and Energy Applications,* 2019pp. 183-187

[25] J. Henkel, B. Chen, M. Liu, and G. Wang, "Analysis, design and testing of an earth contact cooling tube for fresh air conditioning", *Sol. Energy,* pp. 285-290, 2004.

[26] G.N. Tiwari, *Solar Energy-Fundamentals, Design, Modeling and Applications.* Narosa Publications, Centre of Energy Studies IIT: Delhi, 2002.

[27] S. Kashyap, P.K. Chandra, V.K. Gaba, and A.K. Tiwari, "Enviro-economic technical analysis of solar chimney integrated with soil air heat exchanger: Creating passive thermal comfort for hot subtropical regions", *AIP Conf. Proc.,* vol. 2273, p. 050042, 2020.

http://dx.doi.org/10.1063/5.0024381

[28] R.M. Lazzarin, "Solar cooling: PV or thermal? A thermodynamic and economical analysis", *Int. J. Refrig.,* pp. 38-47, 2019.

[29] G.N. Tiwari, and A. Tiwari, *Handbook of Solar Energy Theory, Analysis and Applications.* Springer Science & Business Media: Singapore, 2016.

[30] A. Rabl, "Optical and thermal properties of compound parabolic concentrators", *Sol. Energy,* vol. 18, no. 6, pp. 497-511, 1976.

http://dx.doi.org/10.1016/0038-092X(76)90069-4

[31] A. Rabl, and R. Winston, "Ideal concentrators for finite sources and restricted exit angles", *Applied Optics,* pp. 2880-2883, 1976.

CHAPTER 4

Thermal Modeling of Solar Stills

Desh Bandhu Singh[1],* and **G.N. Tiwari[2]**

[1]Department of Mechanical Engineering, Graphic Era Deemed to be University, Bell Road, Clement Town, Dehradun, Uttarakhand, 248002, India

[2]Bag Energy Research Society, Jawahar Nager (Margupur), Chilkhar-221701, Ballia UP, India

Abstract: The design, analysis and modeling of solar energy-based water purifiers, commonly known as a solar still, which is based on the greenhouse effect, is the requirement of time as there is a scarcity of freshwater throughout the globe. The technology of purifying dirty water using solar energy is a promising solution for simplifying contemporary water scarcity as this technology does not create any bad effect on the surroundings, unlike conventional water purification technology, which creates a lot of polluting elements and ultimately has become problematic for the environment. Most solar energy-based water purifiers are self-sustainable, and they can be installed in remote locations where sunlight and source of impure water are available in abundance. This solar energy-based technology of water purification should perform better in hilly locations as the intensity of light is higher than the intensity of light in fields. The current chapter deals with the thermal modeling of different types of passive and active solar stills, including solar stills loaded with water-based nanofluids, followed by their energy and exergy analyses.

Keywords: Solar still, Thermal modeling, Passive, Active, Nanofluid, Exergy analysis.

1. INTRODUCTION

This chapter discusses the thermal modeling of different types of solar energy-based water purifiers (SEBWP) in passive as well as active modes. Thermal modeling of SEBWP means writing the Eq. for different parts of SEBWP based on making equal input heat/energy to output energy/heat. Eqs obtained in this way are

*Corresponding author Desh Bandhu Singh: Department of Mechanical Engineering, Graphic Era Deemed to be University, Bell Road, Clement Town, Dehradun, Uttarakhand, 248002, India; E-mail: dbsiit76@ gmail.com

Manoj Kumar Gaur, Brian Norton & Gopal Tiwari (Eds.)

subsequently simplified to get a differential Eq. containing unknown parameter, which is to be expressed in terms of known parameters like solar intensity, heat transfer coefficients and some constants. The differential Eq. so obtained is solved under boundary conditions for getting the expression of an unknown parameter in terms of various known parameters like solar intensity, heat transfer coefficients and some known constants.

The first accepted research work on SEBWP was done in the sixteenth century by Arab alchemists [1]. Della Porta in 1589 made use of inverted earthen pots [2]. Further, earthen pots were exposed to the sun's radiation to heat up, due to which water got evaporated, condensed and finally got collected into containers [3]. Talbert *et al.* [4] presented a review of SEBWP on its historical background. Delyannis and Delyannis [5] studied the main SEBWP existing throughout the globe. Malik *et al.* [6] presented work reported on SEBWP in passive mode till 1982 and further, Tiwari [7] updated the work reported on SEBWP till 1992, which consisted of SEBWP operating on both passive and active modes. Delyannis [8] presented a compressive review on SEBWP. Delyannis [9] and Tiwari *et al.* [10] have investigated the various designs of SEBWP for making fresh water available. The first known work on SEBWP in active mode as per available literature was presented by Rai and Tiwari [11] and they concluded that the output (freshwater) obtained from SEBWP in active mode was 24% higher than SEBWP of the same basin area acting in passive mode. Since then, a lot of changes in SEBWP operating in passive as well as active modes have been reported. In this chapter, the development of thermal modeling for such types of SEBWP has been discussed.

2. SEBWP OF SINGLE SLOPE TYPE IN PASSIVE MODE

The schematic diagram of SEBWP of single slope type in passive mode is shown in Fig. (**1**). The work is based on the greenhouse effect. It consists of condensing cover, basin liner and water mass. So, the development of thermal modeling equations for SEBWP of single slope type in passive mode consists of writing equations for the outer surface of condensing cover, an inner surface of condensing cover, water mass and basin liner. The solar energy-based water purifier of single slope type in passive mode is oriented towards the south if the system is being studied for the place in the northern hemisphere and towards the north if the system being studied lies in the southern hemisphere for getting better annual solar energy. Assumptions [12] for writing equations based on balancing input energy to output energy are as follows:

(i)   The vapor leakage in SEBWP is neglected.

(ii) Solar distiller unit's water depth is constant. The change in distilled water yield is very small when the water depth changes thus, change in depth can be neglected.

(iii) The brackish water held in the basin does not develop layers.

(iv) The heat capacity of the bottom and side insulating material, along with condensing glass cover is neglected.

(v)   The condensation with film-type characteristics occurs at the inside plane of the condensing cover. Careful cleaning of the inner surface of the glass ensures film-wise condensation and providing a small angle to the condensing cover favors it. The component of gravity force along the condensing cover will allow the condensate to trickle down along the surface and finally be collected in a measuring jar.

The interaction of heat for the outer surface of the condensing cover has been shown in Fig. (**1a**). Heat reaches the surface on the outer side of the glass cover from the surface on the inner side of the glass cover. The temperature of the surface on the inner side of the glass cover is higher than the temperature of the outer surface of the condensing cover. Heat is lost by the outer surface of condensing cover to the surrounding through convection and radiation. Hence, the Eq. for the surface on outer side of glass cover based on balancing input heat to output heat can be written as [12]:

$$\frac{K_g}{L_g}(T_{gi} - T_{go})A_g = h_{1g}(T_{go} - T_a)A_g \tag{1}$$

Here, h_{1g} stands for total heat transfer coefficient (HTC) from the surface on the outer side of the glass cover to the surrounding. Fourier's law has been used for the heat transfer by conduction mode from the inner surface to an outer surface of condensing cover (expression on the left side of Eq. 1). Newton's law of cooling has been used to write heat transfer by convection from the surface on the outer side of a glass cover to the surrounding (expression on the right side of Eq. 1). The expression for h_{1g} can be written as:

$$h_{1g} = 5.7 + 3.8V \tag{2}$$

Where V represents blowing air velocity.

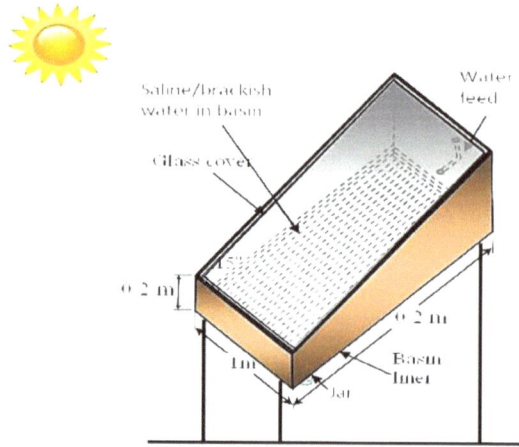

Fig. (1). SEBWP of single slope type operating in passive mode.

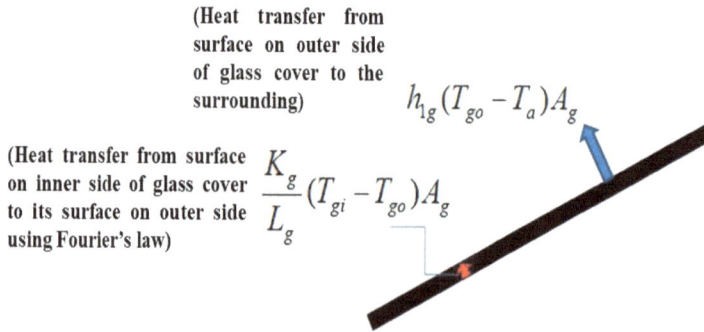

Fig. (1a). Heat interaction for the surface on outer side of glass cover.

The interaction of heat for the inner surface of the condensing cover has been shown in Fig. (**1b**). The inner condensing surface receives heat from the water surface through convective, radiative and evaporative heat transfer mechanisms and heat is lost from inner surface to outer surface of condensing cover through conduction.

Hence, the Eq. for inner surface of condensing cover based on balancing input heat to output heat can be written as:

$$\alpha'_g I(t)A_g + h_{1w}\left(T_w - T_{gi}\right)A_b = \frac{K_g}{L_g}\left(T_{gi} - T_{go}\right)A_g \tag{3}$$

Here, h_{1w} represents total HTC (corresponding to convection, radiation and evaporation) from the water surface to the inner surface of the glass cover. Hence, h_{1w} can be estimated as:

$$h_{1w} = h_{cw} + h_{rw} + h_{ewg} \tag{3a}$$

Here, h_{cw} is known as the convective HTC and can be estimated as:

$$h_{cw} = 0.884 \left[(T_w - T_{gi}) + \frac{(P_w - P_{gi})(T_w + 273)}{268.9 \times 10^3 - P_w} \right] \tag{3b}$$

$$P_w = exp \left[25.317 - \frac{5144}{T_w + 273} \right] \tag{3c}$$

$$P_{gi} = exp \left[25.317 - \frac{5144}{T_{gi} + 273} \right] \tag{3d}$$

h_{rw} is known as the radiative HTC and can be estimated as:

$$h_{rw} = (0.82 \times 5.67 \times 10^{-8}) \left[(T_w + 273)^2 + (T_{gi} + 273)^2 \right] \left[T_w + T_{gi} + 546 \right] \tag{3e}$$

h_{ewg} is known as evaporative HTC and can be estimated as:

$$h_{ewg} = 16.273 \times 10^{-3} h_{cw} \left[\frac{P_w - P_{gi}}{T_w - T_{gi}} \right] \tag{3f}$$

The interaction of heat for basin liner of SEBWP in passive mode has been shown in Fig. (**1c**). The basin liner receives heat from the sun and heat is lost by the basin liner to the surrounding as well as water mass. Hence, the Eq. for basin liner based on balancing input heat to output heat can be written as:

$$\alpha_b' I(t) A_b = h_{bw}(T_b - T_w) A_b + h_{ba}(T_b - T_a) A_b \tag{4}$$

Here, h_{bw} represents HTC from basin liner to water and h_{ba} represents HTC from basin liner to the surrounding. The value of h_{bw} is taken as 100 W/m²-K.

(Heat received from sun due to absorptivity of surface)

$\alpha'_g IA_g$

(Heat transfer from inner surface of condensing cover to its outer surface based on Fourier's Law) $\dfrac{K_g}{L_g}(T_{gi}-T_{go})A_g$

$h_{1wg}(T_w-T_{gi})A_b$

(Heat transfer from water surface to inner surface of condensing cover)

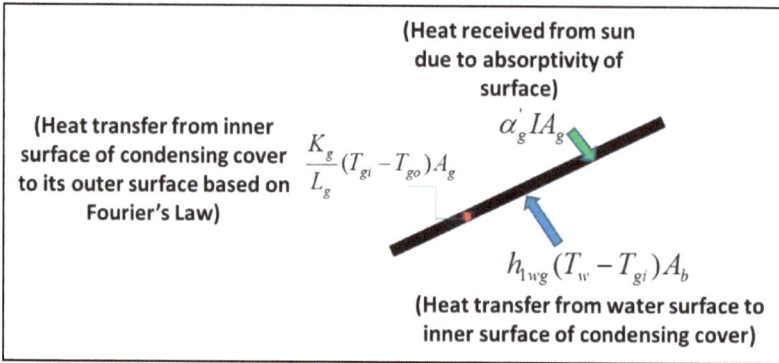

Fig. (1b). Heat interaction for glass cover at its inner surface.

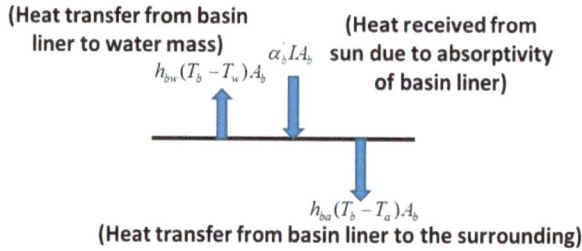

(Heat transfer from basin liner to water mass) $h_{bw}(T_b-T_w)A_b$

(Heat received from sun due to absorptivity of basin liner) $\alpha'_b IA_b$

$h_{ba}(T_b-T_a)A_b$
(Heat transfer from basin liner to the surrounding)

Fig. (1c). Heat interaction for basin liner.

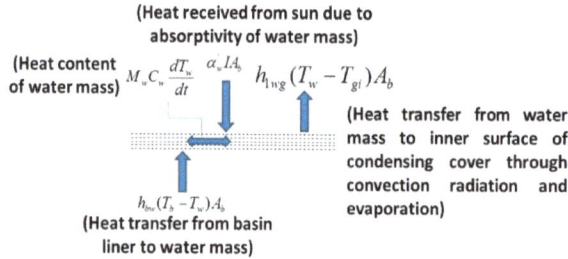

(Heat received from sun due to absorptivity of water mass)

(Heat content of water mass) $M_w C_w \dfrac{dT_w}{dt}$ $\alpha'_w IA_b$ $h_{1wg}(T_w-T_{gi})A_b$

(Heat transfer from water mass to inner surface of condensing cover through convection radiation and evaporation)

$h_{bw}(T_b-T_w)A_b$
(Heat transfer from basin liner to water mass)

Fig. (1d). Heat interaction for water mass.

The interaction of heat for water mass of SEBWP of single slope type in passive mode has been shown in Fig. (**1d**). Water mass receives heat from basin liner and sun and heat lost by water mass to the inner surface of the condensing cover. The difference between heat received and heat lost by water mass is contained by water and the temperature of water increases. Hence, the Eq. for water mass based on balancing input heat to output heat can be written as:

$$(M_w C_w)\frac{dT_w}{dt} = \alpha'_w I(t)A_b + h_w(T_b - T_w)A_b \tag{5}$$
$$- h_{1w}(T_w - T_{gi})A_b$$

After the rearrangement of Eq. 1, one can get:

$$\frac{K_g}{L_g}(T_{gi} - T_{go})A_g = U_c(T_{gi} - T_a \tag{6}$$

Where, $$U_c = \frac{\frac{K_g}{L_g}h_{1g}}{\frac{K_g}{L_g} + h_{1g}} \tag{7}$$

Using Eq. 3 and 6, one can get:

$$h_{1w}(T_w - T_{gi}) = U_{ta}(T_w - T_a) - h'_1\alpha'_g I(t) \tag{8}$$

Where, $$U_{ta} = \frac{h_{1w}U_c A_g}{h_{1w}A_b + U_c A_g} \tag{9}$$

and $$h'_1 = \frac{h_{1w}A_g}{h_{1w}A_b + U_c A_g} \tag{10}$$

Again, Eq. (4) can be rearranged as,

$$h_{bw}(T_b - T_w) = \alpha_{beff} I(t) - U_{bwa}(T_w - T_a) \tag{11}$$

Where, $$\alpha_{beff} = \frac{\alpha'_b h_{bw}}{h_{bw} + h_{ba}} \tag{12}$$

and $$U_{bwa} = \frac{h_{bw}}{h_{bw} + h_{ba}} \tag{13}$$

Using Eq. 5, 8 and 11, one can obtain the differential equation as:

$$\frac{dT_w}{dt} + aT_w = f(t) \tag{14}$$

Where, $$a = \frac{(UA)_{eff}}{M_w C_w} \tag{15}$$

$$f(t) = \frac{(\alpha\tau I)_{eff} + (UA)_{eff} T_a}{M_w C_w} \tag{16}$$

$$(UA)_{eff} = U_{bwa}A_b + U_{ta}A_b) \tag{17}$$

and $$(\alpha\tau I)_{eff} = \left(\alpha_{beff} + \alpha'_w + h'_1\alpha'_g\right)I(t)A_b \tag{18}$$

The differential Eq. 14 can be solved under the following assumptions:

(a) The time interval Δt is small ($0 < t < \Delta t$)
(b) Values of T_a and I(t) can be considered as an average value between 0 and t and hence f(t) will take its average value *i.e.* $\overline{f(t)}$. It means f(t) becomes constant.

Multiplying both sides of Eq. 14 by e^{at} and rearranging, one can get

$$\frac{d}{dt}(e^{at}T_w) = (\overline{f(t)}.e^{at} \tag{19}$$

Now, integrating Eq. 19, one can get

$$e^{at}T_w = \overline{f(t)}\frac{.e^{at}}{a} + c \tag{20}$$

Here, c is a constant and its value can be obtained by applying boundary conditions as $T_w = T_{w0}$ at $t = 0$ to Eq.20 as follows:

$$c = T_{w0} - \frac{\overline{f(t)}}{a} \tag{21}$$

Putting the value of c in Eq. 19 and simplifying it, the expression for water temperature can be expressed as:

$$T_w = \frac{f(t)}{a}(1 - e^{-at}) + T_{w0}e^{-at} \tag{22}$$

From Eq. 1, T_{go} can be expressed in terms of T_{gi} and putting this value of T_{go} in Eq. 2 and rearranging, one can estimate the value of T_{gi} as:

$$T_{gi} = \frac{\dot{\alpha}_g I_s(t)A_g + h_{1w}T_w A_b + U_c T_a A_g}{U_c A_g + h_{1w} A_b} \tag{23}$$

After estimating values of T_w from Eq. 22 and T_{gi} from Eq. 23, the value of hourly yield can be estimated as

$$\dot{m}_{hy} = \frac{h_{ewg}(T_w - T_{gi})}{L} \times 3600 \tag{24}$$

Where, h_{ewg} is HTC of evaporation and can be estimated using the Eq. 3f.

3. SEBWP OF DOUBLE SLOPE TYPE IN PASSIVE MODE

The schematic diagram of solar energy-based water purifier of double slope type in passive mode is shown in Fig. (2). Its work is based on the greenhouse effect. It consists of condensing cover oriented towards East and that towards West, basin liner and water mass. So, the development of thermal modeling equations for solar energy based water purifier of single slope type in passive mode consists of writing equations for outer surfaces of condensing cover oriented towards East as well as West, inner surfaces of condensing cover oriented towards East as well as West, water mass and basin liner. The solar energy-based water purifier of double slope type in passive mode is oriented toward East-West in the northern hemisphere for getting better annual solar energy. Assumptions [12] for writing equations on the basis of balancing input energy with output energy are as follows:

(a) The vapor leakage in SEBWP is neglected.
(b) Solar distiller unit's water depth is constant. The change in distilled water yield is very small when the water depth changes thus change in depth can be neglected.
(c) The brackish water held in the basin does not develop layers.

(d) The heat capacity of the bottom and side insulating material along with condensing glass cover is neglected.

(e) The condensation with film-type characteristics occurs at the inside plane of the condensing cover. Careful cleaning of the inner surface of the glass ensures film-wise condensation and providing a small angle to the condensing cover favors it. The component of gravity force along the condensing cover will allow the condensate to trickle down along the surface and finally be collected in a measuring jar.

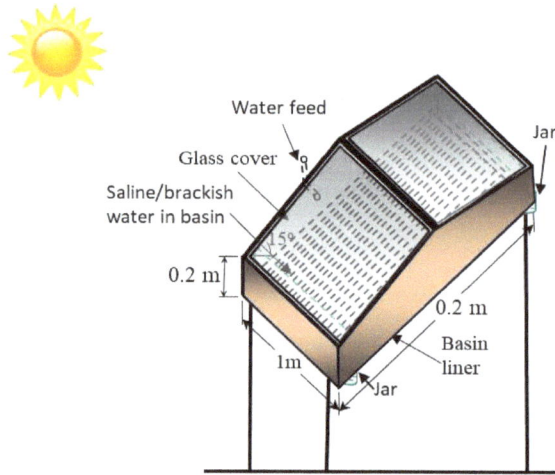

Fig. (2). Solar energy based water purifier of double slope type in passive mode.

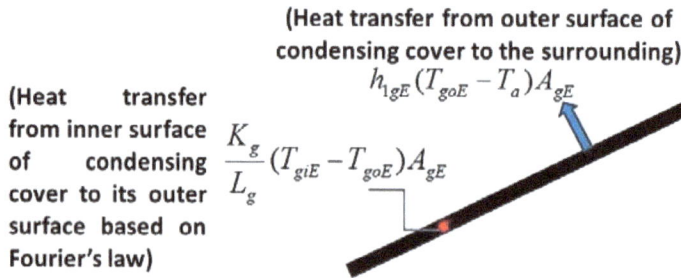

(Heat transfer from inner surface of condensing cover to its outer surface based on Fourier's law)

$$\frac{K_g}{L_g}(T_{giE} - T_{goE})A_{gE}$$

(Heat transfer from outer surface of condensing cover to the surrounding)

$$h_{1gE}(T_{goE} - T_a)A_{gE}$$

Fig. (2a). Heat interaction for outer surface of condensing cover oriented towards East.

The interaction of heat for the outer surface of the condensing cover having East orientation has been shown in Fig. (**2a**). Heat reaches the outer surface of the condensing cover having East orientation from the inner surface of the condensing cover having East orientation through conduction. The temperature of the inner surface of the condensing cover having East orientation is higher than the temperature of the outer surface of condensing cover having East orientation. Heat is lost by the outer surface of the condensing cover having East orientation to the surrounding through convection and radiation. Hence, the equation for outer surface of condensing cover having East orientation based on balancing input heat to output heat can be written as [12]:

$$\frac{K_g}{L_g}\left(T_{giE} - T_{goE}\right)A_{gE} = h_{1gE}\left(T_{goE} - T_a\right)A_{gE} \tag{25}$$

Here, h_{1gE} is total HTC from the outer surface of condensing cover oriented towards East to the surrounding and its value can be estimated using Eq. 2. Fourier's law has been used for conductive heat transfer from the inner surface to an outer surface of condensing cover (expression on the left side of Eq. 25). Newton's law of cooling has been used to write heat transfer from an outer surface of condensing cover to the surrounding (expression on right side of Eq. 25).

The interaction of heat for the inner surface of the condensing cover having East orientation has been shown in Fig. (2b). The inner condensing surface oriented towards East receives heat from water surface through convective, radiative and evaporative heat transfer mechanisms and heat is lost from inner surface to outer surface of condensing cover having East orientation through conduction. Hence, the Eq. for inner surface of condensing cover having East orientation based on balancing input heat to output heat can be written as:

$$\alpha'_g I_{SE}(t)A_{gE} + h_{1wE}\left(T_w - T_{giE}\right)\frac{A_b}{2} - h_{EW}\left(T_{giE} - T_{giW}\right)A_{gE} = \frac{K_g}{L_g}\left(T_{giE} - T_{goE}\right)A_{gE} \tag{26}$$

Here, h_{1wE} is the total HTC (corresponding to convection, radiation and evaporation) from the water surface to inner surface of the glass cover oriented towards the East. Hence, h_{1wE} can be estimated as:

$$h_{1wE} = h_{cwE} + h_{rwE} + h_{ewgE} \tag{26a}$$

Here, h_{cwE} is known as the convective HTC oriented towards the West and can be estimated using Eq. 3b. Summary, h_{rwE} and h_{ewgE} are known respectively as HTC due to radiation and evaporation. They can be estimated using Eq. 3e and 3f respectively.

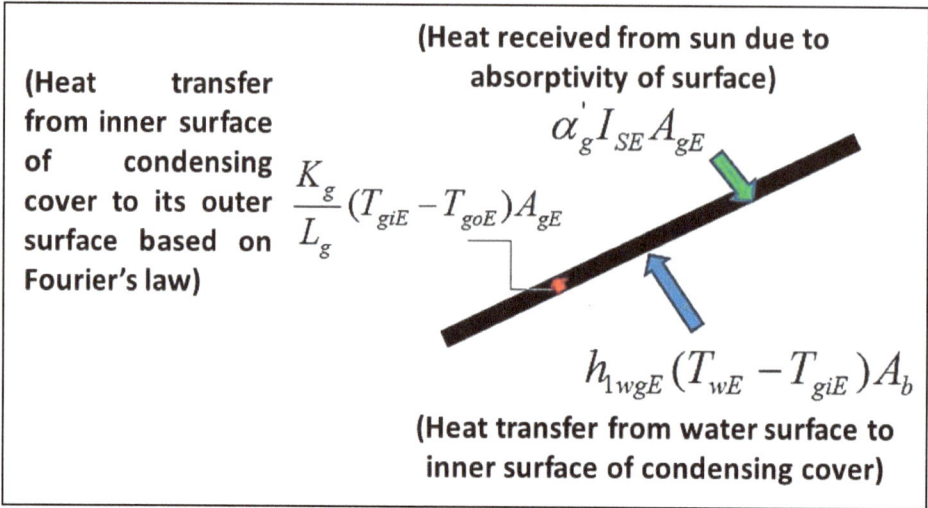

Fig. (2b). Heat interaction for inner surface of condensing cover having East orientation.

Similarly, the equation for outer surface of condensing cover having West orientation based on balancing input heat to output heat can be written as:

$$\frac{K_g}{L_g}(T_{giW} - T_{goW})A_{gW} = h_{1gW}(T_{goW} - T_a)A_{gW} \tag{27}$$

Here, h_{1gW} is total HTC from an outer surface of condensing cover having West orientation to the surrounding and its value can be estimated using Eq. 2. Fourier's law has been used for conductive heat transfer from the inner surface to an outer surface of condensing cover (expression on the left side of Eq. 27). Newton's law of cooling has been used to write heat transfer from an outer surface of condensing cover to the surrounding (expression on right side of Eq. 27).

The equation for inner surface of condensing cover having West orientation based on balancing input heat to output heat can be written as:

$$\alpha'_g I_{SW}(t)A_{gW} + h_{1wW}(T_w - T_{giW})\frac{A_b}{2} + h_{EW}(T_{giE} - T_{giW})A_{gE}$$
$$= \frac{K_g}{L_g}(T_{giW} - T_{goW})A_{gW}$$ (28)

Here, h_{1wW} is the total HTC (corresponding to convection, radiation and evaporation) from the water surface to inner surface of the glass cover oriented towards the West. Hence, h_{1wW} can be estimated as:

$$h_{1wW} = h_{cwW} + h_{rwW} + h_{ewgW}$$ (28a)

Here, h_{cwW} is known as the convective HTC oriented towards the West and can be estimated using Eq. 3b. Summary, h_{rwW} and h_{ewgW} are known respectively as HTC due to radiation and evaporation. They can be estimated using Eq. 3e and 3f respectively.

The interaction of heat for basin liner of SEBWP of double slope type in passive mode has been shown in Fig. (**2c**). The basin liner receives heat from the sun and heat is lost by the basin liner to the surrounding as well as water mass. Hence, the equation for basin liner based on balancing input heat to output heat can be written as:

$$\alpha'_b(I_{SE}(t) + I_{SW}(t))\frac{A_b}{2} = h_{bw}(T_b - T_w)A_b + h_{ba}(T_b - T_a)A_b$$ (29)

Here, h_{bw} is HTC from basin liner to water and h_{ba} is the HTC from basin liner to the surrounding. The value of h_{bw} is taken as 100 W/m²-K.

The interaction of heat for water mass of solar energy-based water purifier of double slope type in passive mode has been shown in Fig. (**2d**). Water mass receives heat from basin liner and sun and heat is lost by water mass to inner surfaces of condensing cover oriented towards East as well as West. The difference of heat received and heat lost by water mass is contained by water and the temperature of water increases. Hence, the equation for water mass based on balancing input heat to output heat can be written as:

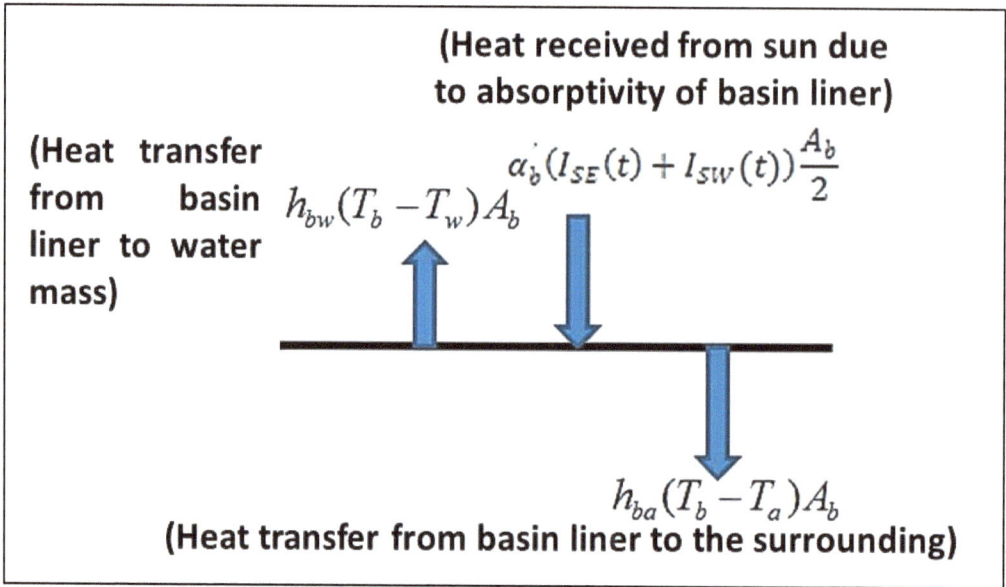

Fig. (2c). Heat interaction for basin liner.

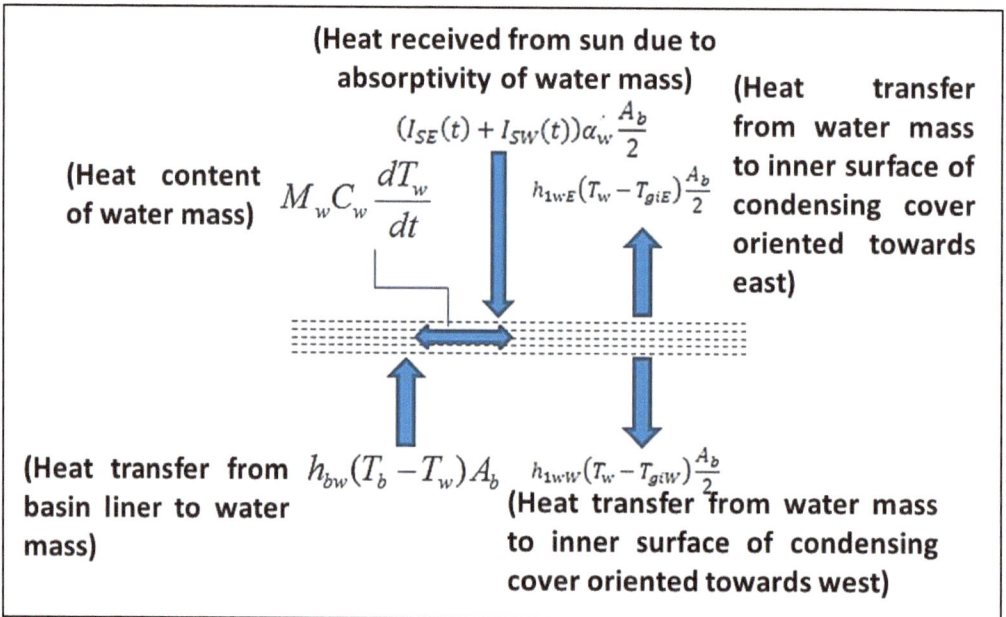

Fig. (2d). Heat interaction for water mass.

$$(M_w C_w)\frac{dT_w}{dt} = (I_{SE}(t) + I_{SW}(t))\alpha'_w \frac{A_b}{2} + h_{bw}(T_b - T_w)A_b - h_{1wE}(T_w - T_{giE})\frac{A_b}{2}$$
$$- h_{1wW}(T_w - T_{giW})\frac{A_b}{2} \tag{30}$$

Eq. 25 can be written as:

$$\frac{K_g}{L_g}(T_{giE} - T_{goE})A_{gE} = U_{cE}(T_{giE} - T_a)A_{gE} \tag{31}$$

Where,

$$U_{cE} = \frac{\frac{K_g}{L_g}h_{1gE}}{\frac{K_g}{L_g} + h_{1gE}} \tag{32}$$

From Eq. 26 and 31, one can obtain

$$\alpha'_g I_{SE}(t)A_{gE} + h_{1wE}(T_w - T_{giE})\frac{A_b}{2} - h_{EW}(T_{giE} - T_{giW})A_{gE} = U_{cE}(T_{giE} - T_a)A_{gE} \tag{33}$$

Similarly, using Eq. 27 and 28, one can get

$$\alpha'_g I_{SW}(t)A_{gw} + h_{1wW}(T_w - T_{giW})\frac{A_b}{2} + h_{EW}(T_{giE} - T_{giW})A_{gw}$$
$$= U_{cW}(T_{giW} - T_a)A_{gw} \tag{34}$$

Where,

$$U_{cW} = \frac{\frac{K_g}{L_g}h_{1gW}}{\frac{K_g}{L_g} + h_{1gW}} \tag{35}$$

Using Eq. 33 and 34, one can get

T_{giW}

$$= \frac{C_1 + C_1' + T_w \left(h_{1wE} \frac{A_b}{2} + h_{1ww} \frac{A_b}{2}\right) - T_{giE}(h_{1wE} \frac{A_b}{2} + U_{cE} A_{gE})}{h_{1ww} \frac{A_b}{2} + U_{cW} A_{gW}} \tag{36}$$

Eq. 36 can further be rearranged using T_w as:

$(T_w - T_{giW})$

$$= \frac{T_w \left(U_{cW} A_{gW} - h_{1wE} \frac{A_b}{2}\right) - C_1 - C_1' + T_{giE}(h_{1wE} \frac{A_b}{2} + U_{cE} A_{gE})}{h_{1ww} \frac{A_b}{2} + U_{cW} A_{gW}} \tag{37}$$

Eq. 36 can also be rearranged using T_{giE} as:

$(T_{giE} - T_{giW})$

$$= \frac{T_{giE} \left(h_{1wE} \frac{A_b}{2} + h_{1ww} \frac{A_b}{2} + U_{cE} A_{gE} + U_{cW} A_{gW}\right) - C_1 - C_1' - T_w(h_{1wE} \frac{A}{2} + h_{1ww} \frac{A}{2})}{h_{1ww} \frac{A_b}{2} + U_{cW} A_{gW}} \tag{38}$$

$$C_1 = \alpha_g' I_{SE}(t) A_{gE} + U_{cE} A_{gE} T_a \tag{39}$$

And,

$$C_1' = \alpha_g' I_{SW}(t) A_{gW} + U_{cW} A_{gW} T_a \tag{40}$$

Using Eq. 34 to 38, T_{giE} can be estimated as

$$T_{giE} = \frac{A + B T_w}{H} \tag{41}$$

$$A = C_1 U_2 + C_2 \tag{41a}$$

$$C_2 = \alpha_g I_{SW}(t) h_{EW} A_{gE} A_{gW} + U_{cW} h_{EW} A_{gE} A_{gW} T_a \tag{41b}$$

$$U_2 = h_{1ww} \frac{A_b}{2} + h_{EW} A_{gW} + U_{cW} A_{gW} \tag{41c}$$

$$B = \left(h_{1wE}U_2 + h_{1ww}h_{EW}A_{gW}\right)\frac{A_b}{2} \tag{41d}$$

$$H = U_1 U_2 - h_{EW}^2 A_{gE} A_{gW} \tag{41e}$$

Putting the value of T_{giE} from Eq. 41 in Eq. 36 and further simplifying different terms, one can estimate T_{giW} as:

$$T_{giW} = \frac{A' + B'T_w}{H} \tag{42}$$

Where,

$$A' = C_1'U_1 + C_2' \tag{42a}$$

$$U_1 = h_{1wE}\frac{A_b}{2} + h_{EW}A_{gE} + U_{cE}A_{gE} \tag{42b}$$

$$C_2' = \alpha_g' I_{SE}(t)h_{EW}A_{gE}A_{gW} + U_{cE}h_{EW}A_{gE}A_{gW}T_a \tag{42c}$$

$$B' = \left(h_{1ww}U_1 + h_{1wE}h_{EW}A_{gE}\right)\frac{A_b}{2} \tag{42d}$$

Eq. 29 can be expressed as:

$$2h_{bw}(T_b - T_w)\frac{A_b}{2} = \alpha_b'h_1 A_b(I_{SE}(t) + I_{SW}(t)) - U_b A_b(T_w - T_a) \tag{43}$$

Where,

$$h_1 = \frac{h_{bw}}{2(h_{bw} + h_{ba})} \tag{43a}$$

$$U_b = \frac{h_{bw}h_{ba}}{2(h_{bw} + h_{ba})} \tag{43b}$$

Using Eq. 30, 41, 42 and 43, the linear differential equation for T_w can be written as:

$$\frac{dT_w}{dt} + aT_w = f(t) \tag{44}$$

Where,

$$a = \frac{1}{M_w C_w} \left[U_b A_b + \frac{h_{1wE}(H-B)A_b}{2H} + \frac{h_{1ww}(H-B')A_b}{2H} \right] \tag{44a}$$

$$f(t) = \frac{1}{M_w C_w} \left[\left(\frac{\alpha'_w}{2} + h_1 \alpha'_b \right) A_b (I_{SE}(t) + I_{SW}(t)) + U_b A_b T_a + \left(\frac{h_{1wE}A + h_{1ww}A'}{H}\right)\frac{A_b}{2} \right] \tag{44b}$$

The differential Eq. 44 can be solved under the following assumptions:

(a) The time interval Δt is small ($0 < t < \Delta t$)
(b) Values of T_a and I(t) can be considered as an average value between 0 and t and hence f(t) will take its average value *i.e.* $\overline{f(t)}$. It means f(t) becomes constant.

Multiplying both sides of Eq. 44 by e^{at} and rearranging, one can get,

$$\frac{d}{dt}(e^{at}T_w) = (\underline{f(t)}.e^{at}) \tag{45}$$

Now, integrating Eq. 19, one can get:

$$e^{at}T_w = \underline{f(t)}\frac{e^{at}}{a} + c \tag{46}$$

Here, c is a constant and its value can be obtained by applying boundary conditions as $T_w = T_{w0}$ at $t = 0$ to Eq. 46 as follows:

$$c = T_{w0} - \frac{\underline{f(t)}}{a} \tag{47}$$

Putting the value of c in Eq. 46 and rearranging, one can obtain,

$$T_w = \frac{\underline{f(t)}}{a}(1 - e^{-at}) + T_{w0}e^{-at} \tag{48}$$

After estimating values of T_w from Eq. 48, T_{giE} from Eq. 41 and T_{giW} from Eq. 42, the value of hourly yield ($\dot{m}_{hy}$) for solar energy based water purifier of double slope type can be estimated as:

$$\dot{m}_{hy} = \frac{\left[h_{ewgE}\left(T_w - T_{giE}\right) + h_{ewgW}\left(T_w - T_{giW}\right)\right]}{L} \times 3600 \qquad (49)$$

Where values of h_{ewgE} and h_{ewgW} are HTCs due to evaporation and can be estimated using Eq. 3f.

4. SEBWP OF BASIN TYPE BY INCORPORATING DIFFERENT TYPES OF COLLECTORS

If SEBWP of basin type in passive mode is integrated with solar collectors for adding heat to the basin, the system is known as basin type active solar still or SEBWP of basin type in active mode. In this section, three types of solar collectors namely flat plate collectors, compound parabolic concentrators and evacuated tubular collectors have been incorporated with SEBWP.

4.1. SEBWP of Single Slope Type Integrated with Partially Covered PVT-FPCs

The schematic diagram of SEBWP of single slope type integrated with partially covered PVT flat plate collectors (PVT-FPC) has been shown in Fig. (**3**). N number of collectors in series has been added to SEBWP of single slope type because the main objective of adding PVT-FPC is to increase the temperature of water. Collectors are connected in series for getting low discharge at high temperature; whereas, they are connected in parallel to get a high discharge at low temperature [13-17].

The equations based on equating input heat to output for SEBWP of single slope type have already been written in section 2. In this section, equations based on equating input heat to output heat have been presented for N alike series-connected PVT-FPC. Assumptions are the same as in Tiwari *et al.* [15].

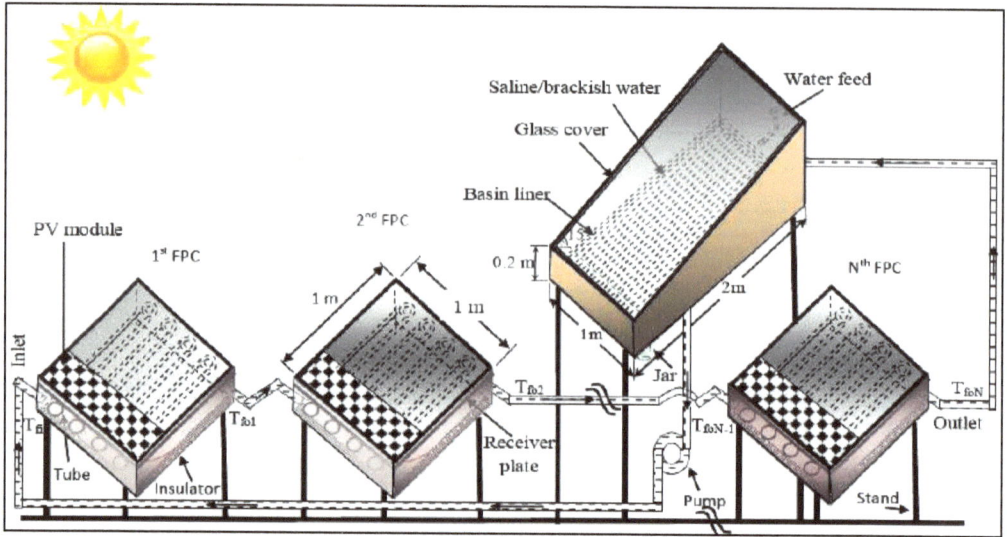

Fig. (3). SEBWP of single slope type integrated with N alike PVT-FPCs.

The mathematical modeling of PVT-FPC means to write equations for different components of PVT-FPC on the basis of equating input heat to output heat for the particular component followed by solving these equations to find the expression for unknown parameters in terms of some known parameters like solar intensity, the surrounding temperature, HTCs and some known constants. The equations based on equating input heat to output heat can be written as Tiwari *et al.* [15]:

For Solar Cells of Semitransparent PV Module

A semi-transparent PV module covers the lower portion of the absorber of FPC, and the upper portion is covered by a glass cover, as shown in Fig. (**3a**). The equation based on equating input heat to output heat for a solar cell of PV can be written as:

$$\alpha_c \tau_g \beta_c I(t) W dx = \left[U_{tc,a}(T_c - T_a) + U_{tc,p}(T_c - T_p) \right] W dx + \eta_c \tau_g \beta_c I(t) W dx \quad \textbf{(50)}$$

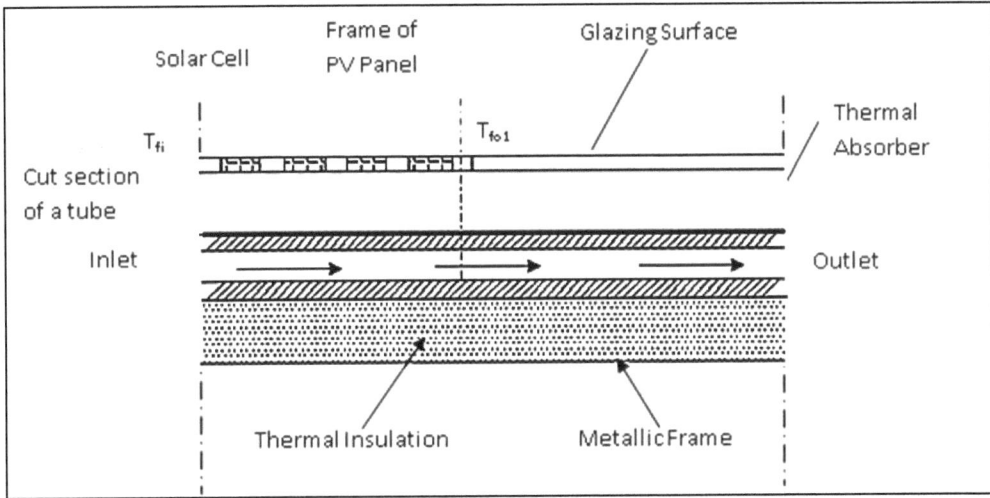

Fig. (3a). Cut section of 1st PVT-FPC.

It should be noted in Eq. 1 that the electrical efficiency of the semitransparent PV module is,

$$\eta_m = \eta_c \tau_g \beta_c \tag{51}$$

The expression for solar cell temperature from Eq. 1 can be found as:

$$T_c = \frac{(\alpha\tau)_{1,eff} I(t) + U_{tca} T_a + U_{tc,p} T_p}{U_{tca} + U_{tcp}} \tag{52}$$

The electrical efficiency of the cell can be written as:

$$\eta_c = \eta_0 \{1 - \beta_0 (T_c - T_0)\} = \frac{\eta_0 \left[1 - \beta_0 \left\{ T_a - T_0 + \dfrac{\alpha_c \tau_g \beta_c I(t) + U_{tc,p}(T_p - T_a)}{U_{tc,a} + U_{tc,p}} \right\}\right]}{1 + \dfrac{\beta_c \beta_0 \tau_g I(t)}{U_{tc,a} + U_{tc,p}}} \tag{53}$$

For Blackened Absorber Plate of FPC

The Eq. based on equating input heat to output heat for absorber plate of FPC can be written as:

$$\alpha_p(1 - \beta_c)\tau_g^2 I(t)Wdx + U_{tcp}(T_c - T_p)Wdx = F'h_{pf}(T_p - T_f)Wdx \tag{54}$$

From Eq. 5, the expression for plate temperature can be expressed as:

$$T_p = \frac{(\alpha\tau)_{2eff}I(t) + PF_1(\alpha\tau)_{1eff}I(t) + U_{L1}T_a + F'h_{pf}T_f}{U_{L1} + F'h_{pf}} \tag{55}$$

For Water Flowing Through An Absorber Pipe

When PV is integrated at the bottom of the collector, the flow pattern of water is shown in Fig. (1). The equation based on equating input heat to output heat for absorber pipe of FPC can be written as:

$$\dot{m}_f C_f \frac{dT_f}{dx} dx = F'h_{pf}(T_p - T_f)Wdx \tag{56}$$

Using Eq. 55, Eq. 56 can be expressed as,

$$\dot{m}_f C_f \frac{dT_f}{dx} dx = F'[PF_2(\alpha\tau)_{meff}I(t) - U_{Lm}(T_f - T_a)]Wdx \tag{57}$$

By rearranging and integrating both sides of Eq. 57 and using boundary conditions, namely, at $T_f\big|_{x=0} = T_{fi}$ and at $T_f\big|_{x=L} = T_{fo}$, one gets,

$$T_{fo} = \left[\frac{PF_2(\alpha\tau)_{m,eff}I(t)}{U_{L,m}} + T_a\right]\left[1 - \exp\left(-\frac{F'A_m U_{L,m}}{\dot{m}_f C_f}\right)\right] + T_{fi}\exp\left(-\frac{F'A_m U_{L,m}}{\dot{m}_f C_f}\right) \tag{58}$$

The Outlet Water Temperature at the End of the Collector

Following Duffie and Beckman [21] and Tiwari [22], an expression for the outlet water temperature at the end of conventional FPC will be,

$$T_{fo1} = \left[\frac{(\alpha\tau)_{cl,eff}I(t)}{U_{L,cl}} + T_a\right]\left[1 - \exp\left(-\frac{F'A_{cl}U_{L,cl}}{\dot{m}_f C_f}\right)\right] + T_{fi1}\exp\left(-\frac{F'A_{cl}U_{L,cl}}{\dot{m}_f C_f}\right) \tag{59}$$

Here, $T_{fi1} = T_{fo}$, the value of final outlet temperature from the 1st PVT-FPC can be estimated as:

$$
T_{fo1} = \left[\frac{(\alpha\tau)_{c1.eff} I(t)}{U_{L.c1}} + T_a\right]\left[1 - \exp\left(-\frac{F'A_{c1}U_{L.c1}}{\dot{m}_f C_f}\right)\right]
$$
$$
+ \left[\begin{array}{l}\left[\dfrac{PF_2(\alpha\tau)_{m.eff} I(t)}{U_{L.m}} + T_a\right]\left[1 - \exp\left(-\dfrac{F'A_m U_{L.m}}{\dot{m}_f C_f}\right)\right] \\[4mm] + T_{fi}\exp\left(-\dfrac{F'A_m U_{L.m}}{\dot{m}_f C_f}\right)\end{array}\right]\exp\left(-\frac{F'A_{c1}U_{L.c1}}{\dot{m}_f C_f}\right) \tag{60}
$$

Eq. 60 can be rearranged as:

$$
T_{fo1} = \frac{(AF_R(\alpha\tau))_1}{\dot{m}_f C_f}I(t) + \frac{(AF_R U_L)_1}{\dot{m}_f C_f}T_a + T_{fi}\left(1 - \frac{(AF_R U_L)_1}{\dot{m}_f C_f}\right) \tag{61}
$$

Similarly, the temperature of the fluid at end of the second PVT-FPC can be estimated as,

$$
T_{fo2} = \frac{(AF_R(\alpha\tau))_2}{\dot{m}_f C_f}I(t) + \frac{(AF_R U_L)_2}{\dot{m}_f C_f}T_a + T_{fi2}\left(1 - \frac{(AF_R U_L)_2}{\dot{m}_f C_f}\right) \tag{62}
$$

Here $T_{fi2} = T_{fo1}$; therefore, Eq. 62 can be rearranged as,

$$
T_{fo2} = \frac{(AF_R(\alpha\tau))_1}{\dot{m}_f C_f}I_{(t)}(1 + K_k) + \frac{(AF_R U_L)_1}{\dot{m}_f C_f}T_{(a)}(1 + K_k) \\ + T_{fi}K_k{}^2 \tag{63}
$$

Proceeding in a similar fashion, the temperature of the fluid at end of Nth PVT-FPC can be estimated as,

$$
T_{foN} = \frac{\{AF_R(\alpha\tau)\}_1}{\dot{m}_f C_f}\left(\frac{1 - K_K^N}{1 - K_K}\right)I(t) + \frac{\{AF_R U_L\}_1}{\dot{m}_f C_f}\left(\frac{1 - K_K^N}{1 - K_K}\right)T_a + T_{fi}K_K^N \tag{64}
$$

Expressions for $(AF_R U_L)_1$, $(AF_R(\alpha\tau))_1$ and K_k have been given in appendix-A.

The hourly heat gain from N alike PVT-FPCs in series connection can be estimated as:

$$Q_{u,N} = \dot{m}_f C_f (T_{foN} - T_{fi}) \tag{65}$$

Substituting the value of T_{foN} from Eq. 64 into Eq. 65 and rearranging, one can express $\dot{Q}_{u,N}$ as:

$$\dot{Q}_{u,N} = N(A_m + A_c)[(\alpha\tau)_{effN} I(t) - U_{LN}(T_{fi} - T_a)] \tag{66}$$

Equations based on balancing input heat to output heat for outer condensing surface, inner condensing surface and basin liner will remain the same as Eq. 1, 3 and 4. The equation for water mass will be different as heat is being provided by collectors to water mass.

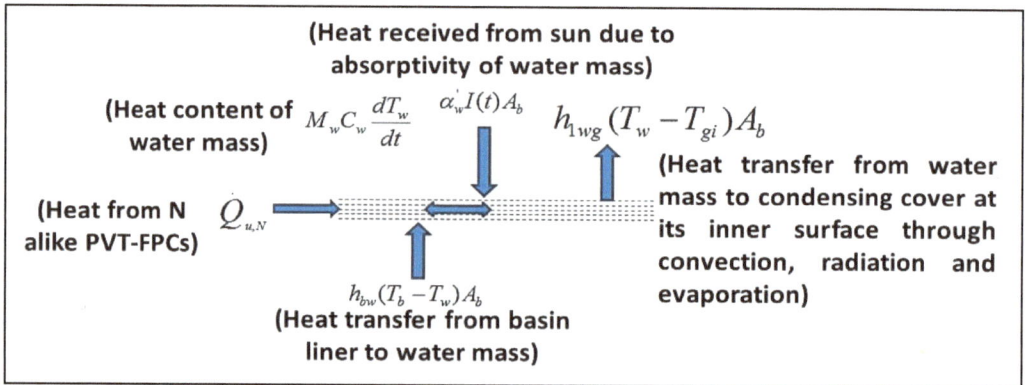

(Heat received from sun due to absorptivity of water mass)

(Heat content of water mass) $M_w C_w \dfrac{dT_w}{dt}$ $\alpha'_w I(t) A_b$ $h_{1wg}(T_w - T_{gi})A_b$

(Heat transfer from water mass to condensing cover at its inner surface through convection, radiation and evaporation)

(Heat from N alike PVT-FPCs) $Q_{u,N}$

$h_{bw}(T_b - T_w)A_b$
(Heat transfer from basin liner to water mass)

Fig. (3b). Heat interaction for water mass.

The interaction of heat for water mass of SEBWP in passive mode has been shown in Fig. (**3b**). Water mass receives heat from basin liner and sun. Heat is lost by water mass to the inner surface of the condensing cover. The difference of heat received and heat lost by water mass is contained by water and the temperature of water increases. Hence, the equation for water mass based on balancing input heat to output heat can be written as:

$$(M_w C_w)\frac{dT_w}{dt} = \alpha'_w I(t) A_b + h_w (T_b - T_w) A_b + \dot{Q}_{u,N} - h_{1w}(T_w - T_{gi})A_b \tag{67}$$

Following the methodology provided in section 2, one can obtain the temperature of water with the help of Eq. 66, 1, 3, 4 and 67 as,

$$T_w = \frac{f(t)}{a}(1 - e^{-at}) + T_{w0}e^{-at} \tag{68}$$

Where,

$$a = \frac{1}{M_w C_w}\left[\dot{m}_f C_f(1 - K_k^N) + U_s A_b\right]; \tag{69}$$

$$f(t) = \frac{1}{M_w C_w}\left[\alpha_{eff}' A_b I_s(t) + \frac{(1 - K_k^N)}{(1 - K_k)}(A\,F_R(\alpha\tau))_1 I_c(t)\right.$$
$$\left. + \left(\frac{(1 - K_k^N)}{(1 - K_k)}(A\,F_R U_L)_1 + U_s A_b\right)T_a\right] \tag{70}$$

Eq. 68 is similar in form to Eq. 22, however; expressions for a and $f(t)$ are different. From Eq. 1, T_{go} can be expressed in terms of T_{gi} and putting this value of T_{go} in Eq. 2 and rearranging, one can estimate the value of T_{gi} as:

$$T_{gi} = \frac{\dot{\alpha}_g I_s(t)A_g + h_{1w}T_w A_b + U_c T_a A_g}{U_c A_g + h_{1w}A_b} \tag{71}$$

Again, the expression for T_{gi} in Eq. 71 is similar in form to Eq. 24; however, the value of T_{gi} will be different as value of T_w is different due to different values of a and $f(t)$. After estimating values of T_w from Eq. 70 and T_{gi} from Eq. 71, the value of hourly yield can be estimated as,

$$\dot{m}_{hy} = \frac{h_{ewg}(T_w - T_{gi})}{L} \times 3600 \tag{72}$$

Where, h_{ewg} is HTC of evaporation and can be estimated using the Eq. 3f.

Eq. 68 can be discussed for the following cases:

Case (i): Putting the value of N = 0 in Eq. 68 to 70 and rearranging equations for a and $\underline{f}(t)$ will be obtained for SEBWP in passive mode and the expression of T_w for SEBWP in passive mode is obtained.

Case (ii): Putting $A_m = 0$ in Eq. 68, the expression of T_w for SEBWP integrated with N alike FPCs is obtained.

4.2. SEBWP OF SINGLE SLOPE TYPE INTEGRATED WITH N ALIKE PARTIALLY COVERED PVT COMPOUND PARABOLIC CONCENTRATORS (PVT-CPC)

The schematic diagram of SEBWP of single slope type integrated with PVT-CPCs has been presented in Fig. (**4**). N number of PVT-CPCs in series has been added to SEBWP of single slope type because the main objective of adding PVT-CPC is to increase the temperature of water. PVT integrated concentrators are connected in series for getting low discharge at high temperature; whereas, they are connected in parallel to get a high discharge at low temperature [18-20].

The equations based on equating input heat to output heat for SEBWP of single slope type in passive mode have already been written in section 2. In this section, equations based on equating input heat to output heat have been presented for N alike series-connected PVT-CPCs. The mathematical modeling of PVT-CPC means to write equations for different components of PVT-CPC on the basis of equating input heat to output heat for the particular component followed by solving these equations to find the expression for unknown parameters in terms of some known parameters like solar intensity, the surrounding temperature, HTCs and some known constants. Assumptions for writing equations based on balancing input energy and output energy for N alike PVT-CPCs are as follows:

(a) The system is considered to be in a quasi-steady state.
(b) Heat loss in PV corresponding to internal losses is negligible.
(c) Amount of heat required for raising the unit temperature for material of glass cover, insulation, absorber and solar cell are negligibly small. Also, the temperature gradient across the thickness of PV module, insulation and glass materials does not exist.

The equations based on equating input heat to output heat can be written as:

For the solar cell of a semitransparent PV module Fig. (**4**).

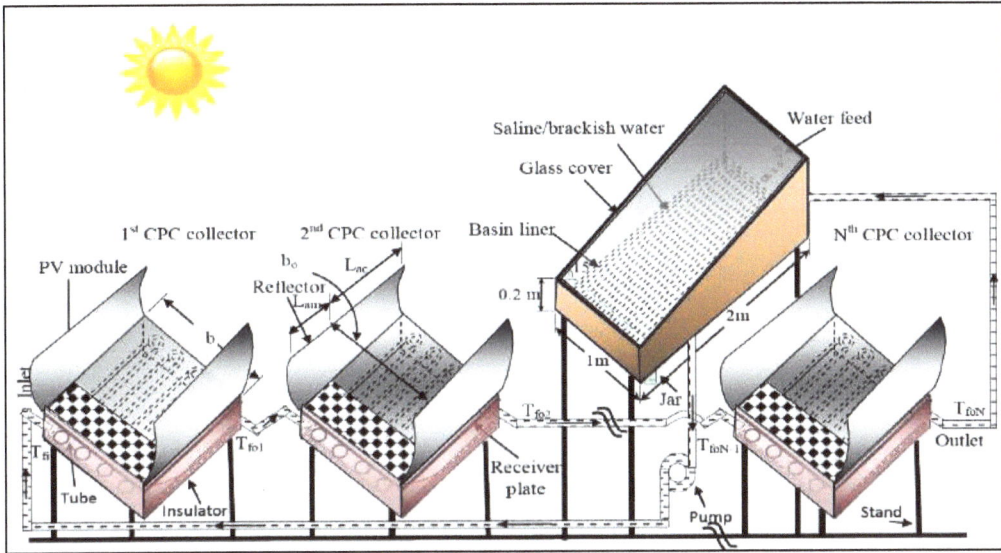

Fig. (4). Schematic diagram of SEBWP of single slope type by incorporating N alike PVT-CPCs.

$$\rho\alpha_c\tau_g\beta_c I_b A_{am} = \left[U_{tc,a}(T_c - T_a) + U_{tc,p}(T_c - T_p)\right]A_{rm} + \rho\eta_m I_b A_{am} \qquad (73)$$

For absorber plate (Fig. **4b**)

$$\begin{aligned}
\rho\alpha_p\tau_g^2(1 - \beta_c)I_b A_{am} + U_{t,cp}(T_c - T_p)A_{rm} \\
= F'h_{pf}(T_p - T_f)A_{rm} + U_{t,pa}(T_p - T_f)A_{rm}
\end{aligned} \qquad (74)$$

Eliminating T_c from Eq. 73 and 74 and rearranging, one can find expression for T_p as,

$$T_p = \frac{[(\alpha\tau)_{2,eff} + PF_1(\alpha\tau)_{1,eff}]I_b + U_{L2}T_a + h_{pf}T_f}{(U_{L2} + h_{pf})} \qquad (75)$$

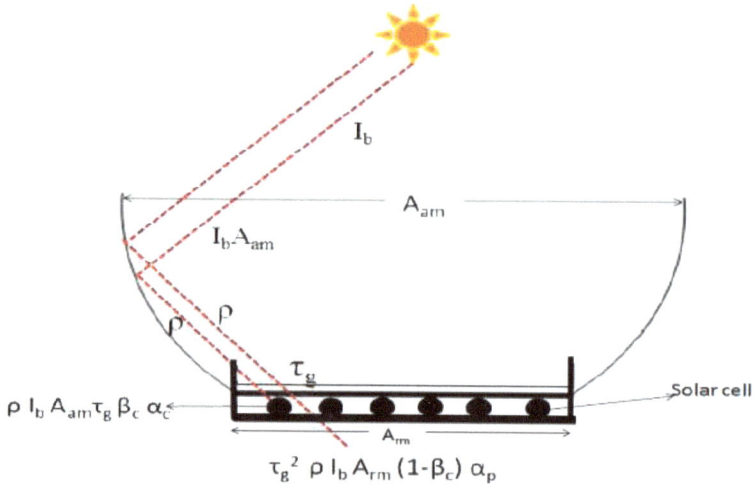

Fig. (4a). XX' view (cut section) of 1st PVT-CPC along width (heat interaction).

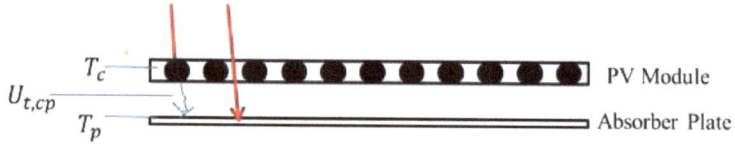

Fig. (4b). Heat interaction for absorber plate.

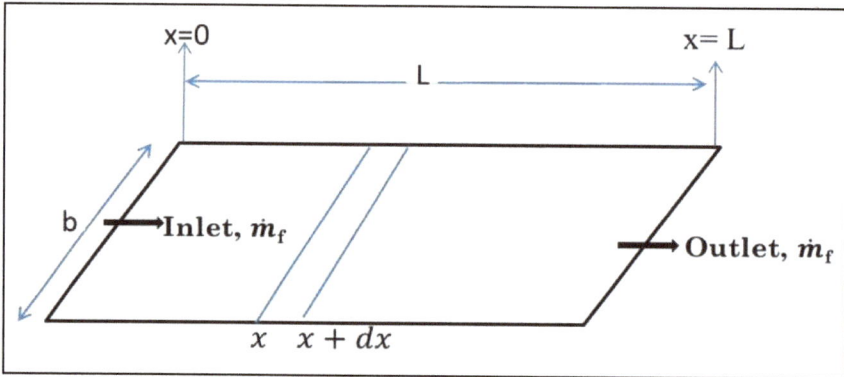

Fig. (4c). Fluid flow through tubes below absorber plate.

For fluid (water) flowing through tubes made of copper (Fig. **4c**),

$$\dot{m}_f c_f \frac{dT_f}{dx} dx = F' h_{pf}(T_p - T_f) b\, dx \tag{76}$$

From Eq. 75 and 76, one can obtain:

$$\dot{m}_f c_f \frac{dT_f}{dx} dx = bF'[I_b PF_2(\alpha\tau)_{m,eff} - U_{l,m}(T_f - T_a)] dx \tag{77}$$

Where, F' $(0 < F' < 1)$ stands for collector efficiency factor.

One can solve Eq. 5 using boundary condition *i.e.* $(T_f|_{x=0} = T_{fi})$ to obtain expression of T_f as:

$$T_f = \left[\frac{PF_2(\alpha\tau)_{m,eff} I_b}{U_{L,m}} + T_a\right]\left[1 - exp\left\{\frac{-bF'U_{L,m}x}{\dot{m}_f c_f}\right\}\right] \\ + T_{fi}\, exp\left[\frac{-bF'U_{L,m}x}{\dot{m}_f c_f}\right] \tag{78}$$

The expression of the temperature of fluid at the outlet where PVT ends can be evaluated as:

$$T_{fom1} = T_f|_{x=L_{rm}} \\ = \left[\frac{PF_2(\alpha\tau)_{m,eff} I_b}{U_{L,m}} + T_a\right]\left[1 - exp\left\{\frac{-F'U_{L,m}A_{Rm}}{\dot{m}_f c_f}\right\}\right] \\ + T_{fi}\, exp\left[\frac{-F'U_{L,m}A_{Rm}}{\dot{m}_f c_f}\right] \tag{79a}$$

After estimating T_{fom1} using Eq. 7; one can obtain the expression for the temperature of water at the outlet of 1st concentrator as:

$$T_{fo1} = \frac{I_b\,(A\,F_R(\alpha\tau))_1}{\dot{m}_f C_f}\frac{(1-K_k{}^N)}{(1-K_k)} + \frac{T_a(A\,F_R U_L)_1}{\dot{m}_f C_f}\frac{(1-K_k{}^N)}{(1-K_k)} + T_{fi}\left(1 - \frac{(A\,F_R U_L)_1}{\dot{m}_f C_f}\right) \tag{79b}$$

Collectors are connected in series, so the outlet from the first collector will be an inlet to the second collector, the outlet of the second collector will be an inlet to the third collector and so on. Applying this condition, the temperature at the end of Nth collector can be estimated as:

$$T_{foN} = \frac{I_b(A\,F_R(\alpha\tau))_1}{\dot{m}_f C_f} \frac{\left(1 - K_k{}^N\right)}{(1 - K_k)} + \frac{T_a(A\,F_R U_L)_1}{\dot{m}_f C_f} \frac{\left(1 - K_k{}^N\right)}{(1 - K_k)} + T_{fi} K_k^N \qquad (80)$$

Heat gain from N alike PVT-CPCs can be written as:

$$\dot{Q}_{u,N} = \dot{m}_f c_f \left(T_{foN} - T_{fi}\right) \qquad (81)$$

Putting the value of T_{foN} in Eq. 81 and rearranging, one can get:

$$\dot{Q}_{uN} = \frac{\left(1 - K_k{}^N\right)}{(1 - K_k)} (A\,F_R(\alpha\tau))_1 I_b + \frac{\left(1 - K_k{}^N\right)}{(1 - K_k)} (A\,F_R U_L)_1 (T_{fi} \\ - T_a) \qquad (82)$$

All known terms of Eq. 73 to 82 are given in Appendix-A.

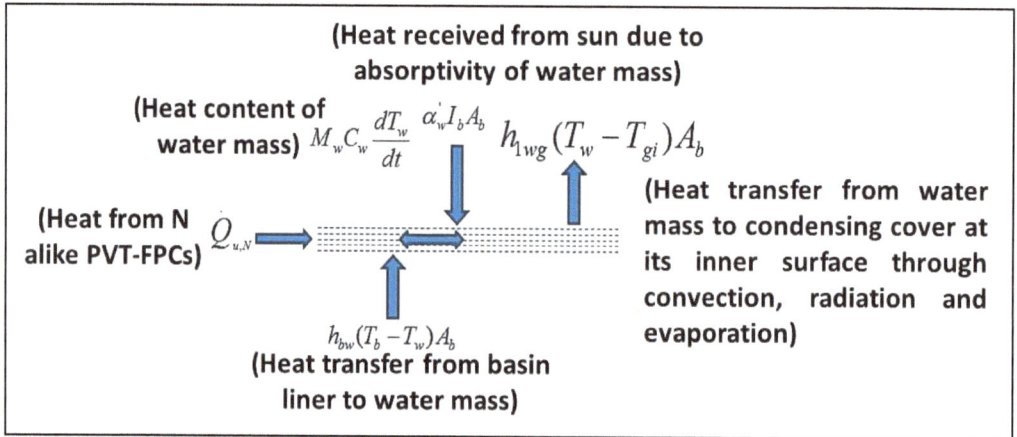

Fig. (4d). Heat interaction for water mass.

Equations based on balancing input heat to output heat for outer condensing surface, inner condensing surface and basin liner will remain the same as Eq. 1, 3 and 4. The equation for water mass will be different as heat is being provided by collectors to water mass.

The interaction of heat for water mass of SEBWP in passive mode has been shown in Fig. (**4d**). Water mass receives heat from basin liner and sun. Heat lost by water mass to the inner surface of the condensing cover. The difference of heat received and heat lost by water mass is contained by water and the temperature of water increases. Hence, the equation for water mass based on balancing input heat to output heat can be written as:

$$(M_w C_w)\frac{dT_w}{dt} = \alpha'_w I_b A_b + h_w(T_b - T_w)A_b + \dot{Q}_{u,N} - h_{1w}(T_w - T_{gi})A_b \quad (83)$$

Following the methodology provided in section 2, one can obtain the temperature of water with the help of Eq. 82, 1, 3, 4 and 83 as,

$$T_w = \frac{f(t)}{a}(1 - e^{-at}) + T_{w0}e^{-at} \quad (84)$$

Where,

$$a = \frac{1}{M_w C_w}\left[\dot{m}_f C_f(1 - K_k^N) + U_s A_b\right] \quad (85)$$

$$f(t) = \frac{1}{M_w C_w}\left[\alpha_{eff}' A_b I_s(t) + \frac{(1 - K_k^N)}{(1 - K_k)}(A\, F_R(\alpha\tau))_1 I_b \right.$$
$$\left. + \left(\frac{(1 - K_k^N)}{(1 - K_k)}(A\, F_R U_L)_1 + U_s A_b\right)T_a\right] \quad (86)$$

Eq. 84 is similar in form as Eq. 22 and 68, however; expressions for a and $\underline{f}(t)$ are different. From Eq. (1), T_{go} can be expressed in terms of T_{gi} and putting this value of T_{go} in Eq. 2 and rearranging, one can estimate the value of T_{gi} as:

$$T_{gi} = \frac{\dot{\alpha}_g I_s(t)A_g + h_{1w}T_w A_b + U_c T_a A_g}{U_c A_g + h_{1w}A_b} \quad (87)$$

Again, the expression for T_{gi} in Eq. 87 is similar in form to Eq. 24 and 71; however, the value of T_{gi} will be different as value of T_w is different due to different values of a and $\underline{f}(t)$. After estimating values of T_w from Eq. 84 and T_{gi} from Eq. 87, the value of hourly yield can be estimated as:

$$\dot{m}_{hy} = \frac{h_{ewg}(T_w - T_{gi})}{L} \times 3600 \tag{88}$$

Where, h_{ewg} is the HTC of evaporation and can be estimated using Eq. 3f.

Eq. 84 can be discussed for the following cases:

Case (i): Putting the value of N = 0 in Eq. 84 to 86 and rearranging, Eq. for a and $\underline{f}(t)$ will be obtained for SEBWP in passive mode and the expression of T_w for SEBWP in passive mode is obtained.

Case (ii): Putting $A_m = 0$ in Eq. 84, the expression of T_w for SEBWP integrated with N alike CPCs is obtained.

Case (iii): Putting $A_m = 0$, $I_b(t) = I_c(t)$, $A_{rm} = A_{am}$ in Eq. 84, the expression of T_w for SEBWP integrated with N alike FPCs is obtained.

Case (iv): Putting $I_b(t) = I_c(t)$, $A_{rm} = A_{am}$ in Eq. 84, the expression of T_w for SEBWP integrated with N alike PVT-FPCs is obtained.

4.3. SEBWP OF SINGLE SLOPE TYPE INTEGRATED WITH N ALIKE EVACUATED TUBULAR COLLECTORS (ETCS)

The schematic diagram of SEBWP of single slope type integrated with N alike ETCs has been shown in Fig. (**5**). N number of ETCs in series has been added to SEBWP of single slope type because the main objective of adding ETCs is to increase the temperature of water. ETCs are connected in series for getting low discharge at high temperature; whereas, they are connected in parallel to get a high discharge at low temperature.

The equations based on equating input heat to output heat for SEBWP of single slope type in passive mode have already been written in section 2. In this section, equations based on equating input heat to output heat have been presented for N alike series-connected ETCs. The mathematical modeling of N alike ETCs means to write equations for different components of ETC on the basis of equating input heat to output heat for the particular component followed by solving these equations to find the expression for unknown parameters in terms of some known parameters like solar intensity, the surrounding temperature, HTCs and some known constants.

The equations based on equating input heat to output heat for different components of N alike ETCs can be written as follows [20, 21]:

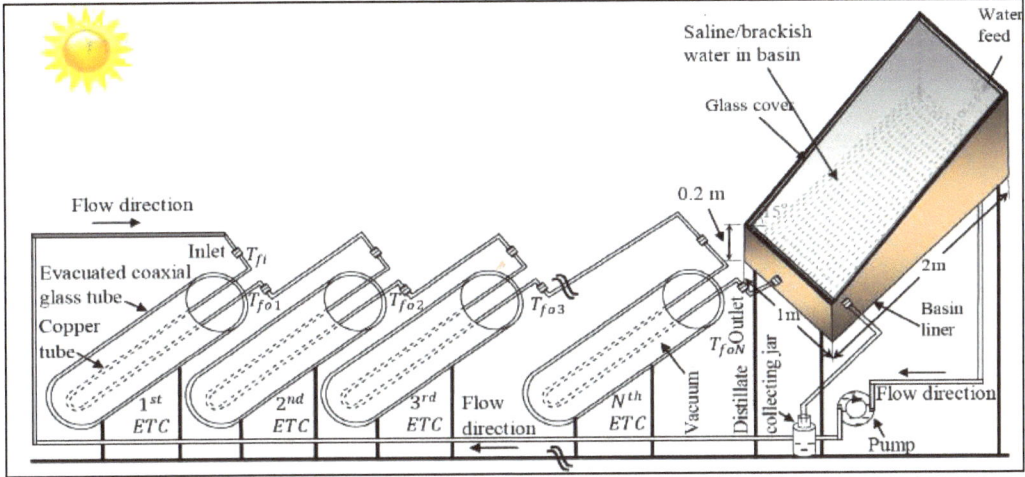

Fig. (5). SEBWP of single slope type integrated with N alike ETCs.

For the Absorber Surface

$$\alpha\tau^2 I(t) 2Rdx = \left[F' h_{pf}(T_p - T_f) + U_{tpa}(T_p - T_a)\right] 2Rdx \tag{89}$$

Where F' denotes collector efficiency factor.

For Fluid Flowing Through Tube

$$\dot{m}_f C_f \frac{dT_f}{dx} dx = F' h_{pf}(T_p - T_f) 2\pi r dx \tag{90}$$

Where r = Radius of copper tube.

Using Eq. 1 and 2, the water temperature at the first collector's outlet can be expressed as:

$$T_{fo1} = \frac{(A\,F_R(\alpha\tau))_1}{\dot{m}_f C_f} I(t) + \frac{(A\,F_R U_L)_1}{\dot{m}_f C_f} T_a + K_k^N T_{fi} \tag{91}$$

Where the value of T_{fi} is equal to T_w.

The temperature at the first collector's outlet will be the same as the temperature at the second collector's inlet; the temperature at the second collector's outlet will be the same as the temperature at the third collector's inlet, and so on. Using this condition, the fluid temperature at the Nth collector's outlet can be calculated as follows:

$$T_{foN} = \frac{(A\,F_R(\alpha\tau))_1}{\dot{m}_f C_f}\frac{\left(1-K_k^{\,N}\right)}{(1-K_k)}I(t) + \frac{(A\,F_R U_L)_1}{\dot{m}_f C_f}\frac{\left(1-K_k^{\,N}\right)}{(1-K_k)}T_a + K_k^{\,N}T_{fi} \qquad (92)$$

The heated fluid (water) available at the outlet of Nth collectors allowed to the basin of SEBWP of single slope type and hence, $T_{wo} = T_{foN}$. After getting the fluid temperature at the outlet of Nth collector, one can obtain the expression for useful heat gain as,

$$\dot{Q}_{uN} = \dot{m}_f C_f\left(T_{foN} - T_{fi}\right)$$
$$= \frac{\left(1-K_k^{\,N}\right)}{(1-K_k)}(A\,F_R(\alpha\tau))_1 I(t) + \frac{\left(1-K_k^{\,N}\right)}{(1-K_k)}(A\,F_R U_L)_1\left(T_{fi} - T_a\right) \qquad (93)$$

Following the methodology provided in section 2, one can obtain the temperature of water with the help of Eq. 93, 1, 3, 4 and 67 as,

$$T_w = \frac{\underline{f}(t)}{a}(1 - e^{-at}) + T_{wo}e^{-at} \qquad (94)$$

Where,

$$a = \frac{1}{M_w C_w}\left[\dot{m}_f C_f(1 - K_k^N) + U_s A_b\right] \qquad (95)$$

$$\underline{f}(t) = \frac{1}{M_w C_w}\left[\alpha_{eff}'A_b\underline{I_s}(t) + \frac{\left(1-K_k^{\,N}\right)}{(1-K_k)}(A\,F_R(\alpha\tau))_1\underline{I_c}(t)\right.$$
$$\left. + \left(\frac{\left(1-K_k^{\,N}\right)}{(1-K_k)}(A\,F_R U_L)_1 + U_s A_b\right)\underline{T_a}\right] \qquad (96)$$

Eq. 68 is similar in form to Eq. 22, however; expressions for a and $\underline{f}(t)$ are different. From Eq. 1, T_{go} can be expressed in terms of T_{gi} and putting this value of T_{go} in Eq. 2 and rearranging, one can estimate the value of T_{gi} as:

$$T_{gi} = \frac{\dot{\alpha}_g I_s(t) A_g + h_{1w} T_w A_b + U_c T_a A_g}{U_c A_g + h_{1w} A_b}$$ (97)

Again, the expression for T_{gi} in Eq. 71 is similar in form to Eq. 24; however, the value of T_{gi} will be different as value of T_w is different due to different values of a and $f(t)$. After estimating values of T_w from Eq. 70 and T_{gi} from Eq. 71, the value of hourly yield can be estimated as:

$$\dot{m}_{hy} = \frac{h_{ewg}(T_w - T_{gi})}{L} \times 3600$$ (98)

Where, h_{ewg} is HTC of evaporation and can be estimated using the Eq. 3f. Different terms used in Eq. 89 to 98 are given in appendix-A.

Eq. 94 can be discussed for the following cases:

Case (i): Putting the value of N = 0 in Eq. 94 to 96 and rearranging, Equations for a and $f(t)$ will be obtained for SEBWP in passive mode and the expression of T_w for SEBWP in passive mode is obtained.

4.4. SEBWP of Double Slope Type Integrated with N Alike PVT-FPCs

The schematic diagram of solar energy-based water purifier (SEBWP) of double slope type integrated with N alike PVT-FPCs has been shown in Fig. (**6**). N number of collectors in series has been added to SEBWP of double slope type because the main objective of adding PVT-FPC is to increase the temperature of water. Collectors are connected in series for getting low discharge at high temperature; whereas, they are connected in parallel to get a high discharge at low temperature.

Equations based on balancing energy input and energy output for different components of series-connected N alike FTCs have been presented in section 4.1 (Eq. 55 to 66). The expression for heat gain by N alike FPCs will be the same as the expression in Eq. 66. Equations for the outer surface of condensing cover having East orientation, an inner surface of condensing cover having East orientation, an outer surface of condensing cover having West orientation, an inner surface of condensing cover having West orientation and basin liner for SEBWP of double slope type integrated with N alike FPCs will be same as Eq. 26, 27, 28, 29 and 30 respectively. The equations based on balancing input energy to output energy for water mass in the basin of the system will be somewhat different and it is being presented here [20,22].

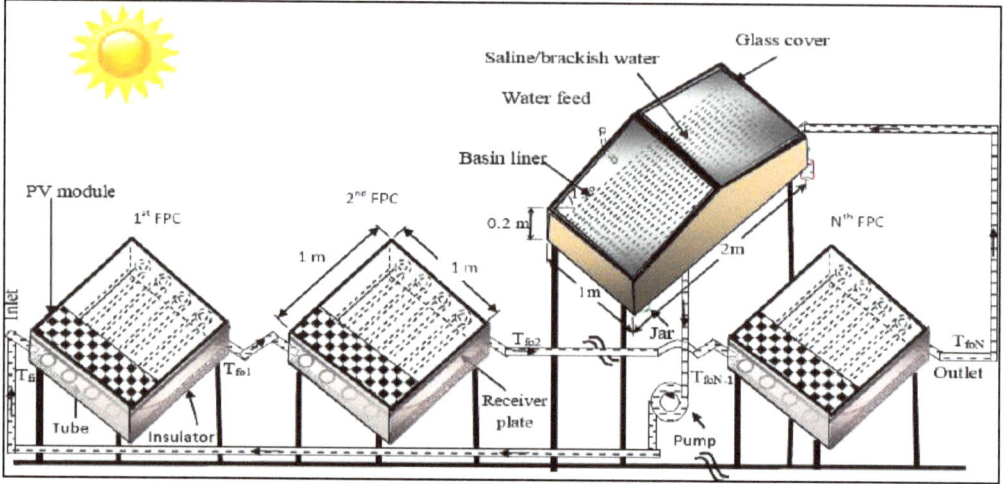

Fig. (6). SEBWP of double slope type integrated with N alike PVT-FPCs.

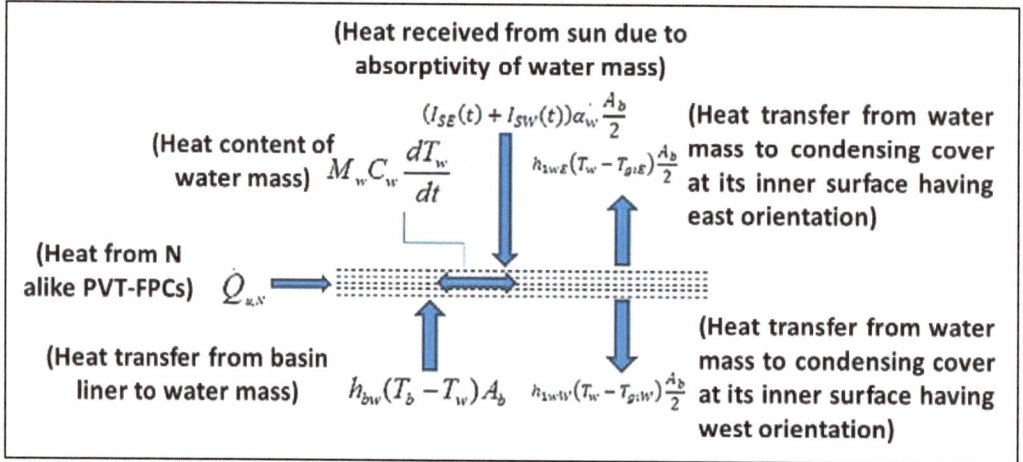

Fig. (6a). Heat interaction for water mass in basin of the system.

The interaction of heat for water mass in the basin of the system is shown in Fig. **(6a)**. It is an equation based on balancing input energy and output energy is as follows:

$$(M_w C_w)\frac{dT_w}{dt} = \dot{Q}_{u,N} + (I_{SE}(t) + I_{SW}(t))\alpha_w' \frac{A_b}{2} + h_{bw}(T_b - T_w)A_b$$
$$- h_{1wE}(T_w - T_{giE})\frac{A_b}{2} - h_{1wW}(T_w - T_{giW})\frac{A_b}{2} \tag{99}$$

Following the methodology of simplifying equations presented in section 3 and using Eq. 66, 26 to 30 and 99, one can get the expression of water temperature, the temperature at inner/outer surface of East oriented condensing cover and temperature at inner/outer surface of West oriented condensing cover as:

$$T_w = \frac{f(t)}{a}(1 - e^{-at}) + T_{w0}e^{-at} \tag{100}$$

$$T_{giE} = \frac{A_1 + A_2 T_w}{P} \tag{101}$$

$$T_{goE} = \frac{\frac{K_g}{L_g}T_{giE} + h_{1gE}T_a}{\frac{K_g}{L_g} + h_{1gE}} \tag{102}$$

$$T_{giW} = \frac{B_1 + B_2 T_w}{P} \tag{103}$$

$$T_{goW} = \frac{\frac{K_g}{L_g}T_{giW} + h_{1gW}T_a}{\frac{K_g}{L_g} + h_{1gW}} \tag{104}$$

After estimating values of T_w from Eq. 100, T_{giE} from Eq. 101 and T_{giW} from Eq. 103, the value of hourly yield ($\dot{m}_{hy}$) for solar energy based water purifier of double slope type can be estimated as:

$$\dot{m}_{hy} = \frac{\left[h_{ewgE}(T_w - T_{giE}) + h_{ewgW}(T_w - T_{giW})\right]}{L} \times 3600 \tag{105}$$

Where values of h_{ewgE} and h_{ewgW} are HTCs due to evaporation and can be estimated using Eq. 3f.

The different unknown terms used in Eq. 100 to 104 are given in appendix-A.

4.5. SEBWP of Double Slope Type Integrated with N Alike Partly Covered PVT-CPCs [23]

The schematic diagram of SEBWP of double slope type integrated with N alike PVT-CPCs has been presented in Fig. (**7**). N number of collectors in series has been added to SEBWP of double slope type because the main objective of adding PVT-CPC is to increase the temperature of water. Collectors are connected in series for getting low discharge at high temperature; whereas, they are connected in parallel to get a high discharge at low temperature.

Equations based on balancing energy input and energy output for different components of series-connected N alike PVT-CPCs have been presented in section 4.2 (Eq. 73 to 82). The expression for heat gain by N alike PVT-CPCs will be the same as the expression in Eq. 82. Equations for the outer surface of condensing cover having East orientation, an inner surface of condensing cover having East orientation, an outer surface of condensing cover having West orientation, an inner surface of condensing cover having West orientation and basin liner for SEBWP of double slope type integrated with N alike PVT-CPCs will be same as Eq. 26-30 respectively. The Eq. based on balancing input energy to output energy for water mass in the basin of the system will be the same as Eq. 99.

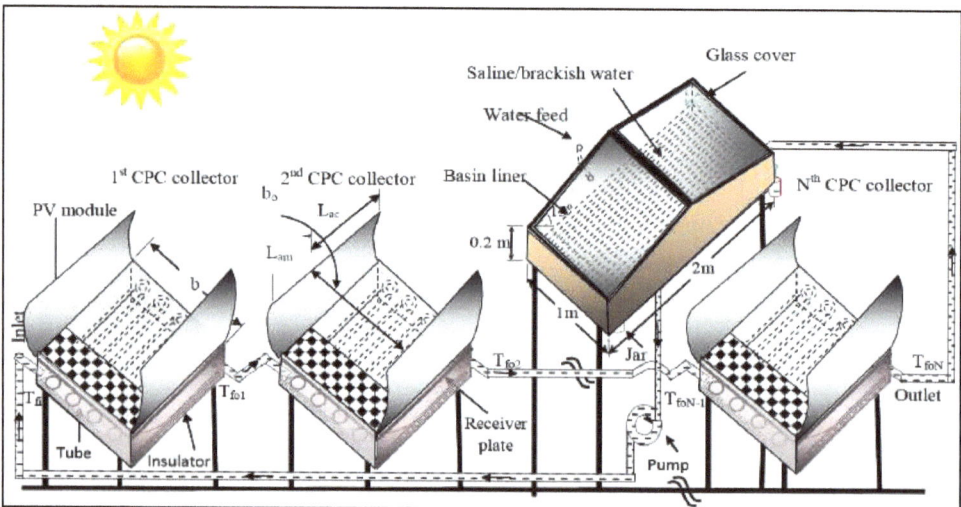

Fig. (7). SEBWP of double slope type integrated with N alike PVT-CPCs.

Following the methodology of simplifying equations presented in section 3 and using Eq. 82, 26 to 30 and 99, one can get the expression of water temperature, the temperature at inner/outer surface of East oriented condensing cover and temperature at inner/outer surface of West oriented condensing cover as:

$$T_w = \frac{f(t)}{a}(1 - e^{-at}) + T_{w0}e^{-at} \tag{106}$$

$$T_{giE} = \frac{A_1 + A_2 T_w}{P} \tag{107}$$

$$T_{goE} = \frac{\dfrac{K_g}{L_g}T_{giE} + h_{1gE}T_a}{\dfrac{K_g}{L_g} + h_{1gE}} \tag{108}$$

$$T_{giW} = \frac{B_1 + B_2 T_w}{P} \tag{109}$$

$$T_{goW} = \frac{\dfrac{K_g}{L_g}T_{giW} + h_{1gW}T_a}{\dfrac{K_g}{L_g} + h_{1gW}} \tag{110}$$

After estimating values of T_w from Eq. 106, T_{giE} from Eq. 107 and T_{giW} from Eq. 109, the value of hourly yield ($\dot{m}_{hy}$) for solar energy based water purifier of double slope type can be estimated as:

$$\dot{m}_{hy} = \frac{\left[h_{ewgE}(T_w - T_{giE}) + h_{ewgW}(T_w - T_{giW})\right]}{L} \times 3600 \tag{111}$$

Where values of h_{ewgE} and h_{ewgW} are HTCs due to evaporation and can be estimated using Eq. 3f.

Expressions for a and $f(t)$ used in Eq. 106 are same as expressions for a and $f(t)$ used in Eq. 100 in which $\underline{I}_c(t) = \underline{I}_b(t)$. Expressions for a and $f(t)$ used in Eq. 100 are given in appendix A. Expressions of different terms used in Eq. 107 to 110 are the same as different terms used in Eq. 100 to 104 respectively, which are given in appendix A.

4.6. SEBWP of Double Slope Type Integrated with N Alike ETCs [17, 18]

The schematic diagram of solar energy-based water purifier (SEBWP) of double slope type integrated with evacuated tubular collectors has been shown in Fig. (**8**). N number of collectors in series has been added to SEBWP of double slope type because the main objective of adding ETC is to increase the temperature of water. Collectors are connected in series for getting low discharge at high temperature; whereas, they are connected in parallel to get a high discharge at low temperature. Equations based on balancing energy input and energy output for different components of series-connected N alike ETCs have been presented in section 4.3 (Eq. 89 to 93). The expression for heat gain by N alike ETs will be the same as the expression in Eq. 93. Equations for the outer surface of condensing cover having East orientation, an inner surface of condensing cover having East orientation, an outer surface of condensing cover having West orientation, an inner surface of condensing cover having West orientation and basin liner for SEBWP of double slope type integrated with N alike ETCs will be same as Eq. 26 to 30 respectively. The Eq. based on balancing input energy to output energy for water mass in the basin of the system will be the same as Eq. 99.

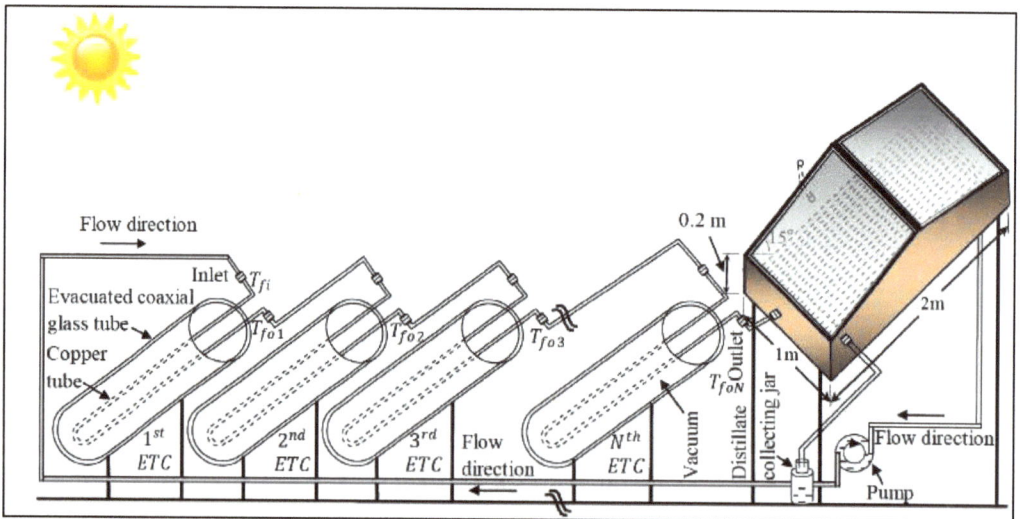

Fig. (8). SEBWP of double slope type integrated with N alike ETCs.

Following the methodology of simplifying equations presented in section 3 and using Eq. 93, 26 to 30 and 99, one can get the expression of water temperature, the temperature at inner/outer surface of East oriented condensing cover and temperature at inner/outer surface of West oriented condensing cover as:

$$T_w = \frac{f(t)}{a}(1 - e^{-at}) + T_{w0}e^{-at} \tag{112}$$

$$T_{giE} = \frac{A_1 + A_2 T_w}{P} \tag{113}$$

$$T_{goE} = \frac{\frac{K_g}{L_g}T_{giE} + h_{1gE}T_a}{\frac{K_g}{L_g} + h_{1gE}} \tag{114}$$

$$T_{giW} = \frac{B_1 + B_2 T_w}{P} \tag{115}$$

$$T_{goW} = \frac{\frac{K_g}{L_g}T_{giW} + h_{1gW}T_a}{\frac{K_g}{L_g} + h_{1gW}} \tag{116}$$

After estimating values of T_w from Eq. 112, T_{giE} from Eq. 113 and T_{giW} from Eq. 115, the value of hourly yield ($\dot{m}_{hy}$) for solar energy based water purifier of double slope type can be estimated as:

$$\dot{m}_{hy} = \frac{[h_{ewgE}(T_w - T_{giE}) + h_{ewgW}(T_w - T_{giW})]}{L} \times 3600 \tag{117}$$

Where values of h_{ewgE} and h_{ewgW} are HTCs due to evaporation and can be estimated using Eq. 3f.

Expressions for a and $\underline{f}(t)$ used in Eq. 112 are same as expressions for a and $\underline{f}(t)$ used in Eq. 100. Expressions for a and $\underline{f}(t)$ used in Eq. 100 are given in appendix

A. Expressions of different terms used in Eq. 113 to 116 are the same as different terms used in Eq. 100 to 104 respectively, which are given in appendix A.

Fig. (9). SEBWP of single slope type by incorporating FPCs and loaded with nanofluid.

5. SEBWP LOADED WITH NANOFLUID [24, 25]

5.1. SEBWP of Single and Double Slope Types in Passive Mode Loaded with Nanofluid

Nanofluids are better agents of heat transfer with the objective of collecting thermal energy for solar energy devices like SEBWP, solar dryers and solar collectors. In recent years, nanofluids have been applied in solar thermal-based investigation throughout the globe in both theoretical and experimental analyses. The thermal efficiency of SEBWPs can be increased by controlling the thermo-physical properties of water/base fluid. Nanofluids are obtained by mixing nanoparticles (1–100 nm) in small amounts to the base fluid. They do not harm environment and hence attract researchers throughout the globe with considerable potential for providing increased heat transfer thermo-physical properties. The thermal modeling of SEBWP loaded with nanofluid will remain the same as the conventional SEBWP using water (base fluid). The thermal modeling of SEBWP of single slope type has been presented in section 2. The thermal modeling of SEBWP of double slope type has been presented in section 3. Nanofluid means mixing of nanoparticles in a very small amount of water (base fluid). So, the properties of nanofluid will be used for performance analysis of SEBWP loaded with nanofluid. Fluid properties of water (base fluid) and nanofluid have been presented in Table **1** to Table **4**. Using these properties, performance parameters of SEBWP in passive mode loaded with nanofluid can be estimated.

Table 1. Thermophysical properties of nanofluids.

Model /Correlation	Quantity	Expression
Pak and Cho [26]	Specific heat	$C_{nf} = \dfrac{[(\varphi_p \rho_p C_p + (1-\varphi_p) \rho_{bf} C_{bf})]}{\rho_{bf}}$; $\varphi_p = \dfrac{V_p}{V_f + V_p}$; where 'V' is the volume.
Pak and Cho [26]	Density	$\rho_{nf} = \varphi_p \rho_p + (1 - \varphi_p)\rho_{bf}$
Khanafer and Vafai [27]	Thermal conductivity	$k_{nf} = k_{bf}\left[1 + (1.0112)\varphi_p + (2.4375)\varphi_p\left(\dfrac{47}{d_p(nm)}\right)\right.$ $\left. - (0.0248)\varphi_p\left(\dfrac{k_p}{0.613}\right)\right]$
Khanafer and Vafai [27]	Viscosity	$\mu_{nf} = -0.4491 + \left(\dfrac{28.837}{T_{nf}}\right) + 0.547\varphi_p - 0.163\varphi_p^2$ $+ (23.653)\left(\dfrac{\varphi_p}{T_{nf}}\right)^2 + (0.0132)\varphi_p^3$ $- (2354.7)\left(\dfrac{\varphi_p}{T_{nf}^3}\right) + (23.498)\left(\dfrac{\varphi_p}{d}\right)^2$ $- (3.018)\left(\dfrac{\varphi_p^3}{d_p^2}\right)$ $1 \leq \varphi_p \leq 9;\ 13 \leq d_p \leq 130nm;\ 20 \leq T \leq 70°C$
Wang *et al.* [28] and Ho *et al.* [29]	Thermal expansion coefficient	$\beta_{nf} = (1 - \varphi_p)\beta_{bf} + \varphi_p\beta_p$

Table 2: Thermal properties of nanoparticles.

Nanoparticle	Density ρ (kg/m³)	Specific Heat C_p (J/kgK)	Thermal Conductivity k (W/mK)
Al₂O₃	3890	880	38.5
TiO₂	4230	697	11.8
CuO	6310	550	17.6

Table 3. Thermophysical properties of vapor [24].

Quantity	Symbol	Expression
Specific heat	C_v	$999.2 + 0.1434 \times (T_v) + 1.101 \times (T_v^2) - 6.7581 \times 10^{-8} \times (T_v^3)$
Density	ρ_v	$353.44/(T_v + 273.15)$
Thermal conductivity	k_v	$0.0244 + 0.7673 \times 10^{-4} \times (T_v)$
Viscosity	μ_v	$1.718 \times 10^{-5} + 4.620 \times 10^{-8} \times (T_v)$
Latent heat of vaporization of fluid	L	$3.1625 \times 10^6 + \left[1 - \left(7.616 \times 10^{-4} \times (T_v)\right)\right]; \quad for\ T_v > 70°C$ $2.4935 \times 10^6 \left[1 - \left(9.4779 \times 10^{-4} \times (T_v) + 1.3132 \times 10^{-7} \times (T_v^2) - 4.7974 \times 10^{-3} \times (T_v^3)\right)\right]; \quad for\ T_v < 70°C$
Partial vapor pressure at condensing cover and fluid temperature	P_{gi}	$P(gi) = exp\left[25.317 - \left(\dfrac{5144}{T(gi) + 273}\right)\right]$ $P(fluid) = exp\left[25.317 - \left(\dfrac{5144}{T(fluid) + 273}\right)\right]$
Thermal expansion coefficient	β_v	$1/(T_v + 273.15)$

Table 4. Thermophysical properties of base fluid [25].

Quantity	Symbol	Expression
Density	ρ_{bf}	$999.79 + 0.0683 \times T_{bf} - 0.0107 \times T_{bf}^2 + 0.00082 \times T_{bf}^{2.5} - 2.303 \times 10^{-5} \times T_{bf}^3$
Specific heat	C_{bf}	$4.217 - 0.00561 \times T_{bf} + 0.00129 \times T_{bf}^{1.5} - 0.000115 \times T_{bf}^2 + 4.149 \times 10^{-6} \times T_{bf}^{2.5}$
Viscosity	μ_{bf}	$\dfrac{1}{\left(557.82 - 19.408 \times T_{bf} + 0.136 \times T_{bf}^2 - 3.116 \times 10^{-4} \times T_{bf}^3\right)}$
Thermal conductivity	k_{bf}	$0.565 + 0.00263 \times T_{bf} - 0.000125 \times T_{bf}^{1.5} - 1.515 \times 10^{-6} \times T_{bf}^2 - 0.000941 \times T_{bf}^{0.5}$

5.2. SEBWP of Double Slope Types in Active Mode Loaded with Nanofluid

SEBWP of double slope type loaded with nanofluid by incorporating FPC has been shown in Fig. (**9**). The characteristic equation of the system has been developed as given below:

The energy balances of different components of the system (A) are expressed as,

(a) East side

$$\alpha_g I_{SE} A_{gE} + h_{1f,E}\left(T_f - T_{giE}\right)\left(\frac{A_B}{2}\right) - h_{EW}\left(T_{giE} - T_{giW}\right)A_{gE}$$
$$= \left(\frac{K_g}{L_g}\right)\left(T_{giE} - T_{goE}\right)A_{gE} \tag{118}$$

$$\left(\frac{K_g}{L_g}\right)\left(T_{giE} - T_{goE}\right)A_{gE} = h_{1gE}\left(T_{goE} - T_a\right)A_{gE} \tag{119}$$

(b) West Side

$$\alpha_g \, I_{SW} \, A_{gW} + h_{1f,W}\left(T_f - T_{giW}\right)\left(\frac{A_B}{2}\right) - h_{EW}\left(T_{giE} - T_{giW}\right)A_{gW}$$

$$= \left(\frac{K_g}{L_g}\right)\left(T_{giW} - T_{goW}\right)A_{gW} \tag{120}$$

$$\left(\frac{K_g}{L_g}\right)\left(T_{giW} - T_{goW}\right)A_{gW} = h_{1gW}\left(T_{goW} - T_a\right)A_{gW} \tag{121}$$

On solving Eq. 1-4, one can get,

$$T_{giE} = \frac{\left[A + T_f B\right]}{H} \tag{122}$$

And

$$T_{giW} = \frac{\left[A' + T_f B'\right]}{H} \tag{123}$$

Unknown terms, $A, A', B, B',$ *and* H in the above equations are given in appendix I.

(c) Basin Liner

$$\alpha_b(I_{SE} + I_{SW}) = 2h_{b,f}\left(T_b - T_f\right) + 2h_{ba}(T_b - T_a) \tag{124}$$

(d) Water Mass

$$M_f C_f \frac{dT_f}{dt} = \alpha_f(I_{SE} + I_{SW})\left(\frac{A_B}{2}\right) + 2h_{b,f}\left(T_b - T_f\right)\left(\frac{A_B}{2}\right)$$

$$- h_{1f,E}\left(T_f - T_{giE}\right)\left(\frac{A_B}{2}\right) - h_{1f,W}\left(T_f - T_{giW}\right)\left(\frac{A_B}{2}\right)$$

$$+ \dot{Q}_{uN} \tag{125a}$$

Where,

$$\dot{Q}_{uN} = A_c F'\left[(\alpha\tau)_{N,eff}I_c - U_{LN}\left(T_f - T_a\right)\right] \tag{125b}$$

Unknown terms are given in appendix-B.

The energy balance of the heat exchanger immersed in the fluid (BF/NF) of the solar still can be expressed as,

$$\dot{m}_f C_f \frac{dT_f}{dx} dx = -(2\pi r_{11} U)(T_{HE} - T_f) dx \qquad (126)$$

Boundary conditions: $T_f(x = 0) = T_{FoN}$ and $T_f(x = L) = T_{fi}$

Solving Eq. 126 using the above boundary conditions, one can get,

$$T_{fi} = T_f \left[1 - exp\left(-\frac{2\pi r_{11} UL}{\dot{m}_f C_f}\right)\right] + T_{FoN}\, exp\left(-\frac{2\pi r_{11} UL}{\dot{m}_f C_f}\right) \qquad (127)$$

Where,

$$U = \left[\frac{1}{h_{bf}} + \left(\frac{r_{11}}{K_1}\right) log\left(\frac{r_{22}}{K_1}\right) + \left(\frac{r_{11}}{r_{22}}\right)\left(\frac{1}{h_{bf}}\right)\right]^{-1} \qquad (128)$$

The value of hourly heat gain from N alike PVT-FPC having series connection and loaded with water can be estimated as;

$$\dot{Q}_{uN} = \dot{m}_f C_f \left(T_{FoN} - T_{fi}\right) \qquad (129)$$

The value of temperature at the end of the Nth PVT-FPC loaded with water can be estimated as:

$$T_{FoN} = \frac{(AF_R(\alpha\tau))_1}{\dot{m}_f C_f}\left(\frac{1 - K_k^N}{1 - K_k}\right) I_c(t) + \frac{(AF_R U_L)_1}{\dot{m}_f C_f}\left(\frac{1 - K_k^N}{1 - K_k}\right) T_a \\ + K_k^N T_{fi} \qquad (130)$$

Unknown terms in the above equation are given in appendix-B.

Substituting T_{FoN} from Eq. 130 in Eq. 127 and rearranging, one can get,

$$T_{FoN} - T_{fi} = (K_k^N - 1)\left(\frac{1 - e^{-z}}{1 - K_k^N e^{-z}}\right)T_f$$

$$+ \frac{(AF_R(\alpha\tau))_1}{\dot{m}_f C_f}\left[1 + \frac{(K_k^N - 1)e^{-z}}{1 - K_k^N e^{-z}}\right]\left(\frac{1 - K_k^N}{1 - K_k}\right)I_c(t) \qquad (131)$$

$$+ \frac{(AF_R U_L)_1}{\dot{m}_f C_f}\left[1 + \frac{(K_k^N - 1)e^{-z}}{1 - K_k^N e^{-z}}\right]\left(\frac{1 - K_k^N}{1 - K_k}\right)T_a$$

Therefore, the rate of useful thermal energy gain (Eq. 129) can be expressed as:

$$\dot{Q}_{uN} = \dot{m}_f C_f\left[(K_k^N - 1)\left(\frac{1 - e^{-z}}{1 - K_k^N e^{-z}}\right)\right]T_f$$

$$+ \left(AF_R(\alpha\tau)\right)_1\left[1 + \frac{(K_k^N - 1)e^{-z}}{1 - K_k^N e^{-z}}\right]\left(\frac{1 - K_k^N}{1 - K_k}\right)I_c(t) \qquad (132)$$

$$+ (AF_R U_L)_1\left[1 + \frac{(K_k^N - 1)e^{-z}}{1 - K_k^N e^{-z}}\right]\left(\frac{1 - K_k^N}{1 - K_k}\right)T_a$$

$$\dot{Q}_{uN} = D_1 T_f + D_2 I_c(t) + D_3 T_a \qquad (133)$$

Where,

$$D_1 = \dot{m}_f C_f\left[(K_k^N - 1)\left(\frac{1 - e^{-z}}{1 - K_k^N e^{-z}}\right)\right] \qquad (134)$$

$$D_2 = \left(AF_R(\alpha\tau)\right)_1\left[1 + \frac{(K_k^N - 1)e^{-z}}{1 - K_k^N e^{-z}}\right]\left(\frac{1 - K_k^N}{1 - K_k}\right) \qquad (135)$$

$$D_3 = (AF_R U_L)_1\left[1 + \frac{(K_k^N - 1)e^{-z}}{1 - K_k^N e^{-z}}\right]\left(\frac{1 - K_k^N}{1 - K_k}\right) \qquad (136)$$

On substituting T_{giE}, T_{giW}, $h_{b,f}(T_b - T_f)$, and $\dot{Q}_{uN}$ from Eq. 122-124, and Eq. 133 respectively in the water mass Eq. 125, one can obtain,

$$\frac{dT_f}{dt} = -T_f \left[h_{1f,E} \left(1 - \frac{B}{H} \right) \left(\frac{A_b}{2} \right) + h_{1f,W} \left(1 - \frac{B'}{H} \right) \left(\frac{A_b}{2} \right) + 2U_b \left(\frac{A_b}{2} \right) \right.$$

$$\left. - D_1 \right] \left(\frac{1}{M_f C_f} \right)$$

$$+ \left[(\alpha_f + 2\alpha_b h_1)(I_{SE} + I_{SW}) \left(\frac{A_b}{2} \right) + h_{1f,E} \left(\frac{A}{H} \right) \left(\frac{A_b}{2} \right) \right. \tag{137}$$

$$+ h_{1f,W} \left(\frac{A'}{H} \right) \left(\frac{A_b}{2} \right) + D_2 I_c(t)$$

$$\left. + \left(2U_b \left(\frac{A_b}{2} \right) + D_3 \right) T_a \right] \left(\frac{1}{M_f C_f} \right)$$

$$\frac{dT_f}{dt} + a_2 T_f = f_2(t) \tag{138}$$

Where,

$$a_2 = \left(\frac{A_b}{2H} \right) (H'_{11} + H''_{33}) \left(\frac{1}{M_f C_f} \right) \tag{139}$$

$$f_2(t) = \left(\frac{A_b}{2H} \right) \left\{ \{ [(\alpha_f + 2\alpha_b h_1)H + K'_{1E}]I_{SE}(t) + [(\alpha_f + 2\alpha_b h_1)H + K'_{1W}]I_{SW}(t) \right.$$

$$\left. + D_2 I_c(t) \} + T_a (H'_{11} + H''_{44}) \right\} \left(\frac{1}{M_f C_f} \right) \tag{140}$$

The solution of first-order differential Eq. 138 can be expressed as,

$$T_f = \frac{f_2(t)}{a_2} [1 - e^{-a_2 \Delta t}] + T_{f0} e^{-a_2 \Delta t} \tag{141}$$

Using thermo-physical properties (Tables **2** and **3**) and HTCs (Table 4), hourly variation of nanofluid (NF) temperature can be obtained from Eq. 18.

Hourly yield from the proposed systems can be estimated from the equation given below,

Where the latent heat of vaporization (L_v) and can be expressed as [26],

$$\dot{M}_{bf} = \frac{\dot{q}_{eg}}{L_v} \times 3600 = \frac{h_{eg}(T_f - T_g)}{L_v} \times 3600 \tag{142}$$

$$
\begin{aligned}
L_v = 3.1625 \times 10^6 \qquad\qquad & for\ T_v > 70°C \\
+ [1 & \\
- (7.616 \times 10^{-4} & \\
\times (T_v))]
\end{aligned} \tag{143}
$$

$$
\begin{aligned}
L_v = 2.4935 \times 10^6 [1 & \\
- (9.4779 \times 10^{-4} \times T_v) & \qquad for\ T_v \\
+ (1.3132 \times 10^{-7} \times T_v^2) - 4.7974 \times 10^{-3} & \qquad < 70°C \\
\times T_v^3)]
\end{aligned} \tag{144}
$$

6. EXERGY ANALYSIS OF SEBWP

The Energy of solar energy-based water purifiers commonly known as solar still can be calculated using the first and second laws of thermodynamics. Exergy can be written as the maximum work output taken from the system in carrying out the system from the current state to the equilibrium state with the surrounding. If the given system is at temperature T and the surrounding temperature is T_a, the maximum work that can be obtained in carrying the system from temperature T to temperature Ta is exergy. Exergy is high-grade energy because complete utilization is possible; whereas, heat is considered low-grade energy because heat can never be completely used. Some portion of the heat is bound to be a loss. Electrical energy, which can also be generated using green technology namely PV module, solar thermal etc., without disturbing the environment from a pollution viewpoint, is high-grade energy because it can be utilized fully. One of the examples of complete utilization is the conversion of electrical energy into heat in heat application processes. Following Nag [30], the expression for exergy can be developed for the solar system as follows:

The change in entropy for the body at temperature T taking mass as unity can be written as (Fig. **10**),

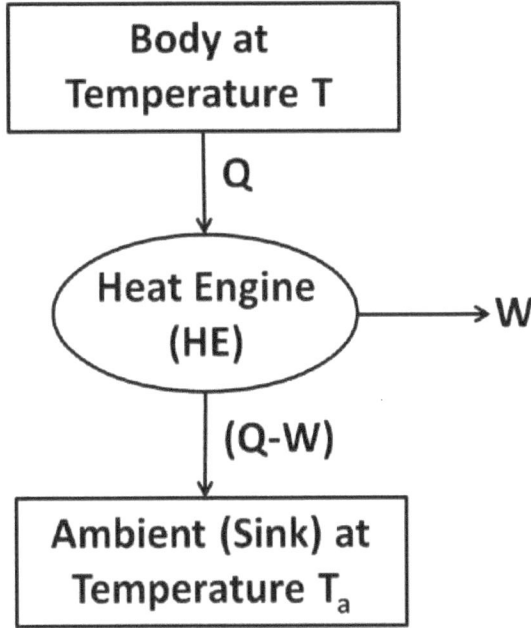

Fig. (10). Heat and work integration between temperature T and Ta.

$$(\Delta S)_{body} = \int_{(T+273)}^{(T_a+273)} \left(\frac{C_p}{T}\right) dT$$

$$= C_p ln \left\{\frac{(T_a + 273)}{(T + 273)}\right\}$$

(145)

Where T is the temperature of the body, Ta is the temperature of surrounding or ambient and Cp is the specific heat capacity at constant pressure.

The entropy change of heat engine (reversible) can be written as,

$$(\Delta S)_{HE} = 0$$

(146)

The entropy change for ambient can be written as,

$$(\Delta S)_{ambient} = \frac{(Q - W)}{(T_a + 273)}$$

(147)

Where Q is the heat supplied to the heat engine and W is the work done by the heat engine. Using the first law of thermodynamics, (Q-W) is heat rejected to the ambient.

Adding Eq. 146, 147 and 148, one gets,

$$(\Delta S)_{body} + (\Delta S)_{HE} + (\Delta S)_{ambient}$$
$$= C_p ln \left\{ \frac{(T_a + 273)}{(T + 273)} \right\} + \frac{(Q - W)}{(T_a + 273)} \tag{148}$$

Here, body, heat engine and ambient together form the universe and hence,

$$(\Delta S)_{Universe} = C_p ln \left\{ \frac{(T_a + 273)}{(T + 273)} \right\} + \frac{(Q - W)}{(T_a + 273)} \tag{149}$$

From entropy principle,

$$(\Delta S)_{Universe} \geq 0 \tag{150}$$

Using Eq. 150 and inequality Eq. 151, one gets,

$$C_p ln \left\{ \frac{(T_a + 273)}{(T + 273)} \right\} + \frac{(Q - W)}{(T_a + 273)} \geq 0 \tag{151}$$

The inequality Eq. 152 can be rearranged as,

$$W \leq Q + (T_a + 273) C_p ln \left\{ \frac{(T_a + 273)}{(T + 273)} \right\} \tag{152}$$

The maximum work W$_{max}$ can be expressed inequality Eq. 153 as,

$$W_{max} = C_p(T - T_a) - (T_a + 273) C_p ln \left\{ \frac{(T_a + 273)}{(T + 273)} \right\} \tag{153}$$

Here, one should always keep in mind that temperatures T and T$_a$ are in °C. Now, maximum work obtainable thermodynamically between glass cover and ambient can be written as,

$$W_{max1} = C_p(T_g - T_a) - (T_a + 273)C_p ln\left\{\frac{(T_g + 273)}{(T_a + 273)}\right\}$$ (154)

Where glass cover temperature (T_g) and surrounding/ambient temperature (T_a) are in °C. Also, maximum work obtainable thermodynamically between water mass in basin and ambient can be written as,

$$W_{max2} = C_p(T_w - T_a) - (T_a + 273)C_p ln\left\{\frac{(T_{w+273})}{(T_a + 273)}\right\}$$ (155)

Subtracting Eq. 155 from Eq. 156, one gets,

$$W_{max2} - W_{max1} = C_p(T_w - T_g) - (T_a + 273)C_p ln\left\{\frac{(T_w + 273)}{(T_g + 273)}\right\}$$ (156)

The maximum work obtainable *i.e.* exergy between water surface in basin and glass cover can be written as (for unit mass flow rate),

$$Exergy = C_p\left[(T_w - T_g) - (T_a + 273)ln\left\{\frac{(T_w + 273)}{(T_g + 273)}\right\}\right]$$ (157)

If the mass flow rate is $\dot{m}_f$, then the exergy between water surface in basin and glass cover can be written as,

$$Exergy = \dot{m}_f C_p\left[(T_w - T_g) - (T_a + 273)ln\left\{\frac{(T_w + 273)}{(T_g + 273)}\right\}\right]$$ (158)

Using $\dot{m}_f C_p = hA$, Eq. 159 can further be written as:

$$Exergy = hA\left[(T_w - T_g) - (T_a + 273)ln\left\{\frac{(T_w + 273)}{(T_g + 273)}\right\}\right]$$ (159)

Here, h is HTC and A is area. Using Eq. 160, exergy gain by the solar energy-based water purifier can be estimated. Eq. 160 can be used to estimate the exergy gain in the case of other solar systems also.

The hourly exergy gain for SEBWP of single slope type can be written as,

$$Hourly\ Exergy\ gain = h_{ewg}A_b\left[(T_w - T_{gi}) - (T_a + 273)ln\left\{\frac{(T_w + 273)}{(T_{gi} + 273)}\right\}\right] \qquad (160)$$

Where T_w is the temperature of water in °C, T_{gi} is the temperature of glass cover at its inner surface in °C, h_{ewg} is evaporative HTC between water and glass cover. A_b is the area of the basin.

The hourly exergy gain for SEBWP of double slope type can be written as,

$$Hourly\ Exergy\ gain = h_{ewgE}\left(\frac{A_{bE}}{2}\right)\left[(T_w - T_{giE}) - (T_a + 273)ln\left\{\frac{(T_w+273)}{(T_{giE}+273)}\right\}\right] +$$
$$h_{ewgW}\left(\frac{A_{bW}}{2}\right)\left[(T_w - T_{giW}) - (T_a + 273)ln\left\{\frac{(T_w+273)}{(T_{giW}+273)}\right\}\right] \qquad (161)$$

Where T_w is the temperature of water in °C and T_{gi} is the temperature of glass cover at its inner surface in °C.

7. RESULTS AND DISCUSSION

The hourly annual yield of SEBWP in passive mode (using Eq. 24 and 49) and in active mode (using Eq. 72, 88 and 98) can be estimated. Further, daily yield can be estimated by adding hourly water yield for 24 h followed by monthly water yield by multiplying daily yield with the number of clear days in a month. The summation of monthly water yield for 12 months gives the estimation of annual water yield. Similarly, hourly exergy for SEBWP can be estimated using Eq. 161 and 162, followed by the estimation of daily thermal exergy by adding hourly thermal exergy for 24 h. The monthly thermal exergy can be estimated by multiplying daily thermal exergy with the number of clear days. The annual thermal exergy can be estimated as the summation of monthly thermal exergy for 12 months.

The comparative study of SEBWP of single slope type on the basis of annual yield and annual thermal exergy has been presented in Fig. (**11**). The annual water yield and exergy both are the smallest for SEBWP of single slope type in passive mode because the external heat is not provided by any source and the water temperature is lower, resulting in lower evaporation of water and finally lower water yield. The annual water yield and annual thermal exergy both are highest for SEBWP of single slope type for the case when SEBWP of single slope type is integrated with N alike PVT flat plate collectors under the optimized condition of mass flow rate and number of PVTFPCs. It happens because the value of N is higher than other cases and higher heat might be added in the case when SEBWP is connected to N alike PVTFPCs. Again, annual water yield and thermal exergy for SEBWP integrated with evacuated tubular collectors are higher than SEBWP integrated with N alike PVT compound parabolic concentrators. It happens because heat loss by convective heat mechanism does not take place as the vacuum is present.

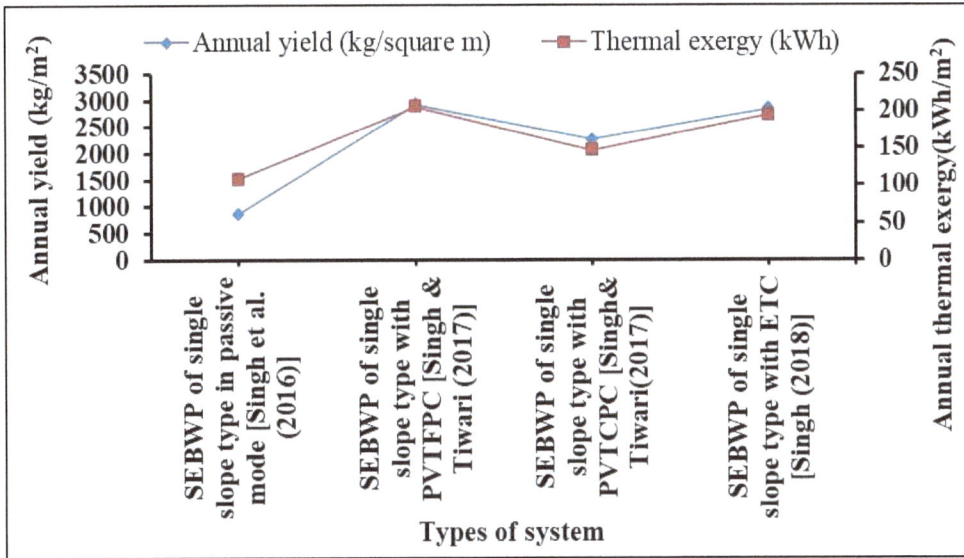

Fig. (11). Comparative study of SEBWP of single slope type on the basis of annual yield and annual thermal exergy.

The comparative study of SEBWP of double slope type on the basis of annual yield and annual thermal exergy has been presented in Fig. (**12**). The annual water yield and exergy both are the smallest for SEBWP of double slope type in passive mode because the external heat is not provided by any source and the water temperature is lower, resulting in lower evaporation of water and finally lower water yield. The annual water yield and annual thermal exergy both are highest for SEBWP of double slope type for the case when SEBWP is integrated with N alike PVT flat plate collectors under the optimized condition of mass flow rate and number of PVTFPCs. It happens because the value of N is higher than other cases and higher heat might be added in the case when SEBWP is connected to N alike PVTFPCs. Again, annual water yield and thermal exergy for SEBWP of double slope type integrated with evacuated tubular collectors are higher than SEBWP integrated with N alike PVT compound parabolic concentrators. It happens because heat loss by convective heat mechanism does not take place as the vacuum is present in the evacuated tubular collector.

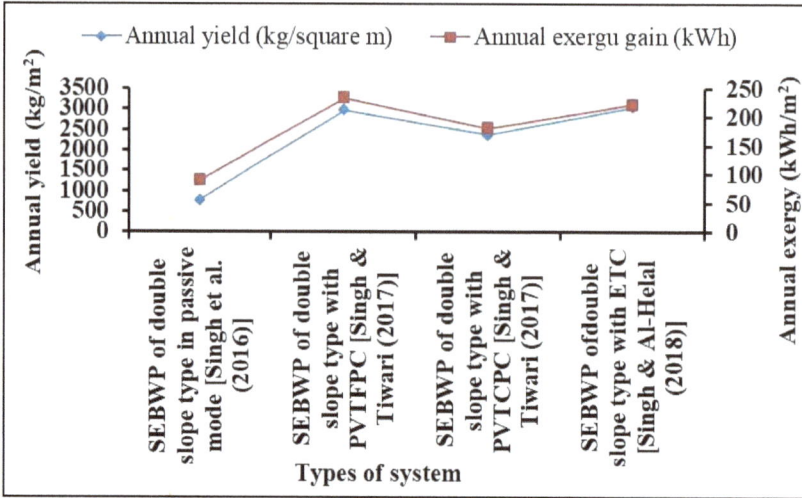

Fig. (12). Comparative study of SEBWP of double slope type on the basis of annual yield and annual thermal exergy.

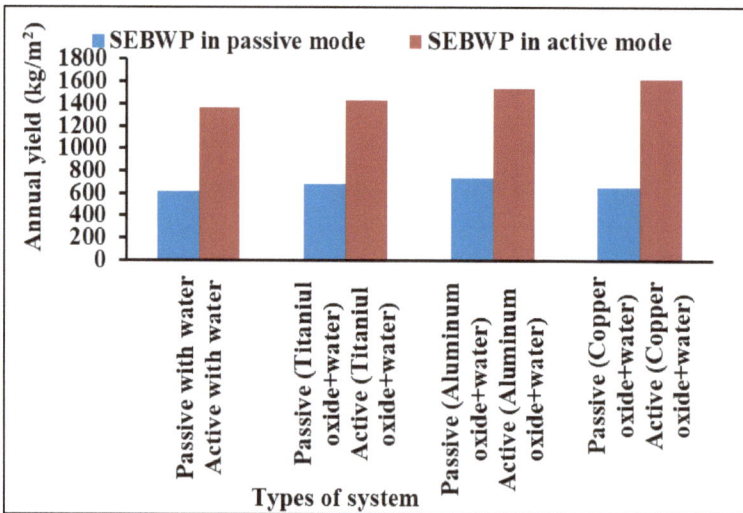

Fig. (13). Comparative study of SEBWP of double slope type in active as well as passive mode with/without nanofluid on the basis of annual yield.

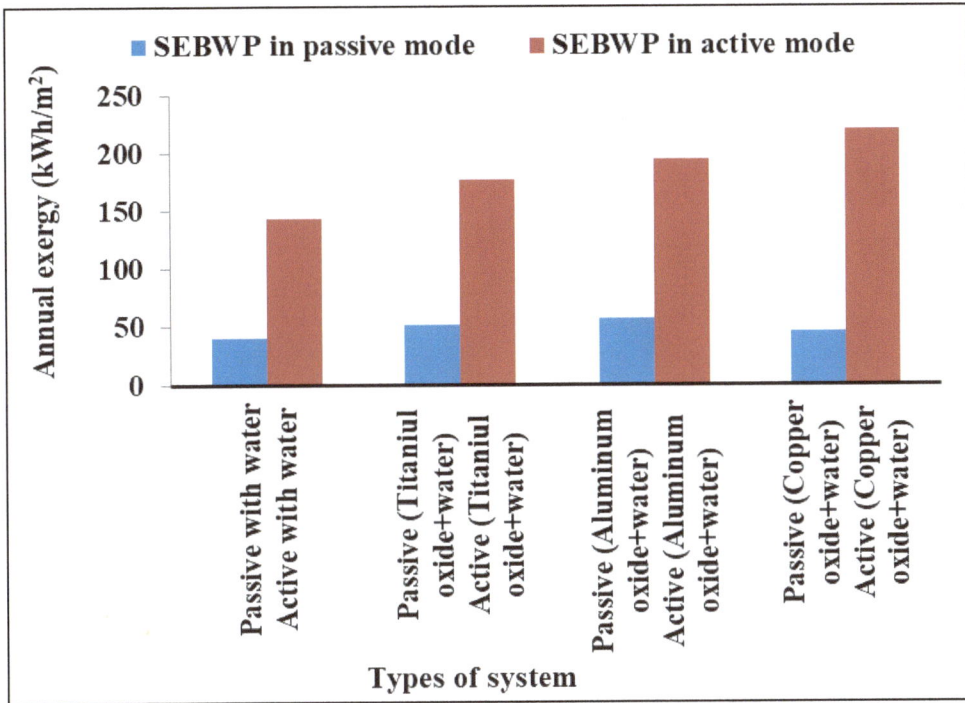

Fig. (14). Comparative study of SEBWP of double slope type in active as well as passive mode with/without nanofluid on the basis of thermal exergy.

The comparative study of SEBWP of double slope type in active as well as passive mode with/without nanofluid on the basis of annual yield is shown in Fig. (**13**). The comparative study of SEBWP of double slope type in active as well as passive mode with/without nanofluid on the basis of thermal exergy has been shown in Fig. (**14**). It has been found that the use of nanofluid increases the performance because of better thermophysical characteristics of nanofluid than the base fluid (water).

CONCLUSION

The thermal modeling of SEBWP of basin type in passive and active modes has been presented, followed by their exergy analyses on the basis of 1st and 2nd laws of thermodynamics. The thermal modeling of SEBWP by incorporating water-based nanofluid has been carried out. The active mode SEBWP performs better than the passive mode SEBWP of the same basin area as external heat is provided in the case of active mode. SEBWP integrated with evacuated tubular collectors performs better than SEBWP integrated with PVT compound parabolic concentrators as the convective heat loss does not take place in the case of evacuated tubular collectors due to the presence of vacuum. The use of nanofluid in SEBWP gives better performance due to better thermophysical characteristics of nanofluid.

NOMENCLATURE

A	Pond surface area, m^2	W	Pond width
C_a	Heat capacity (humid specific), J/kg °C	d	Declination angle, degree
C_p	Heat capacity (specific), J/kg °C	l	Latitude of location, degree
h	Hour angle, degree	A_a	Aperture area
h_s	Hour angle at sunrise and sunset, degree	A_r	Receiver area
h_c	Thermal energy transfer coefficient (convection) among UCZ and air, W/m^2°C	h_{absorb}	Thermal energy exchange coefficient (convection), W/m^2 K
h_e	Thermal energy transfer coefficient (evaporation), W/m^2 °C	$h_{rabsgls}$	Thermal energy exchange coefficient (radiation), W/m^2 K
h_x	Part of solar insolation that available to depth x in m of pond	h_{glass}	Thermal energy exchange coefficient (convection), W/m^2 K
h_1	Thermal energy transfer coefficient (convection) among UCZ and NCZ, W/m^2°C	m_{asc}	Mass passing of air in solar chimney in unit time, kg/s
h_2	Thermal energy transfer coefficient (convection) among NCZ and LCZ, W/m^2°C	S_{absorb}	Solar insolation gains by the absorber, W/m^2

h_3	Thermal energy transfer coefficient (convection) among LCZ with pond, W/m^2°C	S_{glass}	Solar insolation gains by a glass cover, W/m^2
h_4	Thermal energy transfer coefficient (convection) at the groundwater level, W/m^2 °C	q_u	Useful heat energy rate/unit aperture surface area
I	Solar insolation intensity striking the plane of UCZ, W/m^2	T_{ascout}	Temperature of air at solar chimney outlet, K
I_0	Solar insolation intensity entering the pond plane, W/m^2	V	Cubic capacity of the representative room, m^3
k_g	Ground conductivity (thermal), W/m °C	I_b	Incident beam radiation
k_w	Water conductivity (thermal), W/m °C	T_a	Ambient temperature, K
L	Pond dimension (length), m	A_{glass}	Glazing cover area, m^2
m	Layers quantity (number)	A_{absorb}	Absorber plane area, m^2
N	Counting number of day of a year	C_{asc}	Heat capacity (solar chimney air), J/kg K
n_i	Refractive index (air)	C	Concentration ratio
n_r	Refractive index (water)	F'	Collector efficiency factor
P_{vap}	Partial pressure vapour (water) at atmospheric temperature, mm of Hg	U_w	Overall thermal energy exchange coefficient, W/m^2 K
P_{atm}	Atmospheric pressure, mm of Hg	T_{absorb}	Absorber plane temperature, K

P_{UCZ}	Vapour pressure (water) at a temperature of UCZ, mm of Hg	T_i	Air temperature at interior side of the building, K
Q_{cond}	Thermal energy transfer rate (conduction), W	T_{glass}	Glass plane temperature, K
Q_{conv}	Thermal energy transfer rate (convection), W	T_{room}	Air temperature (room), K
Q_{evap}	Thermal energy loss rate (evaporation) from the plane of a pond, W	T_{ascm}	Air average temperature in a solar chimney, K
Q_{rad}	Thermal energy loss rate (radiation) from the surface of a pond, W	T_{ascin}	Temperature of air at solar chimney inlet, K
Q_g	Ground thermal energy loss rate from LCZ, W	F_R	Mass flow–rate factor
Q_{load}	Thermal energy extraction rate from LCZ, W	T_{amb}	Outdoor temperature of air, K
Q_{losses}	Thermal energy loss rate, W	N	Air change per hour
Q_s	Solar radiation absorption rate in a layer, W	W	Image size
R	Reflection coefficient	q	Thermal energy transfer, W
r_t	Ratio of hourly total insolation to total daily insolation	K	Conductivity (thermal), W/m-K
T_a	Atmospheric temperature, °C	***Greek letters***	
T_g	Ground temperature, °C	θ_i	Incident angle, degree
T_{sky}	Sky temperature, K	θ_r	Refraction angle, degree
T	Node temperature, °C	ρ	Mass density, kg/m^3
t	Layer thickness, m	ε_w	Water emissivity
U_g		τ	Present time, s

	Overall thermal energy exchange coefficient of the ground, W/m² °C	dτ	Time increment, s
U_t	Overall thermal energy exchange coefficient of the pond, W/m² °C	λ	Evaporation latent heat, J/kg
		ϕ_h	Air humidity
V_{air}	Velocity wind (m/s)	Ø	Angle formed by optical axis with normal
r	Distance from point of incidence to focus	η	Instantaneous thermal efficiency
S	Absorbed beam radiation rate per unit surface area of un-shaded receiver	*Subscripts*	
T_a	Ambient Temperature	LCZ	Lower convective zone
Tp	Absorber temperature	NCZ	Non-convective zone
T_f	Working fluid temperature	UCZ	Upper convective zone
U_L	Overall heat transfer coefficient		

CONSENT FOR PUBLICATION

Not applicable.

CONFLICT OF INTEREST

The authors declare no conflict of interest, financial or otherwise.

ACKNOWLEDGEMENTS

Declared none.

REFERENCES

[1]   A. Mouchot, *La Chaleur Solavie et ses Applications.* Gauthier- Villars: Paris, 1969.

[2]   G.B. Della Porta, *Magiae naturalis libri XX.* Napoli, 1589.

[3]   G. Nebbia, and G. Mennozi, "A short history of water desalination, Acque Dolce Dal Mare, II Inchiesta Internazionole", *In proceedings of Int Symposium,* 1966 Milano

[4]   S.G. Talbert, and J.A. Eibling, *Manual on Solar Distillation of Saline Water, R & D progress report No. 546, U.S. Department of Interior.* 1970.

[5]   A. Delyannis, and E. Delyannis, "Solar distillation plant of high capacity", *Proceedings of 4th Int. Symposium on Fresh Water from Sea,* 1973

[6]   M.A.S. Malik, G.N. Tiwari, A. Kumar, and M.S. Sodha, *Solar distillation: a practical study of a wide range of stills and optimum design, construction and performance..* Pergamon Press: New York, 1982.

[7]   G.N. Tiwari, "Contemporary Physics-Solar Energy and Energy Conservation", In: *Recent Advances in Solar Distillation.* Wiley Eastern Ltd: New Delhi, India, 1992.

[8]   E.F. Delyannies, "Status of solar assisted desalination: A review", *In proceedings of 3rd world congress on desalination,* vol. 67, 1987 Cannes.
      http://dx.doi.org/10.1016/0011-9164(87)90227-X

[9]   E. Delyannis, "Historic background of desalination and renewable energies", *Sol. Energy,* vol. 75, no. 5, pp. 357-366, 2003.
      http://dx.doi.org/10.1016/j.solener.2003.08.002

[10] G.N. Tiwari, H.N. Singh, and R. Tripathi, "Present status of solar distillation", *Sol. Energy,* vol. 75, no. 5, pp. 367-373, 2003.
      http://dx.doi.org/10.1016/j.solener.2003.07.005

[11] S.N. Rai, and G.N. Tiwari, "Single basin solar still coupled with flat plate collector", *Energy Convers. Manage.,* vol. 23, no. 3, pp. 145-149, 1983.
      http://dx.doi.org/10.1016/0196-8904(83)90057-2

[12] D.B. Singh, G.N. Tiwari, I.M. Al-Helal, V.K. Dwivedi, and J.K. Yadav, "Effect of energy matrices on life cycle cost analysis of passive solar stills", *Sol. Energy,* vol. 134, pp. 9-22, 2016.
      http://dx.doi.org/10.1016/j.solener.2016.04.039

[13] D.B. Singh, and G.N. Tiwari, "Enhancement in energy metrics of double slope solar still by incorporating N identical PVT collectors", *Sol. Energy,* vol. 143, pp. 142-161, 2017.
      http://dx.doi.org/10.1016/j.solener.2016.12.039

[14] D.B. Singh Harendra, S. Kumar, N. Kumar, S.K. Sharma, and A. Mallick, "Effect of depth of water on various efficiencies and productivity of N identical partially covered PVT collectors incorporated single slope solar distiller unit", *Desalination Water Treat.,* vol. 138, pp. 99-112, 2019.
      http://dx.doi.org/10.5004/dwt.2019.23242

[15] G.N. Tiwari, J.K. Yadav, D.B. Singh, I.M. Al-Helal, and A.M. Abdel-Ghany, "Exergoeconomic and enviroeconomic analyses of partially covered photovoltaic flat plate collector active solar distillation system", *Desalination,* vol. 367, pp. 186-196, 2015.
      http://dx.doi.org/10.1016/j.desal.2015.04.010

[16] D.B. Singh, "Energy metrics analysis of N identical evacuated tubular collectors integrated single slope solar still", *Energy,* vol. 148, pp. 546-560, 2018.
      http://dx.doi.org/10.1016/j.energy.2018.01.130

[17] D.B. Singh, and I.M. Al-Helal, "Energy metrics analysis of N identical evacuated tubular collectors integrated double slope solar still", *Desalination,* vol. 432, pp. 10-22, 2018.
http://dx.doi.org/10.1016/j.desal.2017.12.053

[18] D.B. Singh, and G.N. Tiwari, "Exergoeconomic, enviroeconomic and productivity analyses of basin type solar stills by incorporating N identical PVT compound parabolic concentrator collectors: A comparative study", *Energy Convers. Manage.,* vol. 135, pp. 129-147, 2017.
http://dx.doi.org/10.1016/j.enconman.2016.12.039

[19] K. Bharti, S. Manwal, C. Kishore, R.K. Yadav, P. Tiwari, and D.B. Singh, "Sensitivity analysis of N alike partly covered PVT flat plate collectors integrated double slope solar distiller unit", *Desalination Water Treat.,* vol. 211, pp. 45-59, 2021.
http://dx.doi.org/10.5004/dwt.2021.26608

[20] D.B. Singh, "Exergoeconomic and enviroeconomic analyses of N identical photovoltaic thermal integrated double slope solar still", *Int. J. Exergy,* vol. 23, no. 4, pp. 347-366, 2017.
http://dx.doi.org/10.1504/IJEX.2017.086170

[21] D.B. Singh, "Improving the performance of single slope solar still by including N identical PVT collectors", *Appl. Therm. Eng.,* vol. 131, pp. 167-179, 2018.
http://dx.doi.org/10.1016/j.applthermaleng.2017.11.146

[22] D.B. Singh, V.K. Dwivedi, G.N. Tiwari, and N. Kumar, "Analytical characteristic equation of N identical evacuated tubular collectors integrated single slope solar still", *Desalination Water Treat.,* vol. 88, pp. 41-51, 2017.
http://dx.doi.org/10.5004/dwt.2017.21372

[23] D.B. Singh, and G.N. Tiwari, "Effect of energy matrices on life cycle cost analysis of partially covered photovoltaic compound parabolic concentrator collector active solar distillation system", *Desalination,* vol. 397, pp. 75-91, 2016.
http://dx.doi.org/10.1016/j.desal.2016.06.021

[24] L. Sahota, and G.N. Tiwari, "Effect of Al2O3 nanoparticles on the performance of passive double slope solar still", *Sol. Energy,* vol. 130, pp. 260-272, 2016.
http://dx.doi.org/10.1016/j.solener.2016.02.018

[25] L. Sahota, and G.N. Tiwari, "Exergoeconomic and enviroeconomic analyses of hybrid double slope solar still loaded with nanofluids", *Energy Convers. Manage.,* vol. 148, pp. 413-430, 2017.
http://dx.doi.org/10.1016/j.enconman.2017.05.068

[26] B.C. Pak, and Y.I. Cho, "Hydrodynamic and heat transfer study of dispersed fluids with submicron metallic oxide particles", *Experimental Heat Transfer: A Journal of Thermal Energy Generation Transport, Storage, and Conversion,* vol. 11, pp. 151-170, 1998.

[27] K. Khanafer, and K. Vafai, "A critical synthesis of thermophysical characteristics of nanofluids", *Int. J. Heat Mass Transf.,* vol. 54, no. 19-20, pp. 4410-4428, 2011.
http://dx.doi.org/10.1016/j.ijheatmasstransfer.2011.04.048

[28] K.S. Hwang, J.H. Lee, and S.P. Jang, "Buoyancy-driven heat transfer of water-based Al2O3 nanofluids in a rectangular cavity", *Int. J. Heat Mass Transf.,* vol. 50, no. 19-20, pp. 4003-4010, 2007.
http://dx.doi.org/10.1016/j.ijheatmasstransfer.2007.01.037

[29] C.J. Ho, M.W. Chen, and Z.W. Li, "Numerical simulation of natural convection of nanofluid in a square enclosure: Effects due to uncertainties of viscosity and thermal conductivity", *Int. J. Heat Mass Transf.,* vol. 51, no. 17-18, pp. 4506-4516, 2008.

http://dx.doi.org/10.1016/j.ijheatmasstransfer.2007.12.019

[30] P.K. Nag, *Engineering Thermodynamics.* TMH, 2004.

CHAPTER 5

Application and Development in Solar Stills

Vikas Kumar Thakur[1,*] M.K. Gaur[1], G.N. Tiwari[2] and M.K. Sagar[1]

[1]Department of Mechanical Engineering, Madhav Institute of Technology and Science, Gwalior-474005, India

[2]Bag Energy Research Society, Jawahar Nager (Margupur), Chilkhar-221701, Ballia UP, India

Abstract: One-third of the Earth is covered by seawater, yet there is a constant lack of water in many places. A total of 97% of the water is present in the sea as salt water, and only 3% of water is potable, out of which only 1% of clean water reaches the people. Therefore, a device is needed that can convert salt water into clean water. Solar still is a sustainable device through which dirty and salt water can be converted into clear water. Due to the low productivity of conventional solar still; it is not popular in the market. Increasing the productivity of conventional solar still is a major challenge for researchers. Researchers are continuously working on the performance of solar still to increase its productivity. The modifications and designs made by researchers in solar still over the last ten years are encapsulated in this chapter. Solar still with PCM, nanoparticles, reflectors, collectors, external condenser, wick materials, and different angles are studied, and applications of distilled water have also been covered in this chapter.

Keywords: Active solar still, Distilled water, Passive solar still, Solar still.

1. INTRODUCTION

Water is a necessary component of animal and human life. A total of 71% of the Earth's surface is covered with water. 97% is present in the oceans and seas, which is salty and cannot be used for drinking. Only 3% of potable water is available for drinking, of which 2% is found in glaciers in the North and South Pole, and the rest of 1% is available in rivers, lakes, ponds, and groundwater, so a device is required which converts the saline water into potable water.

---

[*]Corresponding author Vikas Kumar Thakur: Department of Mechanical Engineering, Madhav Institute of Technology and Science, Gwalior-474005, India; E-mail: vikasthakur1502@gmail.com

Manoj Kumar Gaur, Brian Norton & Gopal Tiwari (Eds.)

Various water purifier devices are available in the market, but all are operated through electricity, which is not environmentally friendly. Hence, a device is needed that works using renewable energy and can produce fresh water at a low cost. Solar still (SS) is a sustainable device using direct solar radiation to purify impure and saline water [1]. Due to the low production rate, solar still is not popular in the market.

As the population increases, the demand for clean water is also increasing. So solar still is the best alternative to the existing convention water purifiers, but SS is not used on a large scale due to its low productivity. Therefore, researchers constantly use several methods to enhance the yield of SSs. The classification of solar still is given in Fig. (**1**). Solar still is generally of two types; the first is passive SS, and another is active SS. When SS is operated in the natural mode, it is called passive SS, and when an external device is used to heat the water, it becomes active SS. The passive and active SSs are divided into two parts: single slope and double slope. SS with only one inclined condensing cover is called single slope SS, and in the dual-slope SS, two glass covers are installed with opposite facing.

Many researchers have intensively studied the parameters affecting the performance and heat transfer of SS to achieve maximum yield. Various modifications have been made that can increase the productivity of SSs.

A comprehensive review has been prepared by Panchal and Mohan [2]; they studied the performance of fins, heat storage material, and multi-basin. They have studied how to increase the productivity of stills through three effects. Kabeel *et al.* [3] described the various heat exchange mechanisms adopted by the researchers on different modified SSs. External and internal mirror reflectors were used by Tanaka [4] to increase the convective and evaporative heat transfer coefficient of SSs and study their heat and mass transfer. Raju and Narayan [5] performed an experimental study on a SS; they added a different number of flat plate collectors (FPC) to a simple SS. It was found that when a single flat plate collector was added to the setup, its distilled efficiency was 6.82%. When two FPC were added to the setup, the distilled efficiency was 7.29%. As the number of FPCs was added to the SS, the solar radiation receiving area increased, and water got preheated quickly before entering the basin.

Kabeel *et al.* [6] used a separate condenser chamber to increase the productivity of SS and nanoparticles were mixed in the basin water to increase its thermal conductivity and evaporation rate. Productivity increased by 53.2% when an external condenser was added to the setup, and productivity increased by 116% when the nanoparticle was added to the basin water. Refalo *et al.* [7] analyzed the effect of solar chimneys and condensers on the productivity of SS. It was found that when the chimney was added to the SS, the setup produced 5.1 liters/m^2 of distilled water in a day. When the external condenser was added to the setup, the setup productivity was 4.7 liters/m^2 in a single day. The chimney and external condenser increased the surface area of the vapor; hence the productivity increased. Rashidi *et al.* [8] placed a rectangular sponge of rubber inside the SS and did an exergy analysis. The productivity of this modified SS was 17.35% higher than the conventional SS. The sponge produced the capillary effect, broke the water molecules into small sizes, and started to evaporate in less time.

Many modifications have been made to improve the productivity and efficiency of solar still. The development in solar distillation in the past few years is discussed in this chapter. The application of distilled water produced from these SSs is also examined in this chapter.

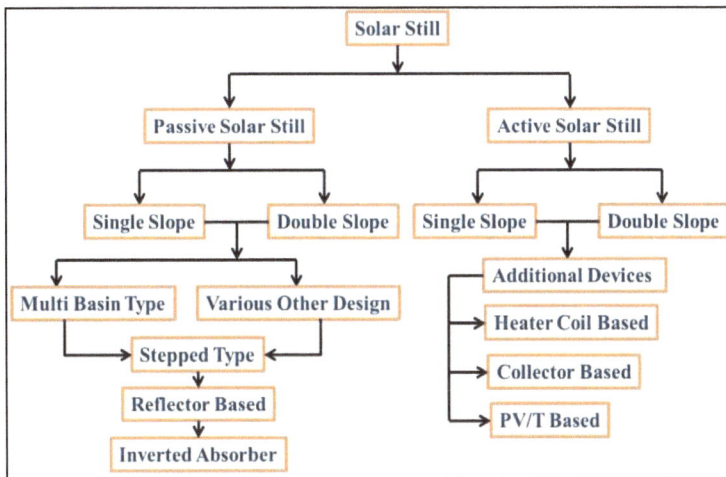

Fig. (1). Classification of solar still.

2. DEVELOPMENTS IN SOLAR STILLS IN THE LAST FEW YEARS

There have been many changes in SSs over time due to their low productivity rate. The authors have used various techniques to improve the yield of SS, which are discussed below.

2.1. Solar Still with Reflector

The reflector has been used to obtain the maximum amount of solar radiation inside the solar. Shows the external and internal reflector-based SS. The external collector increases the radiation collecting area, and hence the more solar radiation goes inside the SS. The internal reflectors reflect the radiation that falls on the SS's sidewall, thus, the evaporation rate increases [9].

2.2. Solar still with Collector

Collectors are used to preheating the water before entering into the basin; hence less heat energy is needed to evaporate the basin water inside the setup. Thus, the evaporation rate increases as more water evaporate simultaneously than in traditional SS. Generally, two types of collectors are integrated into the SS, as shown in Fig. (**2**). The first one is an evacuated tube collector, and another is a flat plate collector. Insulated borosilicate glass tubes are used in evacuated tube collectors. A vacuum is created between the two glass surfaces of the tube to reduce the heat loss through convection. A flat black surface is used in the flat plate collector to increase the absorption capacity, and a pipe is placed inside the flat surface. It is covered by transparent material (Glass, FRP, and Plastic) to create the greenhouse effects. In natural mode, both collectors work on the thermo-siphon effects.

Fig. (2). **(A)** Solar still with evacuated tube collector [12] **(B)** Solar still with flat plate collector [13].

2.3. Solar still with External Condenser

External condensers are used to increase the condensation rate. The temperature of the external condenser is much lower than the internal surface glass of the solar still, due to which the water vapor rapidly releases its latent heat and converts into distilled water. In an external condenser, the water vapor is transferred through a fan or by a natural mode, allowing evaporation inside the SS and outside the external condenser. The condensing cover also obtains potable water along with the external condenser. External condenser integrated SS is shown in Fig. (3).

Fig. (3). Solar still with external condenser [14].

2.4. Solar Still with Ultrasonic Vibrator/Fogger

The ultrasonic transducer vibrates at ultrasonic speeds and causes water molecules to break apart into individual droplets. These droplets get instantly vaporized into the air. Hence, the basin water's evaporation and convection heat transfer coefficient increases compared to convectional SS. Fig. (4) shows the SS with an ultrasonic fogger paced inside it.

Fig. (4). Solar still with ultrasonic vibrator [15].

2.5. Solar Still with Wick Materials

Wick materials have good capillary properties; due to this, the wick materials absorb water molecules, and a thin layer of water is formed at the surface of wick materials. Wick materials increase the surface area of water as compared to conventional SS. Hence, the evaporation rate of water increases, and a higher yield is obtained from the SS. Various wick materials have been used in the SS to increase the evaporation rate. Jute cloth, sponges, and cotton are used as wick materials. Different types of wick material used in SS are shown in Fig. (**5**).

2.6. Solar Still with Different Tilt Angle

The tilt angle of SS is a key factor that affects the production rate. The glass cover of the SS must be in such a position that the solar radiation falls in the normal direction on the glass cover. The maximum radiation intensity is obtained in the normal direction; hence more solar thermal energy is available inside the SS to evaporate the basin water. The reflection of radiation gets reduced at the normal incidence of radiation. If the angle of the glass cover is not perfect, then the solar radiation gets reflected from it, and a sufficient amount of radiation does not reach the basin liner. The sun's position is different every season, such as in summer, the sun travels above the latitudes. Hence, the lower tilt angle of the glass cover gives higher productivity, while a higher angle gives maximum productivity in the winter season because the sun travels lower from the latitude. In yearly performance analysis, it is observed that latitude-based SS gives higher productivity because the sun travels maximum at the latitude. In Fig. (**6a**), the sun position in different seasons is shown and in Fig. (**6b**), the different angle-based SS is shown.

Fig. (5). Different types of wick materials are used in the solar still [16].

Fig. (6). (**a**) Sun position in different seasons (**b**) Solar stills with different angles [17].

2.7. Solar Still with Parabolic Trough Collector (PTC)

PTC preheats the base fluid before entering the basin. PTCs are used as the reflector and concentrator of solar radiation. The water pipe is fixed at a focal point of the PTC, hence receiving an extreme amount of concentrated solar radiation that heats the supply water quickly to high temperatures. The preheated basin water enters the basin, and hence less energy is required to evaporate the water. Fig. (**7**) shows the PTC-based SS.

2.8. Solar Still with Charcoal Cylinder

Charcoal has a good absorption capacity. When placed inside the basin water, it absorbs more solar radiation and transforms it to heat energy; hence, the basin water gets heated in less time and evaporates. Due to its good absorption capacity, charcoal stores the solar thermal energy and releases it after the sunset to keep the SS operational even after the sunset. Hence, the modified SS achieves higher productivity at night than conventional SS. Fig. (**8**) shows the charcoal-based SS.

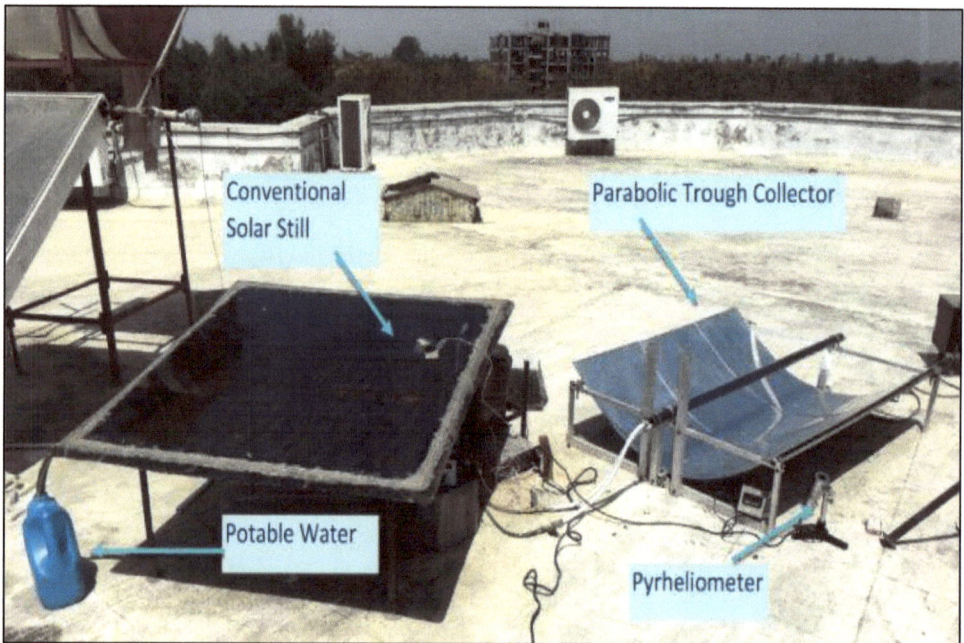

Fig. (7). Parabolic trough collector-based solar still [18].

2.9. Solar Still with Stepped Basin

A stepped SS is shown in Fig. (**9**). A stepped type of basin is constructed to increase the surface area of the basin. There is a thin layer of water in each step of SS. As the amount of water mass in each step of the basin is less than in the flat basin, water gets heated very quickly and evaporates in less time. In stepped SS, the cavity area is also less than the conventional SS. Hence less air is present inside the stepped SS, and the volume of stepped SS gets heated much faster than conventional SS.

Fig. (8). Charcoal-based solar still [19].

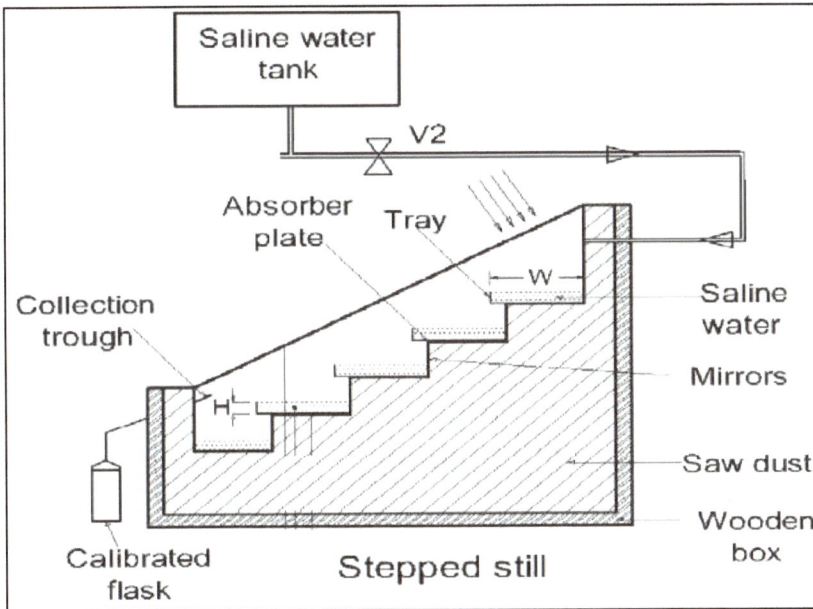

Fig. (9). Stepped solar still [20].

2.10. Solar Still with Nanoparticles

Nanoparticles are very popular among researchers due to their thermophysical properties. They are high thermal conductive materials, which, after being added to water, also change their thermal conductivity. The thermal conductivity of plain water is 0.6 W/m°C, while its thermal conductivity increases after adding nanoparticles. The thermal conductivity of CuO (Copper oxide) nanoparticles is 33 W/m°C [21]; therefore, the thermal conductivity of water also increases when CuO nanoparticles are added to the basin water. The thermal conductivity of nanoparticles and water mixtures depends on the size and concentration of nanoparticles. Nanoparticles absorb solar radiation and transfer heat to water, thus increasing the water temperature. When the thermal conductivity of basin water increases, the internal heat transfer coefficient also increases, which increases the daily yield of the SS.

Fig. (10). Solar still with PCM [22].

2.11. Solar Still with PCM.

The PCM changes its phase after absorbing and releasing heat, from solid to liquid and liquid to solid, respectively. Due to these properties, PCMs store solar thermal energy, and the absorbed energy is used during the off sunshine period to keep the SS operational. PCM absorbs the solar radiation in the daytime and at night; when the basin water temperature is low, the PCM transmits its heat to the basin water, and hence the distilled water is also obtained at night. PCMs used in SSs maintain the basin water temperature even after sunset. Fig. (**10**) shows the PCM-based SS, in which PCM is filled around the still basin.

3. APPLICATIONS OF SOLAR STILL

The application of SS is classified into two parts: domestic applications and industrial applications. Water obtained from the SS is not only used for drinking purposes but there are various industrial applications where distilled water is also used. The distilled water obtained by SS is used in battery industries, automobile industries, hospitals, food technology (preservation), and cosmetic industries. The utility of distilled water is described below in detail.

3.1. Domestic Applications

Solar still is mostly used for drinking water production in the arid and remote zone. Due to the lack of proper access to clean water in the village, consumption of dirty water leads to many fatal diseases, such as cholera, jaundice, fever, *etc.* Fig. (**11**) shows a village where people use contaminated water for drinking purposes. Hence, in remote areas where fresh water is not easily available, solar still can help reduce the drinking water problem. Fig. (**12**) shows the use of solar still by villagers.

Fig. (11). Dirty water used by the villagers [23].

Fig. (12). Solar still used by villagers [24].

3.2. Industrial Applications

Solar still has been used for various applications in many industries. In the past decade, the performance of solar energy has been analyzed by Funken *et al.* [25] to concentrate dilute, hazardous waste. It has also been used to remove organic and inorganic substances through photo-oxidation. Additionally, the feasibility of solar dewatering sludge through a SS has been studied by Haralambopoulos *et al.* [26].

Solar stills were used by Potoglou *et al.* [27] to extract distilled water from olive oil wastage. Then, this distilled water was used to make olive oil. The author used cyclic processes, and the efficiency of solar distillation of olive mill wastewater was examined.

Additionally, an attempt has been made by Velmurugan and Srithar [28] to convert textile industry effluent into potable water using solar desalination methods.

The ability of SSs to remove organic compounds has also been investigated by Hanson *et al.* [29]. The author reports on the performance of single-basin SSs to remove a selected group of inorganic, bacteriological, and organic contaminates. Based on this study, it can be concluded that SS is also used in laboratory work.

Use of Distilled Water in Other Industries

Due to the low productivity, SS is not widely used, but distilled water can be used in various applications.

Battery Industries

Distilled water contains no dissolved minerals, salts, or organic and inorganic compounds that may harm the battery. Therefore, the battery manufacturing company recommends that only distilled water be used for long-lasting and good battery performance because the salt in the normal water spoils the battery quickly.

Medical Field

Instruments used during surgeries need to be sterilized. The zero mineral content of distilled water means sterile instruments will not have any spotting or residue. It will also not leave deposits on the equipment used to sterilize them. Distilled water is also used to clean the wounds due to its purity and prevents infection.

Automobile Field

In the automobile industry, distilled water is used in the coolant because it prevents corrosion and gives long life to the engine.

Food Preservation

Distilled water keeps fruits and vegetables safe for a long time; food spoils quickly when it is cleaned with normal water.

CONCLUSION

The yield of SS depends on solar radiation. The higher the amount of solar radiation received by the setup, the higher the distillate output. Solar collectors, wick materials, stepped basins, *etc.*, use parabolic trough collectors to increase the surface area and receive a large amount of solar radiation. Heat storage materials store solar radiation, such as PCM, charcoal cylinders, *etc.* The higher the amount of solar radiation reaches the basin, the higher the evaporation rate of the basin water, so the researchers have used mirrors and reflectors in their studies. Increasing the evaporation rate alone does not increase productivity; the condensation rate must also be improved. The faster the evaporation, the quicker the latent heat will release, and only then the condensation rate will increase. Hence, the external condenser has been widely used by researchers.

CONSENT FOR PUBLICATION

Not applicable.

CONFLICT OF INTEREST

The authors declare no conflict of interest, financial or otherwise.

ACKNOWLEDGEMENTS

Declared none.

REFERENCES

[1]   V.K. Thakur, M.K. Gaur, and M.K. Sagar, "Role of advance solar desalination technique for sustainable development", In: *Intelligent Computing Applications for Sustainable Real-World Systems,* vol. 13. 2020, pp. 28-38.

http://dx.doi.org/10.1007/978-3-030-44758-8_4

[2]   H. Panchal, and I. Mohan, "Various methods applied to solar still for enhancement of distillate output", *Desalination,* vol. 415, pp. 76-89, 2017.
http://dx.doi.org/10.1016/j.desal.2017.04.015

[3]   A.E. Kabeel, T. Arunkumar, D.C. Denkenberger, and R. Sathyamurthy, "Performance enhancement of solar still through efficient heat exchange mechanism – A review", *Appl. Therm. Eng.,* vol. 114, pp. 815-836, 2017.
http://dx.doi.org/10.1016/j.applthermaleng.2016.12.044

[4]   H. Tanaka, "A theoretical analysis of basin type solar still with flat plate external bottom reflector", *Desalination,* vol. 279, no. 1-3, pp. 243-251, 2011.
http://dx.doi.org/10.1016/j.desal.2011.06.016

[5]   V.R. Raju, and R. Lalitha Narayana, "Effect of flat plate collectors in series on performance of active solar still for Indian coastal climatic condition", *Journal of King Saud University - Engineering Sciences,* vol. 30, no. 1, pp. 78-85, 2018.
http://dx.doi.org/10.1016/j.jksues.2015.12.008

[6]   A.E. Kabeel, Z.M. Omara, and F.A. Essa, "Enhancement of modified solar still integrated with external condenser using nanofluids: An experimental approach", *Energy Convers. Manage.,* vol. 78, pp. 493-498, 2014.
http://dx.doi.org/10.1016/j.enconman.2013.11.013

[7]   P. Refalo, R. Ghirlando, and S. Abela, "The use of a solar chimney and condensers to enhance the productivity of a solar still", *Desalination Water Treat.,* pp. 1-14, 2015.
http://dx.doi.org/10.1080/19443994.2015.1106096

[8]   S. Rashidi, N. Rahbar, M.S. Valipour, and J.A. Esfahani, "Enhancement of solar still by reticular porous media : Experimental investigation with exergy and economic analysis", *Applied Thermal Engineering,* vol. 130, 2017.
http://dx.doi.org/10.1016/j.applthermaleng.2017.11.089

[9]   V.K. Dwivedi, and G.N. Tiwari, "Experimental validation of thermal model of a double slope active solar still under natural circulation mode", *Desalination,* vol. 250, no. 1, pp. 49-55, 2010.
http://dx.doi.org/10.1016/j.desal.2009.06.060

[10] H. Tanaka, "Experimental study of a basin type solar still with internal and external reflectors in winter", *Desalination,* vol. 249, no. 1, pp. 130-134, 2009.
http://dx.doi.org/10.1016/j.desal.2009.02.057

[11] W. Campbell, "Development of a solar furnace with high insulating properties using date palm waste", *The Second International Conference on Green Communications, Computing and Technology,* 2017pp. 1-11

[12] H.N. Panchal, and P.K. Shah, "Enhancement of distillate output of double basin solar still with vacuum tubes", *Front. Energy,* vol. 8, no. 1, pp. 101-109, 2014.
http://dx.doi.org/10.1007/s11708-014-0299-5

[13] E. Deniz, *Solar-Powered Desalination.* Intech, 2015, pp. 90-124.
http://dx.doi.org/10.5772/60436

[14] A.E. Kabeel, Z.M. Omara, and F.A. Essa, "Numerical investigation of modified solar still using nanofluids and external condenser", *J. Taiwan Inst. Chem. Eng.,* vol. 75, pp. 77-86, 2017.
http://dx.doi.org/10.1016/j.jtice.2017.01.017

[15] A.I. Shehata, A.E. Kabeel, D.M.M. Khairat, and A.M. Elharidi, "Enhancement of the productivity for single solar still with ultrasonic humidifier combined with evacuated solar collector: An experimental study", *Energy Convers. Manage.,* vol. 208, p. 112592, 2020.

http://dx.doi.org/10.1016/j.enconman.2020.112592

[16] K. Kalidasa Murugavel, and K. Srithar, "Performance study on basin type double slope solar still with different wick materials and minimum mass of water", *Renew. Energy,* vol. 36, no. 2, pp. 612-620, 2011.
http://dx.doi.org/10.1016/j.renene.2010.08.009

[17] R. Cherraye, B. Bouchekima, D. Bechki, and H. Bouguettaia, "The effect of tilt angle on solar still productivity at different seasons in arid conditions (south Algeria)", *Int. J Ambient Energy,* pp. 1-16, 2020.
http://dx.doi.org/10.1080/01430750.2020.1723689

[18] R. Dev, and G.N. Tiwari, "Characteristic equation of a passive solar still", *Desalination,* vol. 245, no. 1-3, pp. 246-265, 2009.
http://dx.doi.org/10.1016/j.desal.2008.07.011

[19] M.R. Rajamanickam, P. Velmurugan, A. Ragupathy, and E. Sivaraman, "Use of thermal energy storage materials for enhancement in distillate output of double slope solar still", *Materials Today: Proceedings,* vol. 32, no. 2, 2020.
http://dx.doi.org/10.1016/j.matpr.2020.02.203

[20] Z.M. Omara, A.E. Kabeel, and M.M. Younes, "Enhancing the stepped solar still performance using internal and external reflectors", *Energy Convers. Manage.,* vol. 78, pp. 876-881, 2014.
http://dx.doi.org/10.1016/j.enconman.2013.07.092

[21] V.K. Thakur, and M.K. Gaur, "A study on passive solar still with nanoparticles", *Int. J. Energy Technol.,* vol. 2, no. 1, pp. 26-38, 2020.
http://dx.doi.org/10.32438/IJET.203009

[22] D.C. Kantesh, "Design of solar still using Phase changing material as a storage medium", *Int. J. Sci. Eng.,* vol. 3, pp. 1-6, 2012.

[23] Dev. "Water scarcity ravages Rajasthan villages even decades after Independence, Times of India Report."

[24] A.M. Burbano, "Evaluation of basin and insulating materials in solar still prototype for solar distillation plant at Kamusuchiwo community, High Guajira", *Renewable Energy and Power Quality Journal,* vol. 1, pp. 547-552, 2014.
http://dx.doi.org/10.24084/repqj12.395

[25] K.H. Funken, B. Pohlmann, E. Lüpfert, and R. Dominik, "Application of concentrated solar radiation to high temperature detoxification and recycling processes of hazardous wastes", *Sol. Energy,* vol. 65, no. 1, pp. 25-31, 1999.
http://dx.doi.org/10.1016/S0038-092X(98)00089-9

[26] D.A. Haralambopoulos, G. Biskos, C. Halvadakis, and T.D. Lekkas, "Dewatering of wastewater sludge through a solar still", *Renew. Energy,* vol. 26, no. 2, pp. 247-256, 2002.
http://dx.doi.org/10.1016/S0960-1481(01)00114-8

[27] D. Potoglou, A. Kouzeli-Katsiri, and D. Haralambopoulos, "Solar distillation of olive mill wastewater", *Renew. Energy,* vol. 29, no. 4, pp. 569-579, 2004.
http://dx.doi.org/10.1016/j.renene.2003.09.002

[28] V. Velmurugan, and K. Srithar, "Industrial effluent treatment: Theoretical and experimental analysis", *J. Renew. Sustain. Energy,* vol. 3, no. 1, p. 013107, 2011.
http://dx.doi.org/10.1063/1.3558862

[29] A. Hanson, W. Zachritz, K. Stevens, L. Mimbela, R. Polka, and L. Cisneros, "Distillate water quality of a single-basin solar still: Laboratory and field studies", *Sol. Energy,* vol. 76, no. 5, pp. 635-645, 2004.
http://dx.doi.org/10.1016/j.solener.2003.11.010

CHAPTER 6

Potential of Solar Distillation Plant in India

Rajkumar Malviya[1], Veeresh Vishwakarma[1], Prashant V. Baredar[1*] and Anil Kumar[2,3]

[1]*Energy Centre, Maulana Azad National Institute of Technology, Bhopal - 462003, M.P., India*
[2]*Depratment of Mechanical Engineering, Delhi Technological University, Delhi 110042, India*
[3]*Centre for Energy and Environment, Delhi Technological University, Delhi 110042, India*

Abstract: With the rising population and continuous depletion of our natural resources, it has become very tough for everyone to meet their basic needs of food and water. Also, at the rate with which the water-stressed area continues to rise, we soon will be facing a huge water crisis. This chapter specifically talks about India and its potential to make a switch from conventional methods of water usage and switch to a renewable energy-based water desalination unit. This chapter presents an elaborate analysis of the Indian peninsular region and talks about the major cities' comparative performance in the basic design of the solar humidification-dehumidification desalination unit. It can be concluded that the southern-most area has a very large potential for setting up an economically feasible desalination unit. Various parameters are discussed, like humidity ratio, outgoing airstream temperature, and mass rate of evaporated water. As Chennai has the best performance for the particular unit for most of the year, with productivity reaching 44 kg/day, the least favorable site seems to be Puri in Odisha, where productivity remains less and constant at a maximum of 34 kg/day during summers.

Keywords: Desalination, Freshwater, Humidification, Solar.

1. INTRODUCTION

Freshwater is in great demand and is an increasingly vital issue in rural areas of India. Potable water is very scarce in arid areas and the establishment of a human habitat in these areas strongly depends on how much water can be made available. Water is an essential element of life. The issue of supplying potable water can hardly be overstressed. There are abundant resources on earth. The planet's three-fourths of its surface is covered by water. About 97% of the earth's water is

---

*Corresponding author Prashant V. Baredar: Depratment of Mechanical Engineering, Delhi Technological University, Delhi 110042, India; Tel: +919406511666; E-mail: prashant.baredar@gmail.com

Manoj Kumar Gaur, Brian Norton & Gopal Tiwari (Eds.)

saltwater in the oceans. Only 3% is freshwater contained in the poles (in the form of ice), groundwater, lakes, and rivers, which supply most human and animal needs. Nearly 70% of this tiny 3% of the world's fresh water is frozen in glaciers, permanent snow cover, ice, and permafrost. Thirty percent of all freshwater is underground, most of it in deep, hard-to-reach aquifers. Lakes and rivers together contain just a little more than 0.25% of all freshwater; lakes contain most of it.

Salman H. Hamaadi *et al.* (2018) studied the specific effect of temperature change of glass and air along the length of the distiller when it allows for the mass transfer and heat convection between glass cover and air. He stated that the maximum production of freshwater was during the month of March that is 2.2 kg/m^2.day and reached a minimum value during July, which was 0.66 kg/m^{2}day also, the "productivity decreases with increasing air velocity inside the distiller while the evaporation rate increases"; also this study is very much relative to the present study [1].

Ali *et al.* (1993) considered the influence of convection (forced) within the solar still upon the coefficient of mass and heat transfer. Diverse factors are considered, like fluting the surface of the water. The motion of air results in turbulent eddies, an increase of gases that are incondensable, and vapor velocity in the still. He came up with; any rise in the yield of the solar desalination unit is primarily because of the improvement in the coefficients of heat and mass transfer owing to the presence of the vapor-air blend turbulent wave interior of the still [2]. Sartori *et al.* (1996) offered a theoretical evaluation concerning the thermal performance of a basin from a solar still and that of a solar evaporator. The evaporator and the still are fabricated of glass fiber and insulated (thermally) with glass wool of 0.045 m at the sides and bottom. Both surfaces of water have 1m^2 and 0.04 m layers. A conjoint 3 mm impenetrable glass has been engaged to cover the still and established that vaporization in stills was in a smaller amount than in open evaporation regardless of the greater temperatures of water in the previous arrangement. For higher system temperatures, the fraction of evaporation is equal to 50% or even higher than the parallel overall heat transmission [3].

Radhwan *et al.* (2004) offered a transitory study of a solar still in a unique stepped design for humidifying and heating of, particularly agriculture utilized gases. Air dispersed within still present is moistened as well as heat. So the still outcomes indicated produce was 4.92 L for a unit area every day, with a reduction in the flow rate of air having a minor impact on the method output [4]. Mofreh H. Hamed *et al.* (2015) proposed a practical system of humidification-dehumidification desalination and analyzed the operation in two periods, one of which when he calculated the working cycle between 9 am and 5 pm local time; the other begins with preheating the air going for humidification at 1 pm till 5 pm local time. So, the result showed the system ran 4 hr every day, and preheating provides greater production, which is around 22 L/daym², and the overall cost per liter of that unit remains close to Rs. 4 [5].

Zhou *et al.* (2010) introduced another technique relating to the withdrawal of heat and humidity present in a brine solution in the accumulator of the solar vent arrangement regarding electricity production and water treatment. Further, their study resolved that amount produced through water generators and wind turbine generators in the joint plant is lower than that of the standard system because of the discharge of latent heat of vapor following the gases escalating in the chimney [6].

Okati *et al.* (2016) considered a desalination method through solar energy of humidification–dehumidification method comprising a humidification setup with underground condenser apparatus intended to generate clean water. Brine gets heated inside the humidification device, then at that point the steam gets delivered over a fixed set of ducts concealed in the ground, the condensation mechanism then starts and freshwater is produced. The outcomes presented that the production rate of fresh H_2O per concealed duct's size is 3.8120 (kg/m.hr) [7]. Okati*et al.* (2018) offered a work of another solar-assisted desalination unit working on the above principle of humidification-dehumidification (HDH) method, combining the solar still with a concealed dehumidifier working as a condenser. The outcome specified that the water production rate could range above 265 kg/day (approx.) [8].

Abdel Dayem and Fatouh (2009) set up as well as examined a "numerous-effect HDH solar desalination unit. The system involves 2 loops, a water desalination loop and a solar loop. Three systems were analyzed by testing and numerically. It was evident that the unlocked structure with free circulation for solar energy is more effective irrespective of real awkwardness, whereas the locked structure with a supplementary heater has the maximum distilled production of water [9]. Prakash *et al.* (2010) presented a broad analysis with a futuristic "solar humidiifcation dehumidiifcation (HDH) desalination unit." Specific consideration had been set on heaters, with limited modification parameters; also, direct water heating was matched with direct air heating for process calculations. Different methods centered on the HDH theory were also looked over and compared. It was decided that HDH machinery has pronounced potential for decentralized limited water production uses, even though further studies and advances are desirable intended for reducing capital cost and enhancing system efficiency [10]. Design, along with a study regarding various HDD developments, was examined by Ettouney *et al.* (2005).

Providing hands-on equations related to the proposal of an HDH Process in addition to a working outline for a given setup with the base as exit temperature of the air leaving humidification unit, They considered some downside of HDH technique which primarily is caused due to occurrence of air in huge quantity accumulated with the production of vapor. The huge sum needs to dehydrate throughout the dehumidification stage diminishes the procedure's efficiency [11]. Zhani *et al.* (2010) premeditated an innovative solar desalination setup using the HDD technique built at "National Engineering School of Sfax, Tunisia." Its premium feature that cleaned water attained through such novel theory favors its usage to make water to meet irrigation as well as consumer needs. Evaluation of heat performance had been done in the form of gained output ratio (GOR) and also water productivity. Runs of trials with the simulation findings to authenticate the designed replicas as the outcome, the planned replica can be used for dimensioning and inspecting the performance of this type of desalination unit [12]. Nafey *et al.* (2004) examined an HDH desalination unit in which both the air and water are using solar heating in Egypt. A freshwater generation proportion of 10.71 L/day was accomplished in midsummer, but then the rate fell to 5.1 L/day later at the end of the summer. Although many of these former studies have contributed in the direction of the advances of solar-operated HDH desalination, the electrical specific energy exhaustion was not stated [13]. Chafik *et al.* (2003) proposed an innovative kind of seawater desalination plant with the help of solar energy. The method is comprised of multiple steps for air heating, each coupled with a humidification stage [14]. El-Agouz *et al.* (2010) determined by testing the effectiveness of a single-step bubble picket using air bubbles fleeting through seawater. The outcomes presented that the output of the system improves using the upsurge of the water temperature and the reduction in the rate of airflow. The determined system productivity touched the airflow rate of 14 kg/h and 8.22 kg/h at 86°C for water temperature [15]. Orfi *et al.* (2004) considered a solar HDH desalination system ideally and experimentally. To increase the output of the arrangement, what they planned is to utilize the latent heat of condensed vapor in the condenser to warm up the water to be fed [16]. Abdel-Dayem *et al.* (2019) discovered theoretically and by testing the effect of an evolved unit of solar water desalination on the generation capacity of a desalination unit using HD operation under the meteorological circumstances of Cairo, Egypt [17].

2. ROLE OF RENEWABLE ENERGY IN WATER DESALINATION

The evaporation of aquatic solutions using solar energy and the simultaneous condensation of the vapors due to solar thermal energy is called as solar distillation. Humidification-dehumidification takes place in a device called "solar still." From early antiquity, it has been a physical procedure. There exist many brief historical reviews on solar distillation starting from antiquity, where many trials were performed to produce freshwater from salty waters. Many of the descriptions about desalination are perfect ideas, but the primitive knowledge of technology and construction designs was very poor at the time they were expressed for achieving any practical application.

Even though having a share of 70% of the ecosystem, there is an acute shortage of fresh, usable water. Traditional methods for water desalination are complex and exhaustive with an extraordinary state of toxins. A simple desalination unit has been shown in Fig. (**1**). Renewable energy must have been engaged so that we can create the desalination method that is environmentally friendly and cheaper in remote and sunny places. Renewable energies like solar are clean and have been coupled with the humidification & dehumidification method (HDH) has now been broadly used for the best outcome in the desalination process. In such a method, water warmed up by the solar irradiation humidifies the ambient air.

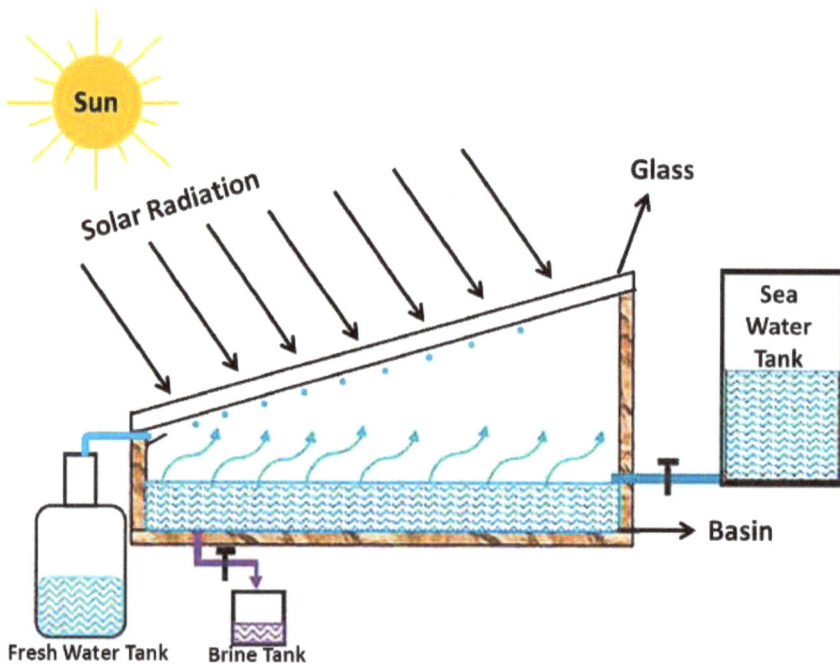

Fig. (1). Simple water desalination unit [18].

Then air that is moist gets dehumidified after contact with a substance whose temperature is relatively much lesser than the humid air's dew point temperature in addition to the vapor advances as desalinated water. With rising demand and consumption of water in the present scenario of global population rise, it is difficult to get access to fresh water at will. This problem coupled with the rise in various water-stressed areas around the globe needs to get to a solution that can help elevate the people in need of clean drinking water and provide them with a healthy lifestyle. A meager 0.014% of all water present on the planet is both fresh and easily reached. Out of that, we have 97% of the total water available is seawater or what is commonly known as saline water. A little less than 3% of it is hard to access, making it completely out of the equation [19].

3. ENERGY TRANSFER PROCESS AND CYCLE IN BASIC SOLAR DISTILLER

The solar distiller is known to be a longitudinal water reservoir enclosed by a clear pane of glass with a dried air intake and a moist air outlet which is described in Fig. (2).

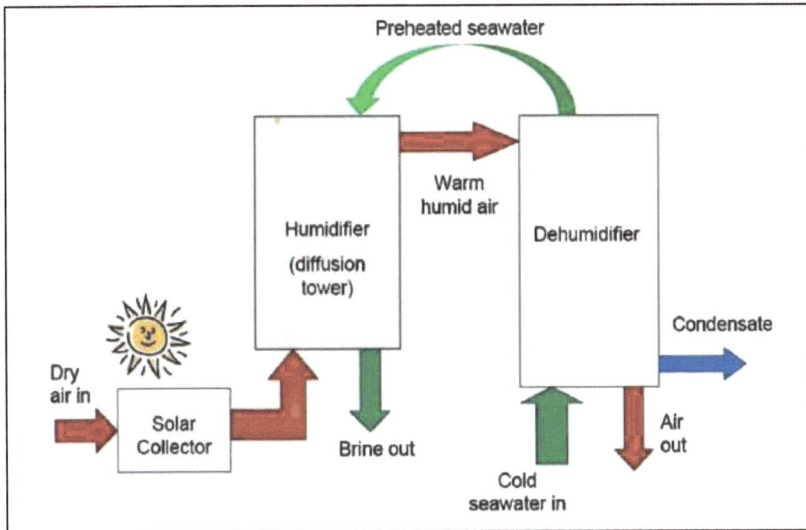

Fig. (2). Basic Humidification-Dehumidification Cycle [10].

Also, heat energy from sunlight warms the seawater and allows it to evaporate through the air stream. Based on the rate of evaporation, the moisture starts to increase along the reservoir to achieve a saturated state at a given period. If the glass shield temperature is well below the humid air dew point, then condensation will follow on the panel. Fig. (3) presented a methodical approach that aims to consider heat and mass transfer between glass cover and air.

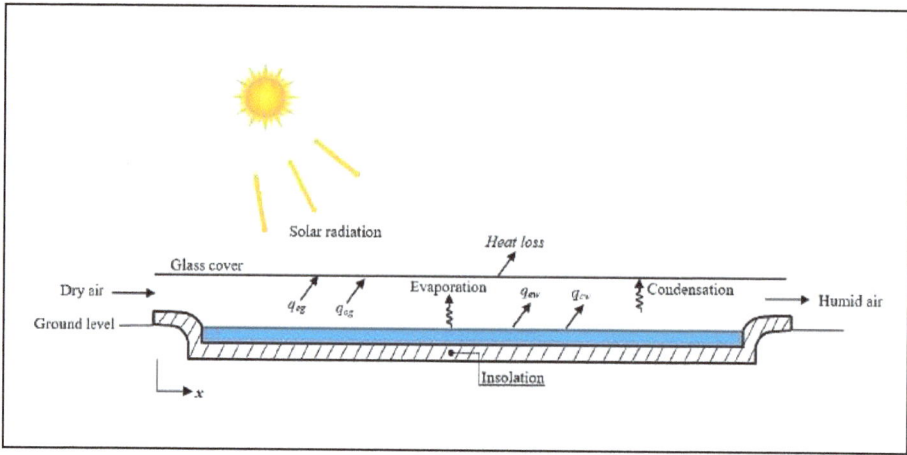

Fig. (3). Energy transfer process [1].

3.1. Energy Balance Equations

In this approach, solar distiller's energy balance is developed:

$$C_{p_f}\left(T_{f_o} - T_{f_i}\right) + w_o h_{v_o} - w_i h_{v_i} = \frac{1}{\dot{m}_f}\left(q_{c_w} + q_{e_w} - q_{c_g} - q_{e_g}\right) \tag{1}$$

The subscripts o, i mean outlet as well as inlet parameters, respectively. Also mentioned expressions q_{e_g}, q_{c_g} denotes the condensation and convection transfer of heat energy within glass cover and airstream also, the terms q_{e_w}, q_{c_w} relate to evaporation and convection transfer of heat energy within airstream and water. These expressions are stated as:

$$q_{c_w} = h_w A\left(T_w - T_f\right) \tag{2}$$

$$q_{e_w} = \frac{h_w}{C_{p_f}} A h_{fg}\left(w_w - w_f\right) \tag{3}$$

$$q_{c_g} = h_f A\left(T_f - T_g\right) \tag{4}$$

$$q_{e_g} = \frac{h_f}{C_{p_f}} A h_{fg}\left(w_f - w_g\right) \tag{5}$$

3.2. Specific Humidity and Humidity Ratio

The specific humidity at T_w for Eq. 3 has been denoted as [21]:

$$w_w = 0.622 \frac{P_{w_s}}{P - P_{w_s}} \tag{6}$$

The specific humidity (w_w) at T_w as well as airstream temperature T_f in context with a division of (dx) length of the distiller is estimated by:

$$w_w = \frac{w_i + w_o}{2} \tag{7}$$

$$T_f = \frac{T_{f_i} + T_{f_o}}{2} \tag{8}$$

$P_{ws,}$ which is the saturated pressure in Eq. 6, will be considered relative to air at saturation with T_w by [22, 23]:

$$P_{w_s} = e^{\left(25.317 - \frac{5144}{T_w + 273}\right)} \tag{9}$$

The inlet air's humidity ratio is given as [7]:

$$P_{f_i} = 0.622 \frac{\emptyset P_{f_{i_s}}}{P - \emptyset P_{f_{i_s}}} \tag{10}$$

Where $P_{f_{i_s}}$ stands for saturation pressure for the temperature of inlet air that has been defined by:

$$P_{f_{i_s}} = e^{\left(25.317 - \frac{5144}{T_{fi} + 273}\right)} \tag{11}$$

Further, in Eq. 5, the humidity ratio w_g is represented by:

$$w_g = 0.622 \frac{P_g}{P - P_g} \tag{12}$$

Where P_g denotes saturation pressure for the temperature at glass surface that will be stated as:

$$P_g = e^{\left(25.317 - \frac{5144}{T_g + 273}\right)} \tag{13}$$

3.3. Heat Transfer

The coefficient of heat transfer for airstream-water h_w as well as glass cover with airstream h_f has been interpreted as [3, 7]:

$$h_w = h_f = 2.8 + 3U_f \tag{14}$$

The water vapors' latent heat can be represented as [10, 24]:

$$h_{f_g} = 1000\left(2500.8 + 2.36T_f + 0.0016T_f^2 - 0.00006T_f^3\right) \tag{15}$$

The saturated vapor's enthalpy for T_f may be obtained by [9, 25]:

$$h_g = 1000\left(2501.3 + 1.82T_f\right), h_v \cong h_g \tag{16}$$

3.4. Mass Transfer, Relative Humidity and Vapor Pressure

The solar distiller is split into tiny (dx) length components. Then, moist air's mass balance involves water film's evaporation as well as vapors' condensation on the glass cover:

$$w_o = w_i + \frac{m_{e_v} - m_{c_o}}{\dot{m}_f} \tag{17}$$

The evaporation and condensation rates m_{e_v}, m_{c_o} are determined as [7, 8, 26]:

$$m_{e_v} = \frac{h_w}{C_{p_f}} A\left(w_w - w_f\right) \tag{18}$$

$$m_{c_o} = \frac{h_w}{C_{p_f}} A\left(w_f - w_g\right) \tag{19}$$

$\emptyset$ is the outlet air's relative humidity, given as:

$$\emptyset = \frac{P_{f_o}}{P_{f_{o_s}}} \tag{20}$$

Where P_{f_o} represents "vapor pressure of outgoing air temperature," which can be calculated by:

$$P_{f_o} = \frac{w_0 P}{0.622 + w_0} \tag{21}$$

The vapors' saturated pressure for the temperature of outgoing air $P_{f_{os}}$ can be calculated as:

$$P_{f_{os}} = e^{\left(25.317 - \frac{5144}{T_{f_o} + 273}\right)} \tag{22}$$

Following the ideal theoretical conditions, there will not be any condensation of the glass cove. So, in case of no condensation $m_{c_o} = 0$ and using Eqs. 19, 18, 17 and 7, the outlet air (w_0) specific humidity can be calculated as:

$$w_o = \frac{\left[\dot{m}_f C_{p_f} - \frac{A}{2} h_w\right] w_i + h_w A w_w}{\dot{m}_f C_{p_f} + \frac{A}{2} h_w} \tag{23}$$

The glass temperature included in Eq. 4 has been calculated using the glass cover heat and mass balance as follows:

$$I(1 - \tau_g)\alpha_g + h_f(T_f - T_g) + \frac{h_f}{C_{p_f}} h_{f_g}(w_f - w_g) = h_a(T_g - T_a) \tag{24}$$

For the case of no condensation, the expression $\left[\frac{h_f}{C_{p_f}} h_{f_g}(w_f - w_g)\right]$ vanishes in Eq. 24 resulting in the glass temperature as:

$$T_g = \frac{I(1 - \tau_g)\alpha_g + h_f T_f + h_a T_a}{h_f + h_a} \tag{25}$$

The "outside heat transfer coefficient h_a" can be written as [27, 28]:

$$h_a = 3.8 U_a + 5.7 \tag{26}$$

3.5. Radiation Energy

The incident solar irradiation at the distillery includes two parts: diffused and direct; direct irradiance may be written as [29]:

$$I_{dr} = I_{sc}\tau^{m_a} \sin \beta \tag{27}$$

Where I_{sc} = 1367 W/m^2 denotes value of solar constant.

Also, diffused irradiation can be calculated by:

$$I_{df} = 0.3 I_{sc}(1 - \tau^{m_a}) \sin \beta \tag{28}$$

Here "τ" denotes transmittance through the atmosphere, whose value stays in a range of 0.65 to 0.75 [29]. For the present study, it is considered 0.7.

Overall incident solar irradiation upon distillery equals the total sum of diffused as well as direct parts:

$$I = I_{dr} + I_{df} \tag{29}$$

3.6. Mean Specific Humidity and Airstream Temperature across Distiller

The air mass (m_a) is expressed as [30]:

$$m_a = \frac{1}{\sin \beta} \tag{30}$$

The temperature of outgoing air (T_{f_o}) has been inferred using Eq. 1 as well as Eqs. 8, 4, and 2, hence:

$$T_{f_o} = \frac{\left[\dot{m}_f C_{p_f} - \frac{A}{2}(h_w + h_f)\right] T_{f_i} + \dot{m}_f (w_i h_{v_i} - w_o h_{v_o}) + h_f A T_g + h_w A T_w + q_{e_w} - q_{e_g}}{\dot{m}_f C_{p_f} + \frac{A}{2}(h_w + h_f)} \tag{31}$$

The mean specific humidity and airstream temperature across distiller can be calculated with means of numeral integration method; the expressions are as follows:

$$\bar{T}_f = \frac{1}{L} \int_0^L T_f \, dx \tag{32}$$

$$\bar{w}_f = \frac{1}{L} \int_0^L w_f \, dx \tag{33}$$

4. ANALYSIS OF BASIC HUMIDIFICATION-DEHUMIDIFICATION PLANT

To keep a constant working platform and analyze the variation of the desalination plant over the period at all the selected sites so that basic comparative results can be obtained. Table **1** Represents here the overall working of the basic humidification-dehumidification plant and will be considered the working cycle for the comparative study. The distiller is considered to be a flat plate still of 1 m wide inlet and covered with glass.

Table 1. Input Parameters.

Air Stream Velocity	0.5 m/s
Cross Sectional Area	0.2 m^2
Width of Distiller	1 m
Inlet Humidity Ratio	0.005
Air Density	1.15 kg/m^3
Air Specific Heat	1005 J/kgK
Water Density	1000 kg/m^3
Water Specific Heat	4200 J/kgK
Water absorptivity	0.9
Glass absorptivity	0.1

The weather and climate data have been collected and validated from a variety of sources, including website data of NREL, IMD, and other private providers. These data are crucial to understanding the flexibility and variations along the coast of India. The input data themselves can be related to the overall working of the unit because the ambient condition takes part in providing the outcome of the system. Hence very exact and practical assumptions have been made to compensate for any irregularities with these deduced variables.

4.1. Solar Irradiance over the Coast of India

Fig. (**4**) shows the Solar Irradiance over the coast of India at various selected locations and it shows that the monsoon has the minimum solar irradiance because of fewer sunlight hours. Also, the highest solar irradiance captured was 696.33 W/m^2 for Surat during winters, while the highest productivity obtained was at Chennai, which is 563.26 W/m^2 during summers. The outlet temperature plays a take part in calculating the productivity of plants as the higher temperature is desirable for a better humidity ratio which in turn improves the water carrying capacity of dry air.

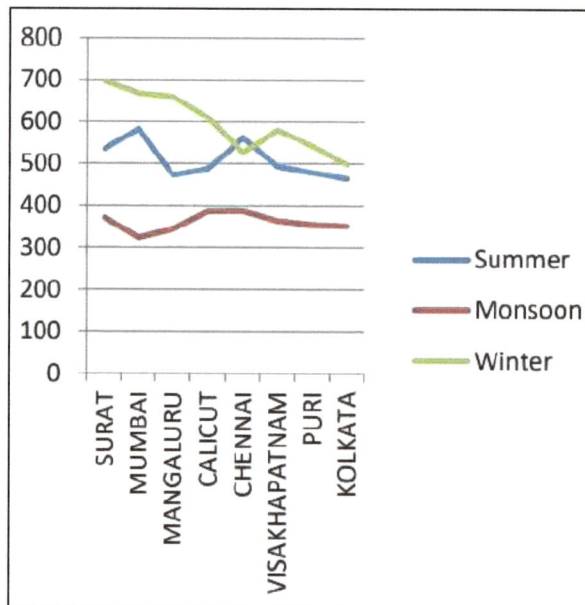

Fig. (4). Irradiation Data.

4.2. Outlet Airstream Temperature

Fig. (**5**) shows the variations in the outlet air temperature, which elaborates the effect of all other parameters taken into account so that the final result could be obtained. Key findings during this study "explained an increase in the evaporation rate with increasing the Solar Irradiance; this is due to the increase of heat and mass transfer coefficient as the solar irradiance increases." The highest temperature obtained was at Surat (62.41 °C) during winter which is basically because of relatively dry ambient conditions. Also, the most unlikely result is by logic, occurs during monsoon only.

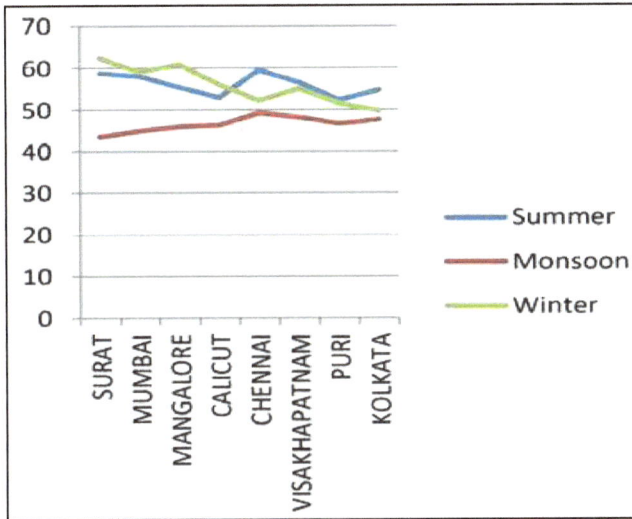

Fig. (5). Outlet Airstream Temperature.

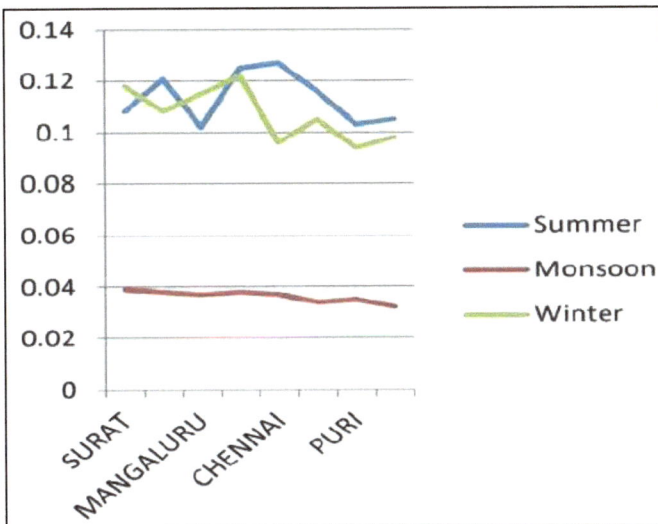

Fig. (6). Humidity Ratio.

4.3. Humidity Ratio of the Air Stream

Fig. (**6**) tells us about the output humidity ratio of the air stream, which in turn will be slightly higher than the inlet humidity ratio, which is kept constant (0.005 kg/kg of dry air) and it is desirable to obtain a high value for this parameter because higher the value of specific humidity the more will be the mass of water vapor in a kg of dry air. The best performing site is in Chennai without going specific humidity equal to 0.125.

4.4. Rate of Mass Flow for Evaporated Water

Fig. (**7**) tells that the mass of evaporated water is very low during the monsoon season all along the coast, which is primarily due to unavailability or limited hours of sunlight resulting in low solar irradiation. The southernmost part of the Indian peninsula happens to be very close to the equator resulting in the ability to capture the highest amount of solar irradiance. Even because of unlikely irradiance, the major development behind such high water evaporation capacity can also be given to higher water inlet temperature. The slop of the evaporation rate moderately decreases with the distiller as enhancing the relative humidity results in decrease in the evaporation capacity.

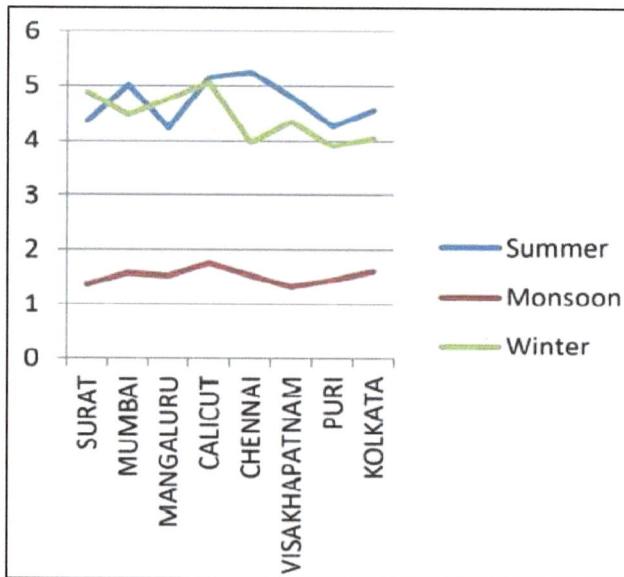

Fig. (7). Rate of Mass Flow for Evaporated Water.

It is noted that the Surat is getting a very much desired result while working in winters; this is mainly because of high Solar Irradiance collected over the place and also a dry environment presents most of the time. Even though having most of the eye-catching results, Surat's productivity is slightly on the lower side; this is because of early condensation inside the distiller chamber. Indeed this problem is less significant during the winters than in summers.

4.5. Productivity of the Desalination Unit

Fig. (**7**) describes the productivity of the desalination unit for all of these sites throughout the year, and it can be deduced that Kozhikode has the maximum productivity for the majority of the time. Also, cities including Chennai, Mumbai, Visakhapatnam, and Surat also have great potential and can be considered while setting up a desalination plant.

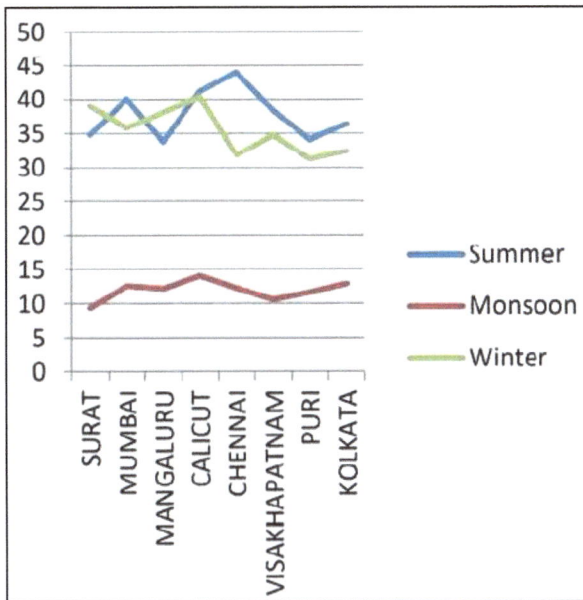

Fig. (8). Productivity.

5. Summary and Outlook

Mumbai has shown some significant data regarding the solar desalination unit to be set up in such an area having good solar energy potential and perfect weather conditions that support setting up one. Even though the output temperature is not quite high for Mumbai, the reason for its high productivity will be a high humidity ratio and high inlet water temperature.

Mangaluru gives a result of exposing that the output for the particular city is not that attractive but still gives a satisfactory result during winters. It is because of the high solar irradiance at the place. Also, the outgoing airstream has quite a high temperature. Hence, it cannot be considered as a favorable place for setting up a desalination unit.

Calicut or Kozhikode is the southernmost city in this study and gave a desirable result as it produced quite a significant amount of freshwater. As the monsoon /hits the coast of the state very early every year, the solar irradiance is quite low comparatively, but the overall production is quite good. Due to the acceptable humidity ratio value, can ensure that the product is satisfactory. Also, the mass of evaporated water is also good because condensation doesn't occur for the majority of the air stream.

Chennai is another city belonging to the southernmost states of the peninsula. Yet the result for the city is a little on the lower side. The only time it is feasible to profitably run the unit is summers that too; because of early monsoon, we get a deficit of what the actual potential of solar irradiance is as well as during winter we see retreating monsoon hitting hard the coast of Tamil Nadu and Andhra Pradesh the productivity is reduced. Also, there is a high risk of damage because lately, it has been one of the severely affected zones.

Visakhapatnam showed that the productivity there was low primarily because of low inlet water temperature and high wind velocity. Apart from that, the retreating monsoon has a significant effect on Andhra Pradesh's coast also. Therefore it affects the output during some parts of the winters.

Puri is situated quite close to one of the largest saltwater lakes in India is a favorable site for seawater desalination. While having a decent sunlight presence over the city, it offers a very favorable location, yet the output suffers for the most part of the year. Also, being around the Bay of Bengal increases the potential risk of cyclones which ultimately affects the working of the unit.

Kolkata has always been one of the major seaports for the country is situated along the coast of the Bay of Bengal. But the solar irradiance obtained is not on the brighter side of data. The outlet temperature itself suggests that the humidity stays quite high. Due to higher ambient temperature, the productivity stays quite positive which is suited for a small-scale desalination unit.

CONCLUSION

A theoretical analysis for a desalination unit design was described to access the working and potential of a solar desalination plant in India along with understanding the effect of weather conditions at that particular site. In this chapter, major coastal cities of the Indian peninsula have been considered which have the capacity and need for solar thermal desalination. Since most of the Indian Peninsula is in the tropic (torrid) region, there is no shortcoming of sunlight over them and that makes it a preferable destination for setting up a solar desalination unit. Out of eight sites that had been considered, five show the most favorable result during the summer season, which has been approximated as the time since the start of March and stays till the end of June; although in June, monsoon starts to arrive in most part of the country and stays till late of October where it is finally seen the retreating monsoon hitting the coast of Tamil Nadu and this factor has been considered in the present chapter because during this season every site has incurred heavy losses in productivity. There is a high solar irradiance present during the remaining of the annual cycle. This can be stated that the potential of setting up a desalination unit in places like Chennai, Kozhikode, Mumbai and Surat is feasible because they provide a viable outcome and favorable climatic conditions to carry out seawater desalination. This is very much obvious that these sites are favorable, considering they are one of the most important cities in their respective states.

It is here concluded that even having a large potential for solar water desalination, the Indian Peninsula is dominatingly performing more efficiently on the south-western coast. And cities like Kozhikode and Chennai are at a natural advantage due to their locations near the equator. Hence the study achieves its aim of analyzing the Indian coast and the results prove that the sites collected can be a preferable choice for setting up a plant or unit, taking into consideration the already mentioned presumptions:

➢ The productivity is maximum in summer for most of the states.
➢ The productivity decreases in winters despite having an increase in solar irradiance because of lower inlet parameters.
➢ Higher humidity gives a natural advantage for productivity.

NOMENCLATURE

A	Area, m^2	m_{co}	Condensation rate, kg/s
P_{fos}	Saturated pressure at outlet air temperature, N/m^2	C_w	Specific Heat of Air, J/kg °C
P_{ies}	Saturated pressure at inlet air temperature, N/m^2	C_{pf}	Specific Heat of Water, J/kg °C
h_a	Heat transfer coefficient between glass and surrounding, W/m^2 °C	h_g	Enthalpy of saturated vapor, J/kg
		T_{max}	Maximum ambient temperature, °C
h_f	Heat transfer coefficient between air and glass, W/m^2 °C	q_{cg}	Convection heat transfer between glass and air, W
P_{ws}	Saturated pressure at the water temperature, N/m^2	q_{eg}	Condensation heat transfer between glass and air, W
q_{cw}	Convection heat transfer between air and water, W	q_{ew}	Evaporation heat transfer between water and air, W
h_v	Enthalpy of vapor, J/kg	I	Total solar radiation, W/m^2
h_{fg}	Latent heat of evaporation, J/kg	T_{min}	Minimum ambient temperature, °C
I_{sc}	Solar constant, W/m^2	T_a	Ambient temperature, °C
I_{df}	Diffused solar radiation, W/m^2	T_f	Airstream temperature, °C
I_{dr}	Direct solar radiation, W/m^2	T_g	Glass temperature, °C
m_a	Air mass, kg	T_w	Water temperature, °C
m_f	Air mass flow rate, kg/s	t	Time, s
m_{ev}	Evaporation rate, kg/s	T_{dp}	Dew point temperature, °C
P_{fo}	Vapor pressure at outlet air temperature, N/m^2	***Greek Letters***	
P	Atmospheric pressure, N/m^2	α_g	Glass absorptivity
P_g	Saturated pressure at inlet air temperature, N/m^2	α_w	Water absorptivity
P_{ws}	Saturated pressure at water temperature, N/m^2	β	Altitude angle (degree)
T_{dp}	Dew point temperature, °C	$\emptyset$	Relative humidity
U_a	Wind velocity, m/s	τ_g	Glass transmissivity

U_f	Airstream velocity, m/s	*Subscript*	
w	Humidity ratio, kg/kg of dry air	f	Air inside the distiller
w_f	Average humidity ratio of air, kg/kg of dry air	g	Glass
x	Horizontal coordinate, m	i	Inlet
y	Day length, hr	o	Outlet
U_a	Wind velocity, m/s	s	Saturation
U_f	Airstream velocity, m/s	w	Water
T_f	Average air temperature, °C		

CONSENT FOR PUBLICATION

Not applicable.

CONFLICT OF INTEREST

The authors declare no conflict of interest, financial or otherwise.

ACKNOWLEDGEMENTS

Declared none.

REFERENCES

[1]   S.H. Hammadi, "Theoretical analysis of humidification – dehumidification process in an open type solar desalination system", *Case Stud. Therm. Eng.,* vol. 12, pp. 843-851, 2018.
http://dx.doi.org/10.1016/j.csite.2018.09.009

[2]   H.M. Ali, "Effect of forced convection inside the solar still on heat and mass transfer coefficients", *Energy Convers. Manage.,* vol. 34, no. 1, pp. 73-79, 1993.
http://dx.doi.org/10.1016/0196-8904(93)90009-Y

[3]   E. Sartori, "Solar still versus solar evaporator: A comparative study between their thermal behaviors", *Sol. Energy,* vol. 56, no. 2, pp. 199-206, 1996.
http://dx.doi.org/10.1016/0038-092X(95)00094-8

[4]   A.M. Radhwan, and M. Radhwan, "Transient analysis of a stepped solar still for heating and humidifying greenhouses", *Desalination,* vol. 161, no. 1, pp. 89-97, 2004.
http://dx.doi.org/10.1016/S0011-9164(04)90043-4

[5]   M.H. Hamed, A.E. Kabeel, Z.M. Omara, and S.W. Sharshir, "Mathematical and experimental investigation of a solar humidification–dehumidification desalination unit", *Desalination,* vol. 358, pp. 9-17, 2015.
http://dx.doi.org/10.1016/j.desal.2014.12.005

[6]   X. Zhou, B. Xiao, W. Liu, X. Guo, J. Yang, and J. Fan, "Comparison of classical solar chimney power system and combined solar chimney system for power generation and seawater desalination", *Desalination,* vol. 250, no. 1, pp. 249-256, 2010.
      http://dx.doi.org/10.1016/j.desal.2009.03.007

[7]   V. Okati, A. Behzadmehr, and S. Farsad, "Analysis of a solar desalinator (humidification–dehumidification cycle) including a compound system consisting of a solar humidifier and subsurface condenser using DoE", *Desalination,* vol. 397, pp. 9-21, 2016.
      http://dx.doi.org/10.1016/j.desal.2016.06.010

[8]   V. Okati, S. Farsad, and A. Behzadmehr, "Numerican analysis of an integrated desalination unit using humidification-dehumidification and subsurface condensation processes", *Desalination,* vol. 433, pp. 172-185, 2018.
      http://dx.doi.org/10.1016/j.desal.2017.12.029

[9]   A.M. Abdel Dayem, M. Fatouh, and M. Fatouh, "Experimental and numerical investigation of humidification/dehumidification solar water desalination systems", *Desalination,* vol. 247, no. 1-3, pp. 594-609, 2009.
      http://dx.doi.org/10.1016/j.desal.2008.12.039

[10] G.P. Narayan, M.H. Sharqawy, E.K. Summers, J.H. Lienhard, S.M. Zubair, and M.A. Antar, "The potential of solar-driven humidification dehumidification desalination for small-scale decentralized water production", *Renew. Sustain. Energy Rev.,* vol. 14, no. 4, pp. 1187-1201, 2010.
      http://dx.doi.org/10.1016/j.rser.2009.11.014

[11] H. Ettouney, "Design and analysis of humidification dehumidification desalination process", *Desalination,* vol. 183, no. 1-3, pp. 341-352, 2005.
      http://dx.doi.org/10.1016/j.desal.2005.03.039

[12] K. Zhani, and H. Ben Bacha, "Experimental investigation of a new solar desalination prototype using the humidification dehumidification principle", *Renew. Energy,* vol. 35, no. 11, pp. 2610-2617, 2010.
      http://dx.doi.org/10.1016/j.renene.2010.03.033

[13] A.S. Nafey, H.E.S. Fath, S.O. El-Helaby, and A. Soliman, "Solar desalination using humidification–dehumidification processes. Part II. An experimental investigation", *Energy Convers. Manage.,* vol. 45, no. 7-8, pp. 1263-1277, 2004.
      http://dx.doi.org/10.1016/S0196-8904(03)00152-3

[14] E. Chafik, "A new type of seawater desalination plants using solar energy", *Desalination,* vol. 156, no. 1-3, pp. 333-348, 2003.
      http://dx.doi.org/10.1016/S0011-9164(03)00364-3

[15] S.A. El-Agouz, "A new process of desalination by air passing through seawater based on humidification–dehumidification process", *Energy,* vol. 35, no. 12, pp. 5108-5114, 2010.
      http://dx.doi.org/10.1016/j.energy.2010.08.005

[16] J. Orfi, M. Laplante, H. Marmouch, N. Galanis, B. Benhamou, S.B. Nasrallah, and C.T. Nguyen, "Experimental and theoretical study of a humidification-dehumidification water desalination system using solar energy", *Desalination,* vol. 168, pp. 151-159, 2004.
      http://dx.doi.org/10.1016/j.desal.2004.06.181

[17] S. Chaudhury, "Causes and Solutions to Water Crisis in India", Available from: https://aidindia.org/causes-and-solutions-to-the-water-crisis-in-india/

[18] F. Alnaimat, J. Klausner, and B. Mathew, "Solar desalination, desalination and water treatment", 2018.

[19] M. Shatat, M. Worall, and S. Riffat, "Opportunities for solar water desalination worldwide: Review", *Sustain Cities Soc.,* vol. 9, pp. 67-80, 2013.
      http://dx.doi.org/10.1016/j.scs.2013.03.004

[20] D.M. Abdel, "A pioneer solar water desalination system: Experimental testing and numerical simulation", *Energy Sci. Technol.,* vol. 1, no. 1, pp. 33-48, 2011.

[21] E.H. Amer, H. Kotb, G.H. Mostafa, and A.R. El-Ghalban, "Theoretical and experimental investigation of humidification dehumidification desalination unit", *Desalination,* vol. 249, no. 3, pp. 949-959, 2009.
http://dx.doi.org/10.1016/j.desal.2009.06.063

[22] S.W. Sharshir, G. Peng, N. Yang, M.A. Eltawil, M.K.A. Ali, and A.E. Kabeel, "A hybrid desalination system using humidification-dehumidification and solar stills integrated with evacuated solar water heater", *Energy Convers. Manage.,* vol. 124, pp. 287-296, 2016.
http://dx.doi.org/10.1016/j.enconman.2016.07.028

[23] C. Elango, N. Gunasekaran, and K. Sampathkumar, "Thermal models of solar still—A comprehensive review", *Renew. Sustain. Energy Rev.,* vol. 47, pp. 856-911, 2015.
http://dx.doi.org/10.1016/j.rser.2015.03.054

[24] R.R. Rogers, and M.K. Yau, *A Short Course in Cloud Physics.* Pergamon Press, 1989.

[25] A.C. Yunus, and A.B. Michael, *Thermodynamics: An Engineering Approach,* 3rd ed Mcgraw-Hill, 2001.

[26] B. Molineaux, B. Lachal, and O. Guisan, "Thermal analysis of five outdoor swimming pools heated by unglazed solar collectors", *Sol. Energy,* vol. 53, no. 1, pp. 21-26, 1994.
http://dx.doi.org/10.1016/S0038-092X(94)90599-1

[27] H. Fath, S. Elsherbiny, and A. Ghazy, "A naturally circulated humidifying/dehumidifying solar still with a built-in passive condenser", *Desalination,* vol. 169, no. 2, pp. 129-149, 2004.
http://dx.doi.org/10.1016/S0011-9164(04)00521-1

[28] S.H. Hammadi, "Tempering of water storage tank temperature in hot climates regions using earth water heat exchanger", *Therm. Sci. Eng. Prog.,* vol. 6, pp. 157-163, 2018.
http://dx.doi.org/10.1016/j.tsep.2018.03.009

[29] Y. El Mghouchi, E. Chham, M.S. Krikiz, T. Ajzoul, and A. El Bouardi, "On the prediction of the daily global solar radiation intensity on south-facing plane surfaces inclined at varying angles", *Energy Convers. Manage.,* vol. 120, pp. 397-411, 2016.
http://dx.doi.org/10.1016/j.enconman.2016.05.005

[30] N. Nijegorodov, and P.V.C. Luhanga, "Air mass: Analytical and empirical treatment; An improved formula for air mass", *Renew. Energy,* vol. 7, no. 1, pp. 57-65, 1996.
http://dx.doi.org/10.1016/0960-1481(95)00111-5

[31] T. Winter, J. Pannell, and M. McCann, *The economics of desalination and its potential application in Australia, Perth WA 6009..* University of Western Australia: Australia, 2005.

[32] R.G. Raluy, L. Serra, J. Uche, and A. Valero, "Life-cycle assessment of desalination technologies integrated with energy production systems", *Desalination,* vol. 167, pp. 445-458, 2004.
http://dx.doi.org/10.1016/j.desal.2004.06.160

[33] S. Bou-H, J.M. Abdel, M. Al-T, and S. Al-S, "Comparative performance analysis of two seawater reverse osmosis plants: Twin hollow fine fiber and spiral wound membranes", *Desalination,* vol. 120, pp. 5-106, 1998.

[34] K. Quteishat, and A. Abu, "Promotion of solar desalination in the MENA region, Middle East Desalination Research Center, Muscat, Oman", Available from: http://shebacss. com/rd/docs/rdresrs003.pdf

[35] O.K. Buros, *The ABCs of desalting,* 2nd edition International Desalination Association: Topsfield, Massachusetts, USA, 2000.

[36] M. Ali Samee, U.K. Mirza, T. Majeed, and N. Ahmad, "Design and performance of a simple single basin solar still", *Renew. Sustain. Energy Rev.,* vol. 11, no. 3, pp. 543-549, 2007.
http://dx.doi.org/10.1016/j.rser.2005.03.003

CHAPTER 7

Design and Thermal Modeling of Solar Cookers

Abhishek Saxena[1]*, Pinar Mert Cuce[2,3] and Erdem Cuce[3,4]

[1]*Department of Mechanical Engineering, Moradabad Institute of Technology, Moradabad 244001, India*
[2]*Department of Energy Systems Engineering, Faculty of Engineering, Recep Tayyip Erdogan University, Zihni Derin Campus, 53100 Rize, Turkey*
[3]*Low/Zero Carbon Energy Technologies Laboratory, Faculty of Engineering, Recep Tayyip Erdogan University, Zihni Derin Campus, 53100 Rize, Turkey*
[4]*Department of Mechanical Engineering, Faculty of Engineering, Recep Tayyip Erdogan University, Zihni Derin Campus, 53100 Rize, Turkey*

Abstract: Solar energy technologies are upgrading day by day in every sunshine-rich region around the globe. These technologies provide a strong platform for humans for high-demand activities like cooking, air heating, power generation, *etc.* Among these activities, solar cooking is much popular due to daily cooking needs. Different designs of solar cookers are available in the market according to the family size. In the present work, the designs of some commonly used solar cookers and their thermal performance evaluation have been discussed. Heat transfer analysis shows that cookers with some potential heat storage materials are better than conventional solar cookers. The design of such cookers is feasible to cook efficiently for long hours, even during off sunshine hours (for a limited period).

Keywords: Design, Heat transfer, Solar cooker, Thermal performance.

1. INTRODUCTION

In recent years, the use of renewable energy sources has been increased to reduce dependence on fossil fuels, and many alternative energy technologies, including solar energy, have become a part of our daily lives. Several studies have been conducted in the last two decades, especially on solar energy, and various environmentally friendly, cost-effective, and sustainable systems related to both electricity and thermal energy generation from solar energy have been developed

***Corresponding author Abhishek Saxena:** Department of Mechanical Engineering, Moradabad Institute of Technology, Moradabad 244001, India; Tel.: +91-97208 14105; E-mail: culturebeat94@yahoo.com

Manoj Kumar Gaur, Brian Norton & Gopal Tiwari (Eds.)

[1]. Solar cookers, one of the promising thermal applications of solar energy, have attracted researchers and people living in rural areas in developing societies due to their simple, cost-effective, environmentally friendly, and reliable features and performance characteristics [2]. Solar cookers are available in numerous designs and applications; therefore, making a general classification for them is not easy. However, it would not be wrong to evaluate solar cookers under three main headings: solar panel cookers, solar box cookers, and solar parabolic/dish cookers, as shown in Fig. (**1**) [3].

When solar cooking technology is examined closely, it is understood that the first applications started with the simplest and cheapest designs, then steady and continuous advancements took place concerning demand due to technological developments and improved income levels. The diversification of performance parameters of solar cookers from year to year has led to the widespread use of scientific studies in the relevant technical field and the introduction of a wide variety of designs on the commercial product side [4].

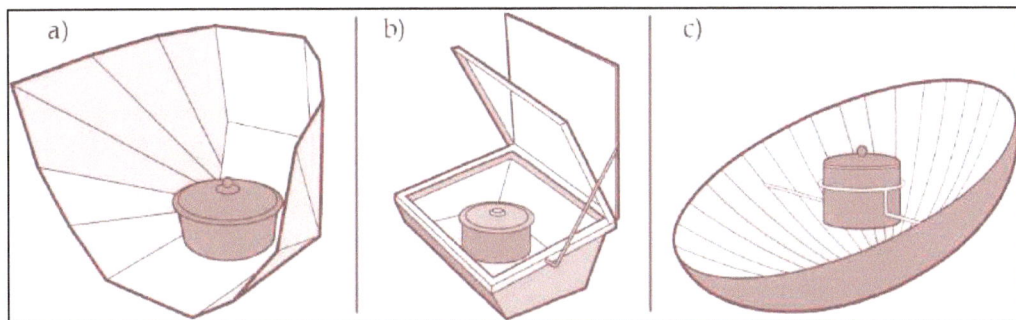

Fig. (1). Typical classification of solar cookers: **a**) solar panel cookers, **b**) solar box cookers, and c) solar parabolic/dish cookers [3].

2. CLASSIFICATION OF SOLAR COOKERS

The aforesaid performance parameters can be listed as efficiency, cooking speed, cost, durability, reliability (attention needed in operation), versatility (being adaptive to cook different foods), portability, heat storage feature (cooking after sunset), performance at cold and windy conditions, performance at low solar radiation conditions, *etc.* Each type of solar oven has many advantages and disadvantages. When the performance mentioned above criteria have been considered, it can be asserted that there is no flawless system. For example, panel-type solar cookers work to reflect the incoming solar radiation on the cooking pots. Therefore, dependence on environmental conditions is relatively high. A similar scenario applies to solar parabolic cookers. However, solar parabolic/dish cookers are driven by the solar concentration phenomenon; thus, their performance is rarely dependent on solar radiation, thermal insulation, or design due to their notably higher cooking power [5].

On the other hand, solar box cookers are highly thermally resistive in most cases. They are appropriate to design with sensible and latent heat storage media for late evening cooking. Phase change materials (PCMs) are often utilized in solar box cookers to enable their use after sunset [6] and enhance the designs' cost-effectiveness and reliability. Solar panel cookers are superior to other cooker types in terms of cost and portability, but the minimal cooking power is their most significant handicap. Solar parabolic/dish cookers are ideal for hard-to-cook foods; however, they have a perspicuous risk of burning food; therefore, someone must be present during cooking. An illustrative, precise, and useful comparison of the solar cooker types concerning the performance mentioned above criteria is given in Table **1**.

Table 1. Classification of solar cooker types by some main performance parameters.

Criteria	Solar Panel Cooker	Solar Box Cooker	Solar Parabolic/ Dish Cooker	Advanced Solar Panel Cooker	Solar Box Cooker With Booster Reflectors And Heat Storage	Advanced Solar Parabolic/Dish Cookers With Solar Tracking And Highly Reflective Coating
Efficiency	Very poor	Fair	Good	Poor	Good	Very good
Cooking Speed	Very poor	Poor	Very good	Poor	Fair	Very good
Cost	Very good	Good	Poor	Good	Fair	Very poor
Durability	Poor	Fair	Good	Fair	Good	Good
Reliability	Good	Good	Poor	Very good	Good	Poor
Versatility	Poor	Poor	Poor	Poor	Fair	Poor
Portability	Very good	Good	Very poor	Very good	Fair	Very poor
Heat Storage	Very poor	Good	Fair	Very poor	Very good	Good

(Table 1) cont.....

Performance in Cold and Windy Conditions	Very poor	Poor	Fair	Poor	Fair	Fair
Performance in Low Solar Radiation Conditions	Very poor	Poor	Fair	Poor	Fair	Good

2.1. Solar Panel Cookers

Solar panel cookers can be considered the simplest, cheapest, and most common solar cooking systems among the available solar cooker types. Solar panel cookers are made of flexible and portable lightweight materials like cardboard, and the external surfaces are coated with highly reflective materials like an aluminum sheet. Efficiency, cooking power, and other performance parameters of solar panel cookers are considerably lower compared to solar box cookers and solar parabolic/dish cookers. Especially their minimal cooking power is the most significant handicap of solar panel cookers.

As illustrated in Fig. (2), the cooking process in solar panel cookers is highly dependent on environmental conditions (incoming solar radiation, ambient temperature, wind speed, *etc.*). Solar radiation falling on the reflective surfaces of the cooker is forwarded onto the surface of the cooking pot, which is usually located in the middle of the solar panel cooker. Therefore, performance is notably affected by the dynamic outdoor conditions.

Solar panel cookers do not have a thermally resistive cooking enclosure like solar box cookers; hence remarkable heat losses occur during the cooking process. To minimize the heat losses, glass covers or transparent plastic bags are utilized, as shown in Fig. (2), to cover the cooking pots/vessels. This also helps provide a greenhouse effect beneath the cover and expedite the cooking process. Solar panel cookers have an expanding usage area yearly following the advancements in coating and material technologies. Advanced solar panel cookers are currently utilized for sterilization, food preservation, and cooking purposes.

2.2. Solar Box Cookers

The history of solar cookers starts with discovering the first solar box cooker in 1767 by Horace de Saussure, a French-Swiss naturalist. The first applications were often used to meet the food needs of French military personnel in North Africa.

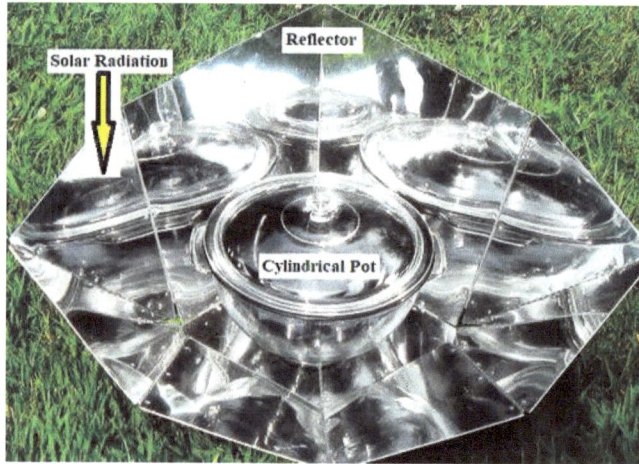

Fig. (2). Typical working principle of solar panel cookers [7].

Since the 20th century, research and development activities related to solar cookers have been accelerated, and various design aspects have been developed with improved performance parameters [8]. Despite the numerous designs of solar box cookers in literature, it is possible to illustrate the system components, as shown in Fig. (**3**) [9]. In a basic manner, a solar box cooker consists of a thermally insulated air medium, a transparent aperture glazing to welcome incoming solar radiation, booster reflectors to maximize solar radiation falling on aperture glazing, and cooking vessels. Thermal insulation in solar box cookers is usually provided by conventional materials, such as EPS, XPS, PUR, glass wool, and rock wool, due to their availability at a global scale, low cost, and good thermal conductivity range [10].

In solar box cookers, cooking pots/vessels are usually placed directly on the ground of the cooker, which is called an absorber surface. The absorber surface is often painted with matte black paint to maximize solar absorption. A similar scenario applies to cooking vessels. Thermal performance parameters of solar box cookers are remarkably affected by design aspects; thus, researchers provide continuous efforts to enhance the current performance. For instance, the height and geometry of the oven part are of vital importance in terms of heat losses. Natural convection-related heat losses exponentially rise with a characteristic length (height of solar box cooker from absorber surface to aperture glazing); hence the volumetric enclosure should be minimized. Cooking in solar box cookers is dominantly driven by the heat transfer mechanism of conduction from the absorber surface to cooking pots. In this respect, cooking pots are integrated with extended surfaces (fins) for heat transfer enhancement [11]. Enhanced heat transfer from the absorber surface to cooking pots reduces cooking time, as clearly illustrated in the results of the thermal performance test.

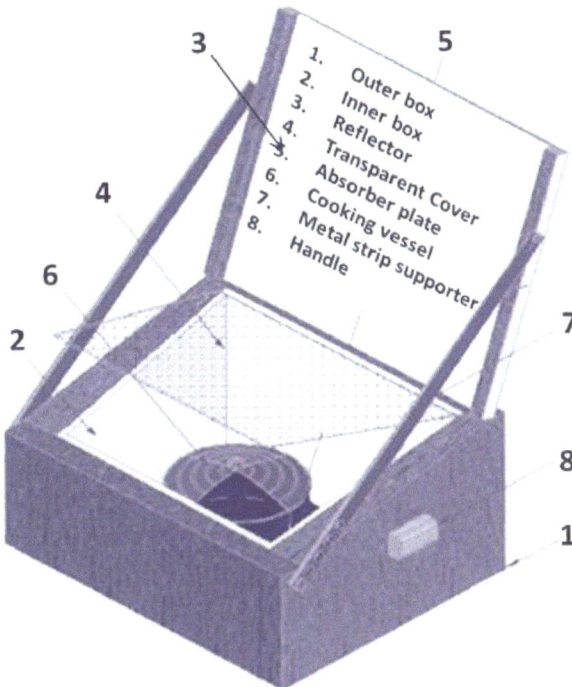

1. Outer box
2. Inner box
3. Reflector
4. Transparent Cover
5. Absorber plate
6. Cooking vessel
7. Metal strip supporter
8. Handle

Fig. (3). A typical solar box cooker with system components and optimum booster reflector angle [9].

Absorber plates are selected from highly thermally conductive metallic materials in solar box cookers to expedite heat absorption and transmission. However, absorber surface area is limited in most cases due to geometric, cost, and portability reasons. Some modifications are also considered for the absorber surface to enhance the solar radiation absorption and accelerate the heat transmission from the absorber surface to the cooking vessel through conduction and natural convection. Considering this, Harmim *et al.* [12] carried out experimental research with ordinary and finned absorber plates, and noticeable enhancements were achieved in thermal performance. The highest absorber surface temperature is 134°C for the finned absorber, whereas it is 129.4°C for the ordinary absorber. Another research conducted by Cuce [13] revealed that micro/nanoporous absorbers in solar box cookers provide notably better absorber plate temperatures than conventional ones. Three different porosity configurations are evaluated by analysing semi-circular, triangular, and trapezoidal porosity. According to the findings, each porous absorber performs better than an ordinary absorber. The ordinary absorber temperature is 110°C, while it is 151.1, 146.6, and 134.1°C for trapezoidal, semi-circular, and triangular cases, respectively. This also reflects the thermodynamic performance (energy and exergy efficiency).

Solar box cookers are appropriate for integrating sensible and latent heat storage systems to continue cooking after sunset. In this respect, a wide range of sensible energy storage materials with high specific heat capacity and numerous phase change materials (PCMs) are utilized in solar box cookers to improve the overall effectiveness of these systems. Low/Zero Carbon Energy Technologies Laboratory at Recep Tayyip Erdogan University conducted many experimental and numerical works. It revealed the appropriateness of heat storage beneath the absorber plate for late evening cooking and its cost-effectiveness.

For instance, Bayburt stone is examined as a potential sensible heat storage material in solar box cookers, and promising performance is achieved from the outdoor tests [14]. In this context, two identical solar box cookers are fabricated, and one is integrated with ground Bayburt stone. Drastic temperature drops are observed in conventional cookers, whereas temperature is kept at preferred ranges in a solar box cooker with sensible heat storage.

2.3. Solar Parabolic/Dish Cookers

Solar panel and solar box cookers usually have a similar disadvantage called limited cooking power. Solar box cookers perform better due to their thermally resistive structure, but their performance under low solar radiation conditions is usually poor. On the contrary to these two types, solar parabolic/dish cookers have an unequivocal advantage of serving even under low solar intensities and windy conditions. Solar parabolic/dish cookers differ from the other types because of the working principle based on solar concentration. The first application of a solar parabolic cooker was developed in India at the National Physical Laboratory in the 1950s, and since then, the technology has shown perspicuous progress. A typical solar parabolic/dish cooker consists of a parabolic or dish reflector and a cooking vessel that is fixed on the mirror's focal point [15, 16].

Both solar parabolic and solar dish cookers can reach very high temperatures on the cooking vessel due to solar concentration. Thus, this solar cooker does not require a special cooking pot. In addition, there is no need to use thermal insulation material around the cooking medium due to the notably higher cooking power characteristics. Mawire *et al.* [17] performed a comprehensive experimental study for two solar dish cookers, as shown in Fig. (**4**), in which one cooking vessel includes sunflower oil as the sensible heat storage material, while the second one has erythritol as PCM.

Fig. (4). Solar dish cookers with sensible and latent heat storage material [17].

When the cooking periods with storage are considered, the cooking pot with the erythritol-based PCM gives better performance due to a decrease in temperature (0.1–9.7°C) from the greatest operation temperatures compared to 8.3 to 34°C for the cooking pot with sunflower oil. A novel design of a solar parabolic cooker equipped with a vacuum medium was developed by El Moussaoui *et al.* [5]. This novel cooker aims to concentrate incoming solar radiation on a vacuum tube to increase the oil's thermal energy content. The said oil is used for the cooking process by rising and heating the cooking vessel. The novel design is appropriate for being utilized in rural and urban areas. It is tested for the climatic conditions of Morocco. The solar parabolic cooker had very promising temperatures for a typical day in October 2019. According to the findings, a vacuum tube-based solar parabolic cooker can reach 250°C for a medium solar intensity range ($G = 550$ W/m^2) [18].

3. POTENTIAL, CHALLENGES, AND ACCESSIBILITY

Cooking is one of the essential requirements of daily food preparation for people worldwide [18]. There is a crucial need to develop alternative cooking systems to decrease the energy demand and hence mitigate global energy consumption, especially fossil fuel-based consumption [19]. On the other hand, firewood is one of the most used energy resources in developing countries. Tremendous demand for firewood for cooking purposes causes deforestation in rural areas. For example, because of the growing urban population, there is a loss of 575000 hectares of forests annually as a consequence of the great demand for charcoal for cooking in Tanzania [20].

Up-to-date data showed a gradual decrease in the number of people who are unable to access clean cooking facilities, as seen in Table **2**. As a result of liquefied petroleum gas (LPG) programs and clean air policies, more than 450 million people have had clean cooking facilities since 2010 in China and India. On the other hand, the situation is not very bright for Sub-Saharan Africa, with only 17% of the total population having access to clean cooking facilities. Over 2.6 billion people still have no access; furthermore, around 2.5 million premature deaths a year are associated with indoor air pollution caused by cooking smoke. Furthermore, the COVID-19 outbreak deprives countries of universal access to clean cooking facilities [21].

Table 2. Access to clean cooking by region [21].

-	The Proportion of the Population With Access to Clean Cooking					Population Without Access (Million)	Population Relying on Traditional Use of Biomass (Million)
	2000	2005	2010	2015	2018	2018	2018
World	52%	55%	58%	63%	65%	2651	2374
Developing Countries	37%	41%	45%	53%	56%	2651	2374
Africa	23%	25%	26%	28%	29%	910	853
North Africa	87%	93%	96%	98%	98%	4	4

(Table 2) cont.....

Sub-Saharan Africa	10%	11%	13%	15%	17%	905	848
Developing Asia	33%	37%	43%	53%	57%	1674	1460
China	47%	51%	55%	67%	72%	399	242
India	22%	28%	34%	44%	49%	688	681
Indonesia	12%	18%	40%	68%	68%	85	55
Other Southeast Asia	36%	42%	48%	54%	58%	164	163
Other Developing Asia	22%	26%	27%	33%	35%	337	318
Central and South America	78%	82%	85%	88%	89%	57	53
Middle East	84%	91%	95%	96%	96%	10	9

It is observed that there has been an increasing trend in energy consumption based on fossil fuels in recent centuries. Reported data shows that the progress in annual energy consumption is 1% and 5% in developed and undeveloped countries, respectively [22, 23]. According to several scenarios that scientists have come up with, it is inevitable to see the rising cost of fossil fuels in the near future, as existing fossil fuel reserves will not meet energy demand [24]. Hence, the interest in renewable energy resources is increasing due to the increasing energy demand, fossil fuel costs, and depleted forests. Today, approximately 20% of the total world energy consumption is produced by using renewable resources (Fig. **1**) [25], and it is expected to be ascended even more in the future [18, 24]. Renewable energy sources play a key role in most applications due to their sustainability and environmental friendliness. Solar energy has a wide use area compared to the other renewables owing to its abundance on the Earth and the feasibility of thermal energy applications. A report released by the International Energy Agency illustrates that solar energy will meet 45% of the world's energy demand by 2050 [26].

In terms of thermal applications of solar energy, solar cooking is one of the most promising solar energy technologies [27]. According to the first law of thermodynamics, energy cannot be destroyed and converted from one form to another. In most of the conventional cooking processes, a significant part of the energy used for cooking is lost due to poor insulation or other reasons. Solar cooking technology provides an energy-efficient way to prepare food while preventing unwanted heat losses. The first experimental study on solar cookers was conducted by a German scientist, namely Tschirnhausen, using a large lens to focus the solar radiation in a clay pot. Nicholas-de-Saussure first fabricated solar cooker dating back history to the 17th century. Thus, he discovered a wooden hotbox to cook the fruit in it. After these studies, Ducurla, a French scientist, provided insulation while reflecting more sunlight by adding mirrors to his previous hot box design [28, 29]. More than 60 main designs and hundreds of solar cookers have been developed [30]. Solar cooker is a low-cost and environmentally friendly device harnessing solar energy. It is a clean technology providing a simple and safe way to cook without consuming any fuel [55].

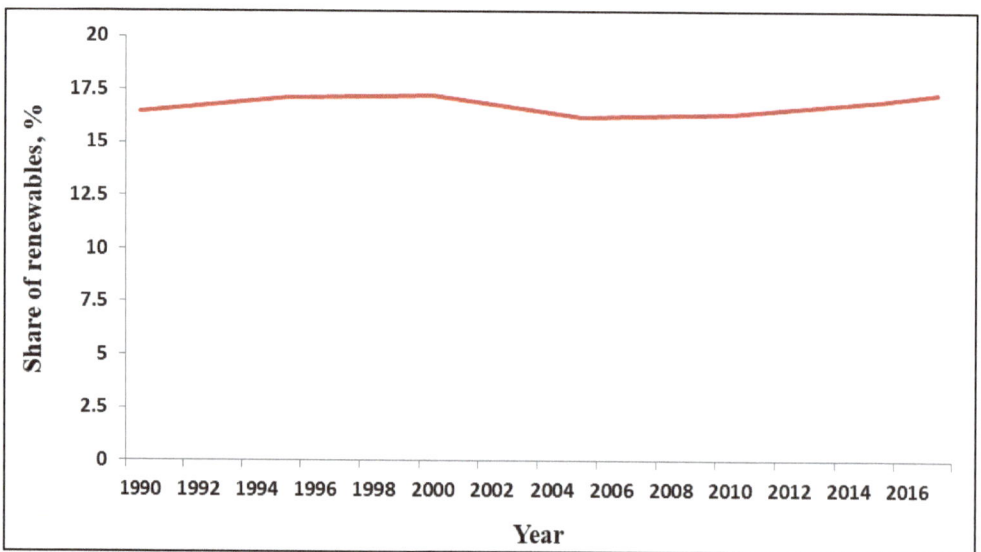

Fig. (5). Renewable share in final energy consumption of the world [26].

In solar cooking technology, sunlight is converted into heat and heat is transferred to the cooking pot. Solar cookers are useful devices to cook food *via* incoming solar radiation. Until today, solar cookers are also used for other processes, such as drying foods, sterilization, and pasteurization [14]. The solar cooking technology is particularly suitable for dry and hot climatic conditions. It provides a cheap and safe way of accessing clean food, especially in low-income countries. There are many reasons to be preferred solar cooking, such as lack of fossil fuels, deforestation, global warming, poverty, and inability to reach firewood. Additionally, acceptable payback periods for solar cookers can be achieved due to fuel savings. Today, researchers are trying to improve solar cookers to make them more efficient and preferable for end-users [21, 31].

Firewood-based stoves are widely used both indoors and outdoors in Africa. Smokes released from firewood damage the eyes and respiratory system. In addition, carbon monoxide gas released from incomplete combustion causes indoor air pollution to be harmful to the lungs. Therefore, stoves are the main reason for lung cancer, asthma, *etc.* Therefore, frequent use of stoves is the main reason for diseases, such as lung cancer, asthma, *etc.* Solar cookers eliminate the smoke and soot released by conventional cooking systems while producing food with a high volume of nutrition [25]. They provide, in brief, the cheapest and cleanest way to prepare quality food. Many studies make solar cookers more accessible and user-friendly for developing and developed countries [32]. Thus, recent studies have shown that usage of such devices has increased in most developing countries due to the abundant solar energy potential [33].

Despite their cost-effectiveness, solar cookers do not frequently use in daily cooking [34, 35]. They are light as long as they are not made of metal. That is why, due to their easy transportation, they have a high potential to be stolen by thieves in rural areas. More portable solar cookers can be designed to eliminate such a problem. Since the position of the sun changes periodically, it is difficult to focus the solar irradiation efficiently on the cooker for every hour of the daytime. Cooking time should be planned as it can take hours due to slow cooking in a solar cooker. In Kenya, for example, people buy food in the morning to prepare dinner. Moreover, they cut the meat into small pieces as it takes longer to cook [9].

As the local cooks claim, it is challenging to prepare two meals simultaneously in a solar cooker. This causes some taste problems in dishes containing more than two components, such as rice and sauce. In addition, solar cookers are not suitable for preparing large quantities of food because of their slow cooking time. In some parts of Africa, people live in extended family units, so they cook once a day. It can be said that standard solar cookers are not adequate in terms of dimensions to meet daily main course requirements [63]. According to different testing standards, the advantages and disadvantages of various solar cookers are illustrated in Table **3** [21].

Table 3. Advantages and disadvantages of various types of solar cookers [21].

Type of Cooker	Advantages	Disadvantages
Solar box cooker **(also known as a solar oven)**	• Uses both direct and scattered solar radiation. • Requires little user intervention. • Very easy and safe to use. • Easy to build. • High acceptance angle. • High tolerance for tracking error.	• Broadly different thermal performance. • Slow cooking.
Panel cooker	• Shows better performance compared to a box cooker.	• Poor performance in cloudy conditions. • More reliance on reflected radiation.
Collector cooker	• Uses both direct and scattered solar radiation. • It can be used as a multi-cooker. • Reaches higher cooking temperatures. • Simple, safe, and comfortable to use.	• Complex to construct. • Expensive to build.
Concentrating cooker	• Quite efficient. • It can reach extremely high temperatures. • Cooks faster.	• More difficult to build and use. • Requires more user attention. • Strong reliance directly on the beam. • Suffers from the wind. • Low acceptance angle. • Relatively high cost of construction. • Higher risk of safety.

4. DESIGN OF SOLAR COOKERS

Designing a solar cooker is an important aspect for efficient cooking performance around the year at different geographical locations. Every element of the cooker has a significant impact on the performance of solar cookers [36]. Therefore, selecting materials for every article or component of a solar cooker becomes important for efficient and uninterrupted cooking performance. However, the size and shape of the solar cooking unit do matter and have a significant effect on cooking timings, especially for cooking hard-boiled stuff, such as meat, fish, gram, *etc.*, [37]. In the present section, the designing procedure of both box and dish cookers has been discussed well.

4.1. Design of a Box Cooker

A solar box cooker (SBC) is a simple solar cooker to design. There is no such universal standard to design an SBC except standard testing. If the cooker satisfies the standard testing, it is a well-designed cooker. However, Indian standard IS 13429 (Indian standard solar cooker- box type- specification): IS 1349 (part 1): 2000) [38] gives some details about the specifications of materials and components of SBC. The main components of an SBC are: a leak-free wooden box, a mirror glass for the reflector, one or two transparent glass sheets for glazing, aluminum-based blackened cooking vessels, an aluminum-based solar absorber/cooking tray (blackened), a gasket, glass wool insulation and lever for adjusting mirror booster [39].

SBCs are available in different shapes and sizes according to the consumers' needs. An SBC for a family size of 05-07 members commonly used a box cooker with the following specifications (Table **4**) recommended by Saxena *et al.*, [40]. This design can fulfill the cooking needs of small and medium-sized households.

Table 4. Specification of a standard box cooker [40].

Specifications of a Standard SBC	
Sizes of the outer box	$645 \times 645 \times 210$ mm^3
Material for outer casing of SBC	Fiberglass
Area of the absorber plate	(i) Top 515×545 mm^2
-	(ii) Bottom 495×525 mm^2
Depth of cooking tray	81 mm
Emissivity of the cooking tray	0.91
Material for cooking absorber plate	*Al* (blackened)
Specific heat of plate material	941 J/kg °C
Thickness of the cooking tray	0.60 mm
Thickness of transparent glass covers	2 mm

(Table 4) cont.....

Spacing between the two glass covers (glazing)	10 mm
Emissivity of the glass cover	0.86
Type of insulation	Glass-wool
Thermal conductivity of insulation	0.06 W/m°C
Thickness of insulation at the bottom	45 mm
Pot height	68 mm
Pot diameter	162 mm
Mirror booster (single glazing)	525 x 550 mm^2

However, if a box cooker is designed simply without following any design standard, it means that it cannot cook the food [41]. The cooker's efficiency becomes low, but it can still cook under high solar radiation conditions. To know the significance of irradiance on the performance of box cookers, many researchers have developed different designs and tested them under different ambient conditions around the globe [42]. Among them, Pande and Thavi [43] designed and tested an SBC for cooking, boiling, and roasting under climatic conditions in India. The system has been placed at an optimum tilt angle at static mode, due to which capturing enormous energy is possible. The system's total cost has been found to be around $10. Nandwani [44] has designed a hybrid SBC and tested it in Costa Rica. This electro-solar cooker has been found suitable for cooking and baking. The system cost is around $110 due to an electric plate of 1500 W. Maximum cooking plate temperature has been achieved at around 145°C. Nahar [45] has designed a large-size family cooker (box type) for 8-10 members and tested it in India. The thermal efficiency of the cooker is about 29.5%, while the payback period (PBP) is around 3.29 years, and the total cost is approximately $27.

Ali [46] has designed a Sudanese solar box type cooker (Fig. **6**) and tested it in India. Results showed that F_1 is about 0.107, F_2 is approximately 0.402, and thermal efficiency is about 22.5% for 3 kg of cooking stuff (water). Nahar [47] has designed a hot box cooker with a double reflector (Fig. **7**) for extra solar radiation gain and reduced cooking timings. Transparent insulation material has been incorporated into the glazing to reduce thermal losses. Improved design efficiency is about 30.5%, and PBP is estimated at 4.33 years with a life cycle of 15 years. Ekechukwu and Ugwuoke [48] discussed the design philosophy and thermal performance of a double-glazed hotbox cooker for Nigerian climatic conditions. The cooker's design is capable of cooking four dishes simultaneously in four different cooking pots of the same capacity (1 kg). The maximum value of T_p has been obtained, which is about 138°C, and 1 kg of water takes only 70 minutes to achieve its boiling temperature.

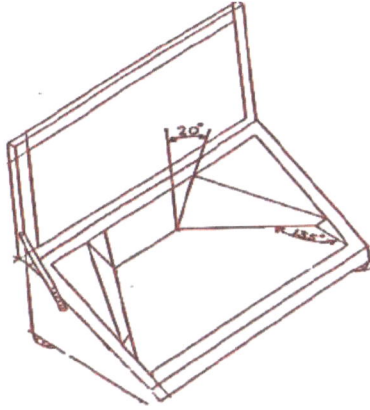

Fig. (6). Sudanese type solar box cooker [46].

I – ALUMINIUM SHEET BOX
(560x560x180)
2 - ALUMINIUM TRAY
3 - REFLECTOR I
4 - REFLECTOR II
5 - CASTOR WHEEL
6 - KAMANI
7 - STAINLESS STEEL
FRAME FOR
TIM MATERIAL
8 GLASS WOOL

Fig. (7). Double reflector type solar box cooker [47].

Mirdha and Dhariwal [49] have optimized the design of conventional box cookers by providing various combinations of reflectors for higher thermal gain. The maximum value of T_p has been obtained at about 140°C in the summer season in

India. Saxena *et al.* [50] have optimized the design of a cooking vessel (Fig. **8**) of the solar cooker by providing the lugs at the bottom side in a curvature form through which convective heat transfer is much improved inside the cooker. The cooking power of SBC is estimated to be about 79.80W in Indian climatic conditions. Saxena *et al.* [51] have designed and tested a hotbox cooker in India. Thermal performance has been evaluated on two different sensible heat storage materials, but the cooker has been best rated on the composition of both materials. The cooker satisfied the conditions of the standard of SBC. Terres *et al.* [52] designed and developed an SBC with an internal reflector mechanism. The mathematical model has been solved for Ranga-Kutta fourth-order equations. Small multiple reflectors inside the cooker enhanced the heat transfer and reduced the cooking time. The maximum value of T_{water} has been obtained at about 106.2°C.

Joshi and Jani [53] have designed a PV/T hybrid box-type cooker in India. The cooker is integrated with 5 PV panels of 15 W each. Cooking efficiency is about 30%, while the total cost is estimated at around $120. The cooker is found feasible to cook almost all Indian dishes. Kumar *et al.* [54] have designed and developed a new solar cooker integrated with evacuated tubes and thermal storage unit.

In comparison to a conventional SBC, the temperature of the cooking tray of a modified cooker has been improved by 20.8%, while cooking time is reduced by 30 minutes. Saxena and Agarwal [55] have designed and developed a hybrid hotbox solar cooker integrated with an air duct to enhance convective heat transfer through a halogen lamp of 200 W placed inside the duct. The thermal efficiency of the SBC has been found to be about 45.11%, and cooking power is estimated at about 60.20W. Saxena *et al.* [56] have designed an SBC integrated with some heat collector copper tubes with an efficient thermal storage capacity (composite of paraffin wax and granular carbon powder). Cooking testing has been done on load and no-load conditions. The thermal efficiency of the SBC has been found to be about 53.81%; cooking power is estimated at about 68.81W, while the total cost is estimated at about $39.11. Shanmugan *et al.* [57] have designed the stepped base SBC using SiO_2/TiO_2 nanoparticles based coating on the cooking/bar plate to enhance the cooker's thermal response by about 15%. Maximum T_p of SBC has been observed at about 167°C in Indian climate conditions. Cooking rice is possible within 105 minutes with this new design.

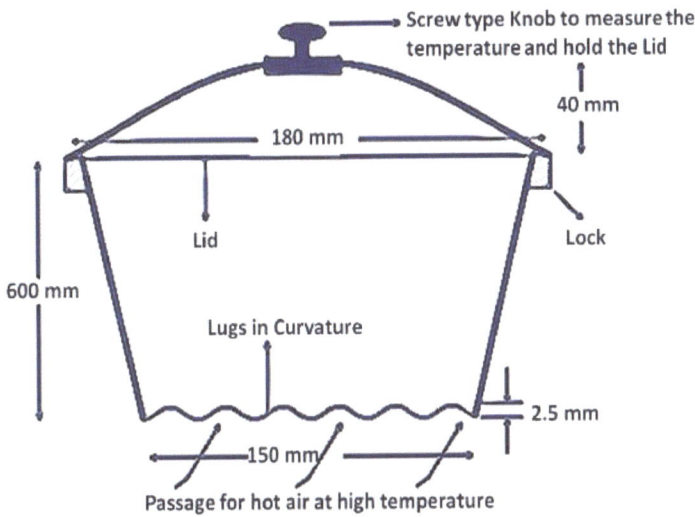

Fig. (8). Modified cooking vessel for fast thermal response [50].

Fig. (9). Various components of a parabolic dish cooker.

4.2. Design of a Parabolic Dish Cooker

The design procedure of the different components of a solar dish cooker (Fig. **9**) is given below [58];

The focal length of the parabolic dish can be estimated as:

$$f = \frac{D}{4R} \tag{1}$$

The parabolic surface can be estimated as follows:

$$A = \frac{8\pi f^2}{3} \left[\left\{ \left(\frac{C_d}{4f} \right)^2 + 1 \right\}^{1.5} - 1 \right] \tag{2}$$

Half of the aperture of a given parabola can be calculated as:

$$\tan \phi = \left(\frac{1}{\dfrac{C_d}{8D} - \dfrac{2D}{C_d}} \right) \tag{3}$$

The volume of the parabola can be estimated as follows:

$$V_p = \frac{\pi}{2}.R^2 D \tag{4}$$

The optical effectiveness factor can be determined by:

$$F'\eta_{opt} = \frac{\dfrac{(F'U_L)A_{pot}}{A_{para}} \left[\left(\dfrac{T_{wf} - T_{amb}}{I_b} \right) - \left(\dfrac{T_{wi} - T_{amb}}{I_b} \right) e^{-\tau/\tau_o} \right]}{1 - e^{-\tau/\tau_o}} \tag{5}$$

Thermal loss factor can be obtained by:

$$F'U_L = \frac{\text{mass of water * specific heat of water}}{A_{pot} * \tau_o} \tag{6}$$

The standard boiling time for the tested solar dish cooker can be obtained by:

$$t_{boil} = \tau_o . In \left[\frac{1}{1 - \left(\dfrac{F'U_L}{F'\eta_{opt}} \right) * \left(\dfrac{A_{pot}}{A_{para}} \right) * \left(\dfrac{100 - T_{amb}}{I_b} \right)} \right] \tag{7}$$

5. THERMAL PERFORMANCE OF SOLAR COOKERS

It is very necessary to evaluate the thermal performance of a solar cooker to estimate the thermodynamic behavior of a solar cooker, cooking timings, cooking capacity (load), and capacity to cook various edibles for a specific region. However, thermal performance depends upon the cooker's design and ambient conditions for significance. However, to finalize or recommend a solar cooker design, thermal performance evaluation is essential under standard conditions.

5.1. Performance Indicators

There are certain standard performance indicators for both the box type and dish type solar cookers. The cooker's design can be recommended for different households or societies to adopt solar cooking based on these parameters.

For box-type solar cookers, Figures of merit (F_1 and F_2), cooking power (P), overall heat loss coefficient (U_L), and thermal efficiency (η_{therm}) are the major performance indicators, while for dish-type cookers, sensible heating of water (I_{boil}), optical efficiency (η_{opt}), cooking power (P), overall heat loss coefficient (U_L), and thermal efficiency (η_{therm}) are the major performance indicators.

5.2. Thermal Modeling of a Box Cooker

5.2.1. Transient Heat Transfer Model

The transient heat transfer model of solar box cookers can be expressed as applying the first law of thermodynamics to each of five system components called glass cover, cooker air, absorber plate, cooking vessel, and vessel fluid as follows [56,61]:

a) Glass cover

$$
\begin{aligned}
\left(mc_p\right)_{gc} \frac{dT_{gc}}{dt} &= \tau_{gc}\alpha_{gc}A_{gc}G + h_{ca-gc}A_{gc}\left(T_{ca} - T_{gc}\right) \\
&\quad + h_{v-gc}A_v\left(T_v - T_{gc}\right) + h_{ap-gc}\left(A_{ap} - A_{vb}\right)\left(T_{ap} - T_{gc}\right) \\
&\quad - h_{gc-amb}A_{gc}\left(T_{gc} - T_{amb}\right) - h_{gc-s}A_{gc}\left(T_{gc} - T_s\right)
\end{aligned}
$$
(8)

b) Cooker air

$$
\begin{aligned}
\left(mc_p\right)_{ca} \frac{dT_{ca}}{dt} &= h_{ap-ca}\left(A_{ap} - A_{vb}\right)\left(T_{ap} - T_{ca}\right) + h_{v-ca}A_v\left(T_v - T_{ca}\right) \\
&\quad - h_{ca-gc}A_{gc}\left(T_{ca} - T_{gc}\right)
\end{aligned}
$$
(9)

c) Absorber plate

$$\left(mc_p\right)_{ap}\frac{dT_{ap}}{dt} = \tau_{gc}\alpha_{ap}(A_{ap}-A_{vb})G - h_{ap-gc}(A_{ap}-A_{vb})\left(T_{ap} - T_{gc}\right)$$
$$- h_{ap-ca}(A_{ap}-A_{vb})\left(T_{ap} - T_{ca}\right) - k_v A_{vb}(T_{ap} - T_v)/w_{vb} \tag{10}$$

d) Cooking vessel

$$\left(mc_p\right)_v\frac{dT_v}{dt} = \tau_{gc}\alpha_v A_v G + h_{ap-v}(A_{ap}-A_{vb})\left(T_{ap} - T_v\right)$$
$$- h_{v-ca}A_v(T_v - T_{ca}) - h_{v-gc}A_v\left(T_v - T_{gc}\right) \tag{11}$$
$$- h_{v-vf}A_{vf}(T_v - T_{vf})$$

e) Vessel fluid

$$(mc_p)_{vf}\frac{dT_{vf}}{dt} = h_{v-vf}A_{vf}(T_v - T_{vf}) \tag{12}$$

Where, $(mc_p)_{gc}$, $(mc_p)_{ca}$, $(mc_p)_{ap}$, $(mc_p)_v$ and $(mc_p)_{vf}$ are the thermal mass of glass cover, cooker air, absorber plate, vessel, and vessel fluid, respectively. τ_{gc} is the transmissivity of the glass cover, α_{gc} is the absorptivity of the glass cover, A_{gc} is the surface area of the glass cover, h_{ca-gc} is the heat transfer coefficient between cooker air and glass cover, h_{ap-ca} is the heat transfer coefficient between absorber plate and cooker air, h_{gc-amb} is the heat transfer coefficient between glass cover and ambient, h_{gc-s} is the heat transfer coefficient between glass cover and sky, T_{gc} is the glass cover temperature, T_{ca} is the cooker air temperature, T_v is the vessel temperature, T_{ap} is the absorber plate temperature, and T_s is the sky temperature.

A similar scenario applies to the rest of the terms in the governing equations in accordance with the subscripts. There are only some surface area terms that need to be clarified. A_v is the surface area of the cooking vessel except for the base area, A_{vb} is the cooking vessel base area, A_{vf} is the surface area of the cooking vessel in which cooking fluid has direct contact. In addition, T_{vf} is the temperature of the fluid inside the vessel, k_v is the thermal conductivity of vessel material, and w_{vb} is the thickness of the vessel base. α_v and α_{ap} are the absorptivity of the vessel and absorber plate, respectively.

5.2.2. Thermodynamic Model

Researchers have widely considered thermodynamic performance assessment of solar box cookers for numerous solar cooker designs. However, the first thermodynamic term for utilizable efficiency was proposed by Khalifa *et al.* [59] in the 1980s. Utilizable efficiency is defined as the ratio of thermal energy stored in the food (Q_f) to the input energy from the sun ($Q_{in,s}$) as follows:

$$\eta_{ut} = \frac{Q_f}{Q_{in,s}} \tag{13}$$

For a typical timeframe (Δt), $Q_{in,s}$ can be calculated by

$$Q_{in,s} = GA_{ag}\Delta t \tag{14}$$

Where G is the incoming solar radiation to the aperture glazing, and A_{ag} is the surface area of aperture glazing. In the thermodynamic performance assessment of solar box cookers, there are two Figures of merit (F_1 and F_2) which are widely utilized and adopted by researchers. First Fig. of merit is determined without load conditions over the average solar intensity value (G) and ambient temperature (T_{amb}) during the test period, and maximum absorber plate temperature ($T_{ap,m}$) as follows [60]:

$$F_1 = \frac{T_{ap,m} - T_{amb}}{G} \tag{15}$$

On the other hand, the second Fig. of merit is calculated with load conditions specifically through a water heating test. F_2 can be determined by the following equation [19]:

$$F_2 = F\eta_o HC_R \tag{16}$$

Where, F is the efficiency factor of heat exchange, η_o is the optical efficiency, and HC_R is the heat capacity ratio. The second Fig. of merit can also be expressed in terms of the first Fig. of merit as follows:

$$F_2 = \frac{F_1 MC_w}{A_{ap}\Delta t} \ln \frac{1 - \frac{1}{F_1}\frac{T_{w1} - T_{amb}}{G}}{1 - \frac{1}{F_1}\frac{T_{w2} - T_{amb}}{G}} \tag{17}$$

Where, MC_w is the product of water mass, A_{ap} is the surface area of the absorber plate and Δt is the time interval. Solar box cookers' energy and exergy efficiencies are also widely used in thermal performance assessment. Energy efficiency (η_{en}) is given by the following equation:

$$\eta_{en} = \frac{m_w c_{pw} \Delta T / \Delta t}{G A_{ag}} \tag{18}$$

Where m_w is the water mass in the cooking pot, c_{pw} is the specific heat capacity of water, and ΔT is the temperature difference achieved in water over the period of Δt. On the other hand, exergy efficiency deals with the available part of the energy to do work, and it is defined as follows [55]:

$$\eta_{ex} = \frac{m_w c_{pw} [(T_{wf} - T_{wi}) - T_{amb} \ln \frac{T_{wf}}{T_{wi}}] / \Delta t}{G_{ex} A_{ag}} \tag{19}$$

Where, G_{ex} is the exergetic part of incoming solar radiation. There are different methods in the literature to calculate G_{ex}. However, Petela's approach [56] is the most common one, which can be given as follows:

$$G_{ex} = G A_{ag} \Delta t [1 + \frac{1}{3} \left(\frac{T_{amb}}{T_{sun}}\right)^4 - \frac{4}{3} \left(\frac{T_{amb}}{T_{sun}}\right) \tag{20}$$

T_{sun} is the outer surface temperature of the sun and is often taken to be 5800 K.

5.2. Thermal Modeling for Dish Cooker

A thermal model for the dish cooker has been prepared by considering thermal heat storage material integration to the base of the cooking vessel. As Fig. (**11**) shows, the rate of solar energy instances on the dish receiver is equal to the rate of useful heat extended through the wind plus the rate of thermal losses from the collector. Therefore, the energy balance equation of the collector can be expressed as follows:

$$Q_{abs} = Q_{useful} + Q_{loss} \tag{21}$$

The energy on the receiver can also be expressed as a function of aperture beam radiation and concentration efficiency:

$$Q_{abs} = \eta_{opt} * A_{aper} * I_b \tag{22}$$

Where the useful heat can be estimated by:

$$Q_{useful} = (\dot{m}C_p)_a *(T_{a,out} - T_{a,in})$$ (23)

And thermal losses from the solar absorber:

$$Q_{loss} = A_{ab}.U_L.(T_{ab} - T_{amb})$$ (24)

Now, the net energy gain can be obtained by substituting equations (23) and (24) into (21);

$$Q_{useful} = \eta_{opt}.A_{aper}.I_b - A_{ab}U_L(T_{ab} - T_{amb}) - \frac{\rho_{ab}.V_{ab}.C_{p-ab}.dT_{ab}}{dt}$$ (25)

For thermal performance characterization of solar concentrating type collector concept of thermal efficiency is essential, which can be estimated as:

$$\eta_{therm} = \eta_{opt} - \frac{A_{ab}U_L(T_{ab} - T_{amb})}{A_{aper}.I_b} - \frac{\rho_{ab}.V_{ab}.C_{p-ab}.dT_{ab}}{A_{aper}.I_b.\dot{m}.C_p.dt}$$ (26)

Where, the overall heat loss coefficient can be calculated as follows:

$$U_L = \left(\frac{1}{h_{conv} + h_{rad}}\right)^{-1}$$ (27)

Where, the value of h$_{conv}$ and h$_{rad}$ can be estimated by the equations (28) and (29):

$$h_{conv} = (K_{air}/d_{rec})*Nu$$ (28)

$$h_{rad} = \sigma.\varepsilon_{rad}.(T_{rec}^2 - T_{amb}^2)(T_{rec} + T_{amb})$$ (29)

Now, assuming that the cooking vessel is surrounded by a thermal heat storage material or completely packed with a suitable phase change material (PCM) jacket. Therefore, the total stored energy is:

$$Q_{total} = m_{pcm}C_{p(pcm-sol)}.(T_{melt} - T_{in}) + m_{pcm}Q_{lat} + m_{pcm}C_{p(pcm-liq)}(T_{final} - T_{melt})$$ (30)

Q$_{total}$ for boiling type (above 100°C) of cooking:

$$Q_{total>100} = m_{pcm} C_{p(pcm-sol)} \cdot (T_{melt} - 100) + m_{pcm} Q_{lat} + m_{pcm} C_{p(pcm-liq)} (T_{final} - T_{melt}) \qquad \textbf{(31)}$$

Now, the overall cooking efficiency of the cooker can be calculated by:

$$\eta_{over} = \frac{\text{net heat gained during storage}}{\text{net heat gained during cooking}} = \frac{Q_{use-st}}{Q_{use-co}} \qquad \textbf{(32)}$$

CONCLUSION

Solar cookers offer cheap and clean technology while being used in various applications. Most non-profit organizations have promoted using solar cookers to mitigate fuel costs, environmental problems, desertification, and deforestation [62]. Additionally, this technology requires little effort to prepare delicious food, given that people in underdeveloped countries need to collect firewood for cooking. It is difficult and time-consuming to collect combustible material in deforested areas. In 1999, around 2 billion people worldwide suffered from a shortage of firewood [63]. An example from the countryside of Tibet shows that locals rely on dried yak dung, as there are very few trees in the area. Moreover, collecting dung from distant pastures from time to time and then carrying it to the mountain top to dry is considered a woman's job. According to a study, women spend nearly 35 hours weekly on this hard work [63].

Since the solar cooker does not require eye tracking, it provides a free cooking environment. In this way, one can take time to cope with other tasks or needs. The smoke caused by open fires can irritate the lungs and lead to many ailments. An estimation by the World Health Organization reports that children younger than five die mostly from acute respiratory infections in the developing world. In this context, annual deaths are approximately 1.6 million [63]. On the other hand, open fire can cause physical damage to the human body, from mild to severe burns.

Deforestation is one of the major threats in the world, especially in Africa. In Sub-Saharan Africa, for example, firewood is still used by 90 percent of the population for heating and cooking. That is why around 90 percent of the forests in West Africa have already been destroyed, and most importantly, food security in most African countries is under threat. One study demonstrated that cooking fires are responsible for 42 percent of the total soot in the South Asian atmosphere. Considering all the aspects, solar cookers are promising technologies to deal with deforestation, food shortages, and global warming.

Solar ovens are also used for drying foods and purifying water. Furthermore, rural South Africans are even utilizing these for ironing [63]. The literature review shows that solar cookers do not have any limitations on the usage area due to their cheapness, lack of harmful effects, such as smoke, environmental friendliness, *etc.*

NOMENCLATURE

D	Depth of parabola	I_b	Beam radiation
R	Radius of parabola	A_{para}	Area of parabola
f	Focal length	A_{ab}	Area of absorber
C_d	Concentrator diameter	A_{pot}	Area of the cooking pot
C_p	Specific heat	η_{opt}	Optical efficiency
m	Mass	Q_{useful}	Useful energy
V_p	Volume of parabola	Q_{total}	Total stored energy
η_{opt}	Optical efficiency	Q_{lat}	Latent heat of PCM
T_a	Temperature of the air	A_{aper}	Aperture area
T_{wf}	Final temperature of the water	ι	Cooking pot transmissivity
T_{wi}	Initial temperature of the water	ι_o	Cooling time constant
T_{rec}	Temperature of receiver	ι_{boil}	Boiling time
T_{amb}	Ambient temperature	U_L	Overall heat loss function
T_{in}	Initial temperature of PCM	h_{conv}	Convective heat loss coefficient
T_{final}	Final temperature of PCM	h_{rad}	Radiative heat loss coefficient
T_{melt}	Melting temperature of PCM	pcm	Phase change material

CONSENT FOR PUBLICATION

Not applicable.

CONFLICT OF INTEREST

The authors declare no conflict of interest, financial or otherwise.

ACKNOWLEDGEMENTS

Declared none.

REFERENCES

[1]   E. Cuce, E.K. Oztekin, and P.M. Cuce, "Hybrid photovoltaic/thermal (HPV/T) systems: From theory to applications", *Energy Research Journal,* vol. 9, no. 1, pp. 1-71, 2018.
      http://dx.doi.org/10.3844/erjsp.2018.1.71

[2]   E. Cuce, and P.M. Cuce, "A comprehensive review on solar cookers", *Appl. Energy,* vol. 102, pp. 1399-1421, 2013.
      http://dx.doi.org/10.1016/j.apenergy.2012.09.002

[3]   "Top solar cookers compared (infographic). Solar cookers, ovens, and stoves: Comparing the options", https://gosun.co/blogs/news/top-solar-cookers-compared

[4]   U.C. Arunachala, and A. Kundapur, "Cost-effective solar cookers: A global review", *Sol. Energy,* vol. 207, pp. 903-916, 2020.
      http://dx.doi.org/10.1016/j.solener.2020.07.026

[5]   N. El Moussaoui, S. Talbi, I. Atmane, K. Kassmi, K. Schwarzer, H. Chayeb, and N. Bachiri, "Feasibility of a new design of a Parabolic Trough Solar Thermal Cooker (PSTC)", *Sol. Energy,* vol. 201, pp. 866-871, 2020.
      http://dx.doi.org/10.1016/j.solener.2020.03.079

[6]   A.A.M. Omara, A.A.A. Abuelnuor, H.A. Mohammed, D. Habibi, and O. Younis, "Improving solar cooker performance using phase change materials: A comprehensive review", *Sol. Energy,* vol. 207, pp. 539-563, 2020.
      http://dx.doi.org/10.1016/j.solener.2020.07.015

[7]   https://en.wikipedia.org/wiki/File:HotPot_solar_cooker_with_panel_reflector_(5_liter_capacity,_front_view).png

[8]   E. Cuce, and P.M. Cuce, "Theoretical investigation of hot box solar cookers having conventional and finned absorber plates", *Int. J. Low Carbon Technol.,* vol. 10, no. 3, pp. 238-245, 2015.
      http://dx.doi.org/10.1093/ijlct/ctt052

[9]   A. Weldu, L. Zhao, S. Deng, N. Mulugeta, Y. Zhang, X. Nie, and W. Xu, "Performance evaluation on solar box cooker with reflector tracking at optimal angle under Bahir Dar climate", *Sol. Energy,* vol. 180, pp. 664-677, 2019.
      http://dx.doi.org/10.1016/j.solener.2019.01.071

[10] E. Cuce, P.M. Cuce, C.J. Wood, and S.B. Riffat, "Toward aerogel based thermal superinsulation in buildings: A comprehensive review", *Renew. Sustain. Energy Rev.,* vol. 34, pp. 273-299, 2014.
      http://dx.doi.org/10.1016/j.rser.2014.03.017

[11] A. Harmim, M. Boukar, and M. Amar, "Experimental study of a double exposure solar cooker with finned cooking vessel", *Sol. Energy,* vol. 82, no. 4, pp. 287-289, 2008.
      http://dx.doi.org/10.1016/j.solener.2007.10.008

[12] A. Harmim, M. Belhamel, M. Boukar, and M. Amar, "Experimental investigation of a box-type solar cooker with a finned absorber plate", *Energy,* vol. 35, no. 9, pp. 3799-3802, 2010.
      http://dx.doi.org/10.1016/j.energy.2010.05.032

[13] E. Cuce, "Improving thermal power of a cylindrical solar cooker via novel micro/nano porous absorbers: A thermodynamic analysis with experimental validation", *Sol. Energy,* vol. 176, pp. 211-219, 2018.
      http://dx.doi.org/10.1016/j.solener.2018.10.040

[14] P.M. Cuce, "Box type solar cookers with sensible thermal energy storage medium: A comparative experimental investigation and thermodynamic analysis", *Sol. Energy,* vol. 166, pp. 432-440, 2018.
      http://dx.doi.org/10.1016/j.solener.2018.03.077

[15] M. Hosseinzadeh, A. Faezian, S.M. Mirzababaee, and H. Zamani, "Parametric analysis and optimization of a portable evacuated tube solar cooker", *Energy,* vol. 194, p. 116816, 2020.
http://dx.doi.org/10.1016/j.energy.2019.116816

[16] B.A. Mekonnen, K.W. Liyew, and M.T. Tigabu, "Solar cooking in Ethiopia: Experimental testing and performance evaluation of SK14 solar cooker", *Case Stud. Therm. Eng.,* vol. 22, p. 100766, 2020.
http://dx.doi.org/10.1016/j.csite.2020.100766

[17] A. Mawire, K. Lentswe, P. Owusu, A. Shobo, J. Darkwa, J. Calautit, and M. Worall, "Performance comparison of two solar cooking storage pots combined with wonderbag slow cookers for off-sunshine cooking", *Sol. Energy,* vol. 208, pp. 1166-1180, 2020.
http://dx.doi.org/10.1016/j.solener.2020.08.053

[18] N.L. Panwar, S.C. Kaushik, and S. Kothari, "State of the art of solar cooking: An overview", *Renew. Sustain. Energy Rev.,* vol. 16, no. 6, pp. 3776-3785, 2012.
http://dx.doi.org/10.1016/j.rser.2012.03.026

[19] P.A. Funk, and D.L. Larson, "Parametric model of solar cooker performance", *Sol. Energy,* vol. 62, no. 1, pp. 63-68, 1998.
http://dx.doi.org/10.1016/S0038-092X(97)00074-1

[20] C.Z.M. Kimambo, "Parametric model of solar cooker performance", *J. Energy South. Afr.,* vol. 18, no. 3, pp. 41-51, 2007.
http://dx.doi.org/10.17159/2413-3051/2007/v18i3a3384

[21] International Energy Agency SDG7: Data and Projections, https://www.iea.org/reports/sdg7-data-and-projections/access-to-clean-cooking#abstract

[22] F. Yettou, B. Azoui, A. Malek, A. Gama, and N.L. Panwar, "Solar cooker realizations in actual use: An overview", *Renew. Sustain. Energy Rev.,* vol. 37, pp. 288-306, 2014.
http://dx.doi.org/10.1016/j.rser.2014.05.018

[23] A.B. Stambouli, and H. Koinuma, "A primary study on a long-term vision and strategy for the realisation and the development of the Sahara Solar Breeder project in Algeria", *Renew. Sustain. Energy Rev.,* vol. 16, no. 1, pp. 591-598, 2012.
http://dx.doi.org/10.1016/j.rser.2011.08.025

[24] A. Herez, M. Ramadan, and M. Khaled, "Review on solar cooker systems: Economic and environmental study for different Lebanese scenarios", *Renew. Sustain. Energy Rev.,* vol. 81, pp. 421-432, 2018.
http://dx.doi.org/10.1016/j.rser.2017.08.021

[25] International Energy Agency Data and Statistics, https://www.iea.org/data-and-statistics

[26] S. Mekhilef, R. Saidur, and A. Safari, "A review on solar energy use in industries", *Renew. Sustain. Energy Rev.,* vol. 15, no. 4, pp. 1777-1790, 2011.
http://dx.doi.org/10.1016/j.rser.2010.12.018

[27] T.J. Hager, and R. Morawicki, "Energy consumption during cooking in the residential sector of developed nations: A review", *Food Policy,* vol. 40, pp. 54-63, 2013.
http://dx.doi.org/10.1016/j.foodpol.2013.02.003

[28] M. Wentzel, and A. Pouris, "The development impact of solar cookers: A review of solar cooking impact research in South Africa", *Energy Policy,* vol. 35, no. 3, pp. 1909-1919, 2007.
http://dx.doi.org/10.1016/j.enpol.2006.06.002

[29] S. Singh, A. Saroj, and N. Singh, "An overview of solar cooking, with the help of phase changing material and solar tracker", *International Journal for Scientific Research & Development,* vol. 3, no. 6, pp. 407-411, 2015.

[30] A. Kundapur, and C.V. Sudhir, "Proposal for new world standard for testing solar cookers", *Journal of Engineering Science and Technology,* vol. 4, no. 3, pp. 272-281, 2009.

[31] E. Biermann, M. Grupp, and R. Palmer, "Solar cooker acceptance in South Africa: Results of a comparative field-test", *Sol. Energy,* vol. 66, no. 6, pp. 401-407, 1999.
http://dx.doi.org/10.1016/S0038-092X(99)00039-0

[32] F. Riva, M.V. Rocco, F. Gardumi, G. Bonamini, and E. Colombo, "Design and performance evaluation of solar cookers for developing countries: The case of Mutoyi, Burundi", *Int. J. Energy Res.,* vol. 41, no. 14, pp. 2206-2220, 2017.
http://dx.doi.org/10.1002/er.3783

[33] J.M.F. Mendoza, A. Gallego-Schmid, X.C. Schmidt Rivera, J. Rieradevall, and A. Azapagic, "Sustainability assessment of home-made solar cookers for use in developed countries", *Sci. Total Environ.,* vol. 648, pp. 184-196, 2019.
http://dx.doi.org/10.1016/j.scitotenv.2018.08.125 PMID: 30114589

[34] R. Wimmer, C. Pokpong, M.J. Kang, and M. Ardeshir, "Analysis of user needs for solar cooking stove acceptance", *Sustainability in Energy and Buildings: Research Advances,* vol. 3, pp. 43-51, 2012.

[35] P.P. Otte, "Analysis of user needs for solar cooking stove acceptance, Solar cooking in Mozambique-an investigation of end-user's needs for the design of solar cookers", *Energy Policy,* vol. 74, no. 3, pp. 366-375, 2014.
http://dx.doi.org/10.1016/j.enpol.2014.06.032

[36] A. Saxena, V. Goel, and M. Karakilcik, "Solar food processing and cooking methodologies", In: *Applications of Solar Energy.* Springer, 2018, pp. 251-294.

[37] A. Saxena, and S. Lath, "Solar cooking by using PCM as thermal heat storage", *MIT Int. J. Mech. Eng.,* vol. 3, no. 2, pp. 91-95, 2013.

[38] *Indian Standard Solar Cooker- Box Type- Specification): IS 1349 (Part 1).* 2000.

[39] A. Saxena, A. Shrotriya, and D. Srivastava, "Thermal performance and financial feasibility of a box type", *TIDEE,* vol. 13, no. 3, pp. 301-309, 2014.

[40] A. Saxena, Varun, S.P. Pandey, and G. Srivastav, "A thermodynamic review on solar box type cookers", *Renew. Sustain. Energy Rev.,* vol. 15, no. 6, pp. 3301-3318, 2011.
http://dx.doi.org/10.1016/j.rser.2011.04.017

[41] A. Saxena, and M. Karakilcik, "Performance evaluation of a solar cooker with low cost heat storage material", *Int. J. Sustain. Green Energy,* vol. 6, no. 4, pp. 57-63, 2017.
http://dx.doi.org/10.11648/j.ijrse.20170604.12

[42] A. Saxena, and G. Srivastava, "Performance studies of a multipurpose solar energy system for remote areas", *MIT Int. J. Mech. Eng.,* vol. 3, pp. 21-33, 2013.

[43] P.C. Pande, and K.P. Thanvi, "Design and development of a solar cooker for maximum energy capture in stationary mode", *Energy Convers. Manage.,* vol. 27, no. 1, pp. 117-120, 1987.
http://dx.doi.org/10.1016/0196-8904(87)90062-8

[44] S.S. Nandwani, "Design, construction and experimental study of electric cum solar oven—II", *Solar Wind Technol.,* vol. 6, no. 2, pp. 149-158, 1989.
http://dx.doi.org/10.1016/0741-983X(89)90024-6

[45] N.M. Nahar, "Design, development and testing of a novel non-tracking solar cooker", *Int. J. Energy Res.,* vol. 22, no. 13, pp. 1191-1198, 1998.
http://dx.doi.org/10.1002/(SICI)1099-114X(19981025)22:13<1191::AID-ER445>3.0.CO;2-V

[46] B.S. Mohamed Ali, "Design and testing of Sudanese solar box cooker", *Renew. Energy,* vol. 21, no. 3-4, pp. 573-581, 2000.
http://dx.doi.org/10.1016/S0960-1481(00)00089-6

[47] N.M. Nahar, "Design, development and testing of a double reflector hot box solar cooker with a transparent insulation material", *Renew. Energy,* vol. 23, no. 2, pp. 167-179, 2001.
http://dx.doi.org/10.1016/S0960-1481(00)00178-6

[48] O.V. Ekechukwu, and N.T. Ugwuoke, "Design and measured performance of a plane reflector augmented box-type solar-energy cooker", *Renew. Energy,* vol. 28, no. 12, pp. 1935-1952, 2003.
http://dx.doi.org/10.1016/S0960-1481(03)00004-1

[49] U.S. Mirdha, and S.R. Dhariwal, "Design optimization of solar cooker", *Renew. Energy,* vol. 33, no. 3, pp. 530-544, 2008.
http://dx.doi.org/10.1016/j.renene.2007.04.009

[50] A. Saxena, "Performance study of a modified cooking vessel for solar box type cooker", *TIDEE,* vol. 9, no. 3, pp. 93-98, 2010.

[51] A. Saxena, N.A. Varun, and G. Srivastava, "A technical note on performance testing of a solar box cooker provided with sensible storage material on the surface of absorbing plate", *Int. J. Renew. Energy Technol.,* vol. 3, no. 2, pp. 165-173, 2012.
http://dx.doi.org/10.1504/IJRET.2012.045624

[52] H. Terres, A. Lizardi, R. López, M. Vaca, and S. Chávez, "Mathematical model to study solar cookers box-type with internal reflectors", *Energy Procedia,* vol. 57, pp. 1583-1592, 2014.
http://dx.doi.org/10.1016/j.egypro.2014.10.150

[53] S.B. Joshi, and A.R. Jani, "Design, development and testing of a small scale hybrid solar cooker", *Sol. Energy,* vol. 122, pp. 148-155, 2015.
http://dx.doi.org/10.1016/j.solener.2015.08.025

[54] S. Kumar, A. Kumar, and A. Yadav, "Experimental investigation of a solar cooker based on evacuated tube collector with phase change thermal storage unit in Indian climatic conditions", *Int. J. Renew. Energy Technol.,* vol. 9, no. 3, pp. 310-336, 2018.
http://dx.doi.org/10.1504/IJRET.2018.093007

[55] A. Saxena, and N. Agarwal, "Performance characteristics of a new hybrid solar cooker with air duct", *Sol. Energy,* vol. 159, pp. 628-637, 2018.
http://dx.doi.org/10.1016/j.solener.2017.11.043

[56] A. Saxena, E. Cuce, G.N. Tiwari, and A. Kumar, "Design and thermal performance investigation of a box cooker with flexible solar collector tubes: An experimental research", *Energy,* vol. 206, p. 118144, 2020.
http://dx.doi.org/10.1016/j.energy.2020.118144

[57] S. Shanmugan, S. Gorjian, A.H. Elsheikh, F.A. Essa, and Z. Mohamed, "Investigation into the effects of SiO2/TiO2 nanolayer on the thermal performance of solar box type cooker", *Energy Sources A Recovery Util. Environ. Effects,* pp. 1-14, 2020.
http://dx.doi.org/10.1080/15567036.2020.1859018

[58] M. Ouannene, B. Chaouachi, and S. Gabsi, "Design and realization of a parabolic solar cooker", *Int. Symp. On convective Heat and Mass Transfer in Sustainable Energy,* 2009pp. 1-9

[59] A.M.A. Khalifa, M.M.A. Taha, and M. Akyurt, "Solar cookers for outdoors and indoors", *Energy,* vol. 10, no. 7, pp. 819-829, 1985.
http://dx.doi.org/10.1016/0360-5442(85)90115-X

[60] P.J. Lahkar, and S.K. Samdarshi, "A review of the thermal performance parameters of box type solar cookers and identification of their correlations", *Renew. Sustain. Energy Rev.,* vol. 14, no. 6, pp. 1615-1621, 2010.
http://dx.doi.org/10.1016/j.rser.2010.02.009

[61] F. Yettou, A. Gama, B. Azoui, A. Malek, and N.L. Panwar, "Experimental investigation and thermal modelling of box and parabolic type solar cookers for temperature mapping", *J. Therm. Anal. Calorim.,* vol. 136, no. 3, pp. 1347-1364, 2019.
http://dx.doi.org/10.1007/s10973-018-7811-9

[62] M.S. Abd-Elhady, A.N.A. Abd-Elkerim, S.A. Ahmed, M.A. Halim, and A. Abu-Oqual, "Study the thermal performance of solar cookers by using metallic wires and nanographene", *Renew. Energy,* vol. 153, pp. 108-116, 2020.
http://dx.doi.org/10.1016/j.renene.2019.09.037

[63] J. MacClancy, "Solar Cooking", *Food Cult. Soc.,* vol. 17, no. 2, pp. 301-318, 2014.
http://dx.doi.org/10.2752/175174414X13871910532060

CHAPTER 8

Application and Development in Solar Cooking

A.K. Dhamneya[1*], M.K. Gaur[1], Vikas Kumar Thakur[1], Pushpendra Singh[1]

[1]Department of Mechanical Engineering, Madhav Institute of Technology & Science, Gwalior-474005, India

Abstract: The consumption of conventional energy has increased exponentially due to the ever-increasing population of the world. Studies revealed that cooking activities contribute majorly to the overall energy consumption throughout the globe, further accounting for an increasing global warming potential. Being an enormous, virtually unlimited, and expandable source, solar energy turns out to be a favorable solution to the situation. Solar energy's widespread availability and processing technologies make the thermal energy conversion process easily accessible. Hence, solar energy has emerged as a 'natural solution' to the energy crisis and the adverse environmental impact, such as the greenhouse effect. This chapter outlines the various solar cooker fundamentals and development in different types of solar cookers, namely box type, panel, funnel type, parabolic type, and indirect type, along with the application of different solar cookers.

Keywords: Application, Box type, Development, Solar cooker, Solar thermal.

1. INTRODUCTION

After absorption and dispersion, the solar flux reaching the Earth's surface is about 1.08108 GW, equivalent to 3,400,000 EJ of energy reaching the Earth's surface each year, contributing to 7000-8000 times the world's annual primary energy consumption [1]. In general, energy management depends upon two key parameters, energy-saving and maximum use of non-conventional sources of energy. Studies revealed that the energy crisis could be reduced by moving towards non-conventional sources of energy [2]. The application of solar energy has a broad array of choices, including utilization as thermal power and photovoltaic conversions. Thermal utilization of solar energy has undergone extensive research

*Corresponding author **A.K. Dhamneya:** Department of Mechanical Engineering, Madhav Institute of Technology & Science, Gwalior-474005, India; Tel+91-7512409313; E-mail: amratkumardhamneya@ mitsgwalior.in

Manoj Kumar Gaur, Brian Norton & Gopal Tiwari (Eds.)

compared to photovoltaic or other forms because it is not limited to a particular solar radiation region. Although medium and high-temperature utilization of solar energy dominates the energy sector, the critical importance of low-temperature necessities in human life, such as cooking, water heating, conventional air dehumidification by using desiccant, *etc.*, cannot be ignored because of their potential possibilities at various scales and forms ranging from our homes, societies, and townships to industries and businesses. Of all the practiced utilities, cookers based on solar energy are the simplest, feasible and lucrative choice that offers an extremely efficient and tidy cooking method. It is widely suited to urban and rural living in all worlds, developed, developing or underdeveloped regions experiencing a scarcity of energy [3]. The expected high demand for conventional fuel for cooking is due to the fast-changing work culture, modernization and global warming. Solar energy is considered a potential way to lower the demand for a conventional energy sources.

Solar cookers are recognized as one of the most successful solar energy applications. Due to their simple design, reachability, cost-efficiency, reliability, and performance, they have become a favorite subject of both scientists and consumers in the rural areas of developing countries [4]. The general classification of solar cookers may be difficult because they are available in numerous designs with various applications. However, scientifically, we can still evaluate solar cookers under the following three major heads based on their shapes: solar box cookers, solar panel cookers, and solar parabolic/dish cookers [5]. Most NGOs around the world are encouraging the utilization of solar cookers to minimize the conventional fuel expenses (an enthusiastic step for below the poverty line people), reduce the emission of greenhouse gases and, slow down deforestation, overcome land degradation due to mining coal and chopping wood for cooking, and increase the status of women [6].

The first solar cooker was developed and tested for cooking by a Swiss Naturalist in 1767 [7]. Until the discovery of solar energy-based advanced cooking devices (*i.e.*, box-type solar cookers), most of their predecessors suffered from numerous deficiencies [8]. However, another truth is that, despite their benefits, solar energy-based devices have not been broadly utilized, especially for domestic purposes, although many inexpensive designs exist [3, 9, 10]. Few studies have been carried out to make solar cookers more user-friendly and practicable for developing and developed countries [11]. In many developing countries with abundant annual solar radiation, the employment of such devices has recently increased [12]. Researchers have compared solar energy-based cookers with conventional energy-based cooking devices in India and revealed that, after the conventional gas (LPG) and naphtha-based cookers, solar box type cooking devices are in the third position, followed by solar concentrating type cookers in the order of practical adoption [13].

2. CLASSIFICATION OF SOLAR COOKERS

A variety of energy sources are used for cooking on Earth. Although, most of the rural areas of developing countries are mainly dependent on wood and coal for cooking. Conventional sources of energy are decreased with the increase in population. Therefore, around 350 NGOs worldwide are promoting a non-conventional source for cooking. Solar Cooker International (SCI, 2020) is one of the top NGOs in the world that is determined to promote pollution-free sustainable development and, more importantly, to make the end-users aware of the benefits and impact.

Solar Cooker International (SCI) has introduced many types of solar cookers, like box cookers, panel cookers, parabolic cookers, trough cookers, evacuated tube cookers, fresnel solar cooker designs, and solar wall oven designs. Currently, many types of solar cookers are available. Therefore, solar cookers can be classified under various distinctions, such as their shapes and operability. The detailed classification of solar cookers is shown in Fig. (**1**).

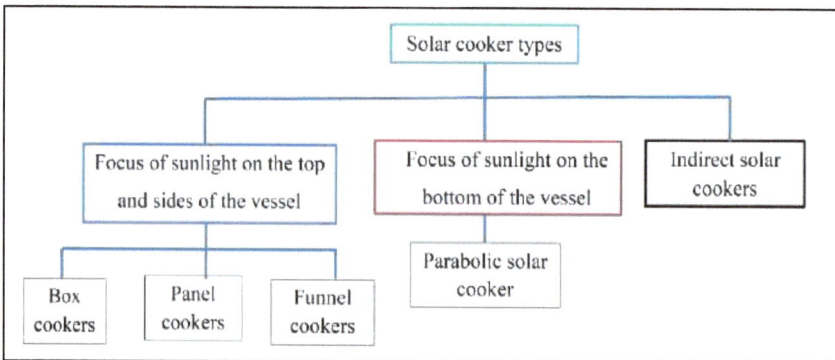

Fig. (1). Solar cooker classification.

A way to describe the performance of solar energy-based cooking devices is through attributes, such as effectiveness, cooking time, payback period, the required operational care, versatility (capable of cooking a variety of food), portability, thermal energy storage function (ability to cook after sunset), performance in severe weather conditions like cold and windy, performance when clouds are obstructing the sunlight or condition with low solar radiation, *etc.* Considering the above-mentioned performance factors, each solar cooker has its own merits and demerits as no system is flawless and has its own limitations and benefits. As we can understand with an example, an evacuated Tube type solar cooking device highly depends on the environmental conditions because it functions by reflecting solar energy on the cooking vessels. However, a concentrating type solar cooker works differently and depends on the sunlight concentration ratio, intensity of incoming solar energy that hardly impacts its effectiveness, and the heat lost to the surroundings or the aspects of its design, constituting its significantly higher cooking potential [14].

Solar box-type cookers generally offer a greater degree of thermal resistance, and hence their design is suitable for storing solar thermal energy for nighttime after sunset. Solar box cookers achieve heating capability after the sunset by utilizing Phase Change Materials (PCMs) [15,16], as this also enhances the reliability of the design along with its cost-effectiveness. Regarding portability and cost, the solar panel cookers are superior to the rest of the designs, but the limited thermal energy/power of cooking is their biggest restrain. Solar concentrating dish cookers have a higher cooking capacity and seem ideal for high-temperature cooking food; however, there is more likelihood of burning food in this type of solar cooker. Hence, it requires someone's presence while cooking to oversee and prevent burning.

3. DEVELOPMENT IN SOLAR COOKERS

3.1. Box Type Cooker

A French-Swiss naturalist, Horace de Saussure, invented the box-type solar cooker in 1767. The initial applications involved meeting the food requirements of French military personnel in North Africa. Research and development on solar cookers have accelerated since the 20th century, and we have noticed improved performance parameters because of enhanced design [16]. A solar box cooker generally consists of a thermally insulated air medium, a transparent aperture glazing to intake solar radiations, booster reflector(s) to maximize the solar radiation fall on the aperture, and enclosed cooking vessels. The thermal insulation material is generally used to prevent heat loss from the solar cookers XPS, EPS, PUR, glass wool, and rock wool, as these are of low cost and have a good thermal conduction range [17]. In box-type cookers, the cooking pots are kept at the base of the enclosed structure, which is usually referred to as an absorber surface. The absorber surface and pots are painted with matt black paint to absorb maximum radiation.

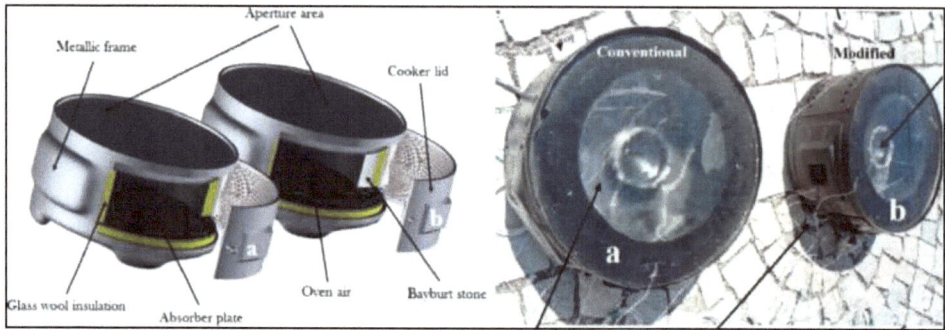

Fig. (2). (a) Modified box type solar cooker without stone, **(b)** Stone base modified solar cooker [18].

Cuce [18] developed a modified solar cooker, in which Bayburt stone has been used to enhance the energy storage capacity of the solar cooker, as shown in Fig. **(2)**. The performance of the modified solar cooker was experimentally analyzed and compared with the conventional solar cooker. Due to heat storage capacity, a modified solar cooker gives a maximum of 35.3% energy efficiency, while a convectional one gives 27.6% energy efficiency.

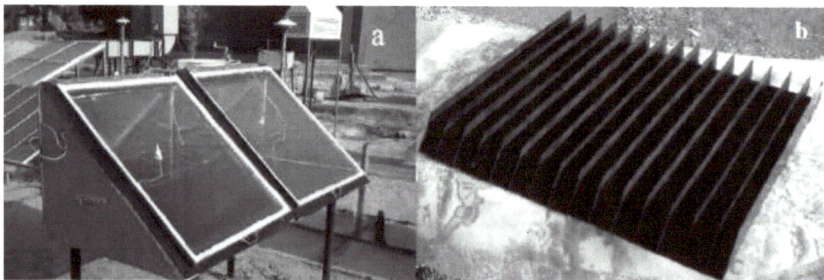

Fig. (3). (a) Experimental setup of box-type solar cooker, **(b)** Fin-type absorber plate [19].

Harmim *et al.* [19] carried out a performance evaluation of box-type cookers with ordinary and finned absorber plates, as shown in Fig. **(3)**. In a cooker with an ordinary absorber plate, the maximum temperature was 129.4°C, while in the case of a finned absorber, it was 134°C. Hence finned absorber cooker gives a lesser cooking time due to the high inside temperature.

Harmim *et al.* [20] studied the performance of a parabolic concentrator integrated box type solar cooker shown in Fig. **(4)**. A concentrator was used in the solar cooker to increase solar radiation concentration and collect the solar radiation at a point. The modified stationary cooker works as a sun-tracking solar cooker without any tracking system. The modified solar cooker is designed for all seasons.

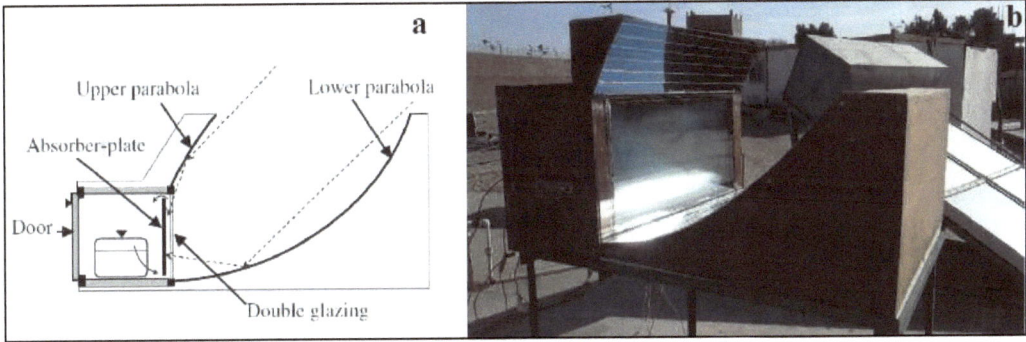

Fig. (4). (a) Schematic diagram of parabolic concentrator integrated box type solar cooker, **(b)** Experimental setup of modified solar cooker [20].

Cuce [21] used micro/nanoporous absorbers in solar box cookers and found significantly higher absorber plate temperatures than conventional ones. The experiment was performed with the absorbers' triangular, semi-circular, and trapezoidal configurations. The highest temperature achieved for the ordinary absorber during the experiment was 110°C, while it was 134.1°C, 146.6°C, and 151.1°C for the triangular, semi-circular, and trapezoidal absorbers, respectively.

To improve the overall effectiveness and cooking time after sunset, several phase change materials (PCMs) and a combination of sensible energy storage materials with high specific heat capacity were utilized in box-type cookers.

3.2. Panel Type Cooker

These are the simple, economical, and common solar cookers available commercially. Portable and flexible lightweight materials like cardboard are used to make its casing, and the external surfaces are coated with highly reflective material like aluminum. The performance parameters of the solar panel cookers, such as efficiency, cooking speed *etc.*, are considerably lower than the solar parabolic/dish cookers and the solar box cookers. Limited cooking capacity/temperature is the major limitation of panel cookers.

Unlike solar box cookers, the solar panel cookers do not have any thermally resistive cooking enclosure, and hence a considerable part of the heat is lost during the cooking process. To minimize such heat losses, glass covers or transparent plastic sheets are often used to enclose the cooking vessel/pot. With the ongoing improvements in surface coating and material technologies, the usage of solar panel cookers is increasing yearly. Advanced solar panel cookers can perform delicate processes like sterilization, food preservation, and unconventional cooking in light of these developments.

Kerr B. *et al.* [22] developed a panel-type solar cooker and used this modified cooker to operate the medical steam sterilizer, as shown in Fig. (**5**). Reflective panels had been integrated into the solar cooker to collect the radiation at a focal point. The sterilizer was used in the medical field to generate the steam.

Fig. (5). Panel solar cooker (**a**) Experimental setup with proper insulated sterilizer, (**b**) Sterilizer [22].

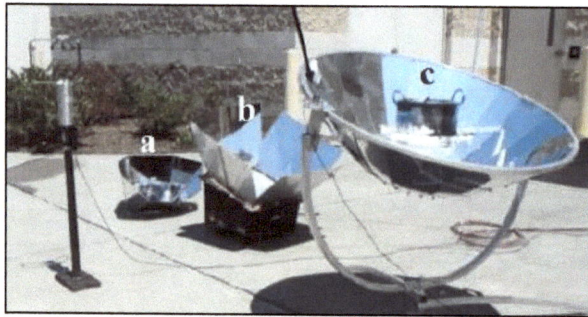

Fig. (6). Three different types of solar cookers (**a**) Panel type, (**b**) Box type, and (**c**) Parabolic type [24].

Pankaj K.G *et al.* [23] constructed and compared the performance of panel type and box type solar cookers. It was concluded that the panel-type solar cooker achieved a low-temperature range compared to the box-type solar cooker. Due to the large open area in the panel-type solar cooker, a large amount of heat is transferred to the atmosphere, reducing the cooking unit's performance. It was suggested that the panel-type solar cooker is not perfect for cooking.

Ebersvillera and Jeter [24] built three different types of solar cookers (Fig. **6**) and compared their performance. The author constructed a panel, box, and parabolic-type solar cooker and operated it in the same climatic conditions. When the temperature difference between ambient and cooking unit was 50°C, the cooking unit of panel type, box type, and parabolic-type solar cooker achieved 13%, 18%, and 33% thermal efficiency.

3.3. Funnel Cookers

Solar Funnel Cookers are harmless, cheap, easy to manufacture, and perform very well in capturing solar radiation for cooking and heating water. It is a sort of hybrid between a parabolic and a box-type cooker. It has a large, deep funnel that focuses the radiation along a line on the base of the funnel, and it incorporates the best features of the box and parabolic cookers.

Chepkurui and Biira investigated solar funnel cooker performance and reported that the maximum temperature of 90°C was observed at 90 cm funnel length [25]. The solar funnel cookers have a temperature range of 90-150°C. It depends on the design of the funnel. The solar funnel cooker is shown in Fig. (**7**).

Fig. (7). Solar Funnel Cooker [25].

A funnel solar cooker was constructed and tested by Hassan [26], as shown in Fig. (**8**). The experiment was performed at different funnel positions and then suggested the optimum angle of the funnel. It was concluded that the funnel at 30° to 90° angle received maximum solar radiation from morning to noon. In contrast, a 110° to 140° funnel angle is suitable for an afternoon.

Chepkurui and Biira [27] studied the effect of different funnel lengths on the performance of funnel solar cookers. The funnel solar cooker of four different lengths of funnels (50 cm, 42.6 cm, 32 cm and 23.3 cm) was constructed, as shown in Fig. (**9**). It was observed that the higher length of the funnel received maximum solar radiation; hence it gives better performance than other funnels. The higher funnel length (50cm) cooker achieved a maximum temperature of 93°C, while 42.6cm, 32 cm and 23.3 cm achieved a maximum temperature of 84 °C, 68 °C and 58 °C, respectively.

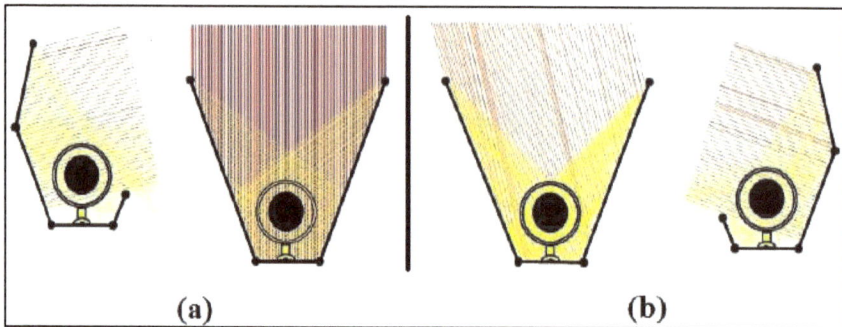

Fig. (8). Funnel positions from morning to afternoon, (**a**) Sunray incident in the morning time, (**b**) Sunray incident in the afternoon [26].

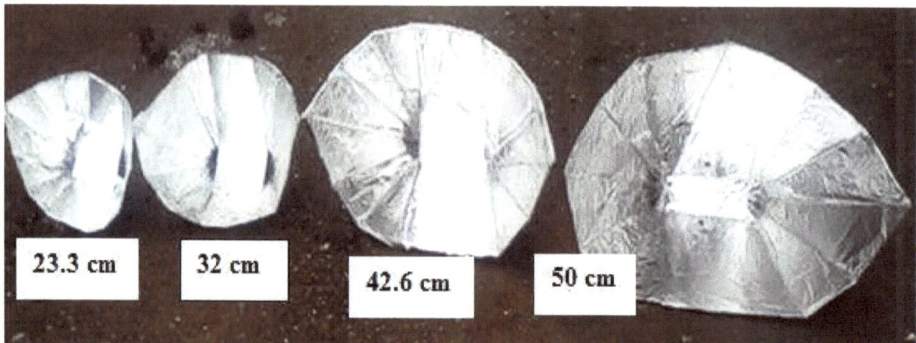

Fig. (9). Funnel solar cooker with different funnel lengths [27].

Andres *et al.* [28] designed a multi-face reflector integrated funnel solar cooker, as shown in Fig. (**10**). A transparent glass cover was used to cover the cooking unit for the greenhouse effect. It was observed that the modified system gives a maximum of 35% thermal efficiency.

3.4. Solar Parabolic/Dish Cookers

The Solar box cookers perform better due to their thermally insulative design; however, their performance severely deteriorates under low solar radiation conditions. Solar parabolic/dish cookers perform better under low solar intensities and windy conditions. These dish cookers work on the principle of concentrated solar radiation. In the 1950s, the first solar parabolic cooker was developed in India at the National Physical Laboratory, and since then, this technology has traveled a long path of progress. A parabolic/dish reflector and a cooking vessel placed on the focal point constitute a typical solar parabolic cooker [29]. The solar parabolic/dish cooker can attain very high temperatures on the cooking vessel placed at the focal point of the parabola.

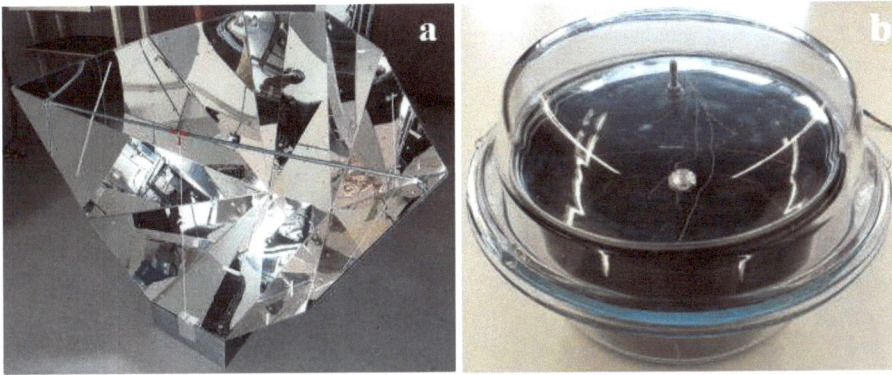

Fig. (10). Funnel solar cooker with multi-face reflector and transparent encloser [28].

El Moussauoui *et al.* [15] developed a novel solar parabolic/dish cooker fortified with a vacuum medium. This new cooker design concentrates the incident solar radiation on a vacuum tube to increase the thermal energy content of the sunflower oil circulating inside the big tube, which is further used for raising the temperature and heating the cooking pot. Both the urban and rural areas found this new design suitable, supported by tests performed in the climatic conditions of Morocco. On a typical day in October 2019, the solar parabolic cooker achieved promising temperatures reaching up to 250°C in the case of a vacuum tube-based cooker for a solar intensity range of 550 W/m^2.

Ahmed *et al.* [30] studied the performance of a parabolic solar cooker with different reflective materials, as shown in Fig. (**11**). The developed cooker used aluminum foil, stainless steel, and mylar tape as reflective materials. It was observed that the mylar tape gives better performance than the other two reflectors. The cooking unit achieved a maximum 93.7°C temperature when mylar tape was used. In the case of aluminum foil and stainless steel, the cooking unit achieved a maximum temperature of 77.1°C and 73.1°C, respectively.

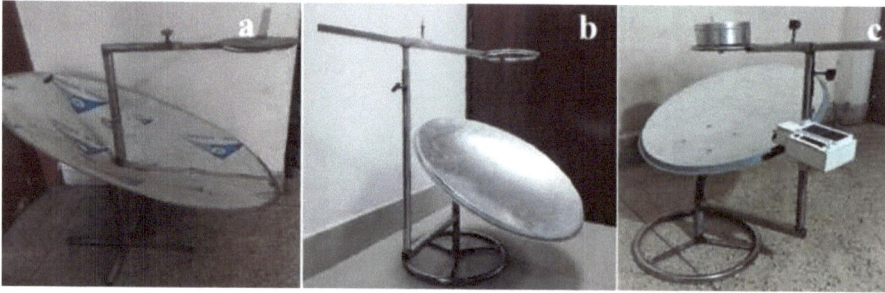

Fig. (11). Parabolic solar cooker using three different reflecting materials, **(a)** Stainless Steel, **(b)** Aluminum Foil, and **(c)** Mylar Tape [30].

A sun tracker system was integrated into a parabolic solar cooker by Al-Soud *et al.* [31], as shown in Fig. (**12**). PLC system had been used to control the sun tracker automatically. The sun tracking system helps to receive the maximum solar radiation on the cooking unit, which improves the performance of the solar cooker. The authors concluded that the modified system achieved a maximum of 90°C water temperature at 36°C atmospheric temperature.

The PCMs, namely paraffin and erythritol, were used as heat storage material inside the cooking unit by Lecuona *et al.* [32], and the performance of the modified system was investigated experimentally. Due to heat storage capacity, PCM stores solar energy in the daytime and releases it at night and during the off-sunshine period. The developed parabolic cooker with PCM is shown in Fig. (**13**).

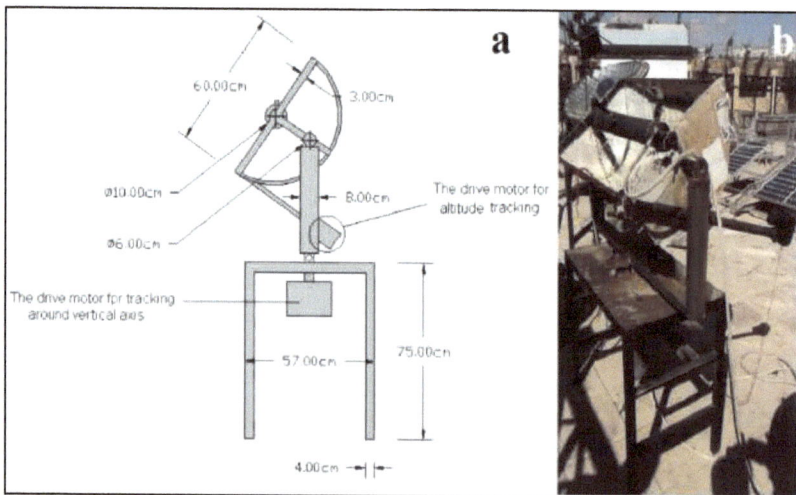

Fig. (12). (a) Schematic diagram of parabolic solar cooker using automatic sun tracking system **(b)** Experimental setup [31].

Fig. (13). (a) Parabolic solar cooker **(b)** Box for PCM integrated cooking unit [32].

3.5. Indirect Type Solar Cooker

Many researchers have suggested indirect type solar cookers that can trap solar thermal energy exterior of the house to cook inside the indoor kitchen. Hosseinzadeh M. *et al.* [33] used the nanoparticles inside the heat pipe, and a multi-layer carbon tube was used to connect the cooking unit from the parabolic collector, as shown in Fig. **(14)**. Due to high thermal conductivity, nanoparticles have been used as a base fluid, which increases the performance of the cooking unit.

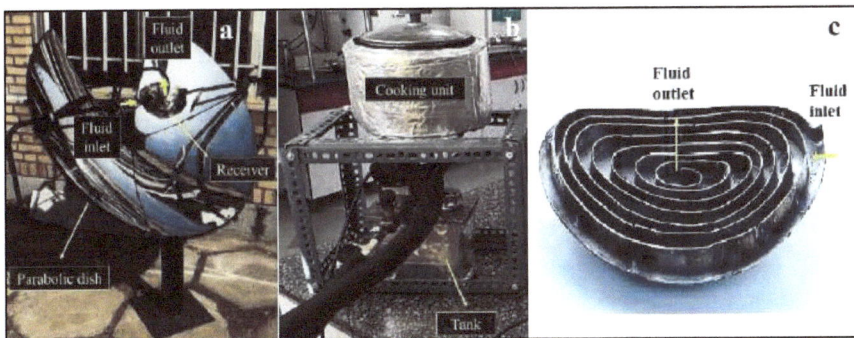

Fig. (14). (a) Radiation collector, **(b)** Cooking unit, **(c)** Receiver unit [33].

A flat plate collector, reflector and PCM were used in the indirect solar cooker by Hussein *et al.* [34]. A flat plate collector and reflector were used for the radiation collection to improve the radiation intensity. At the same time, the PCM layer has been used inside the cooking unit to keep the food warm at night and in the early morning. The schematic diagram of the modified solar cooker is shown in Fig. **(15)**.

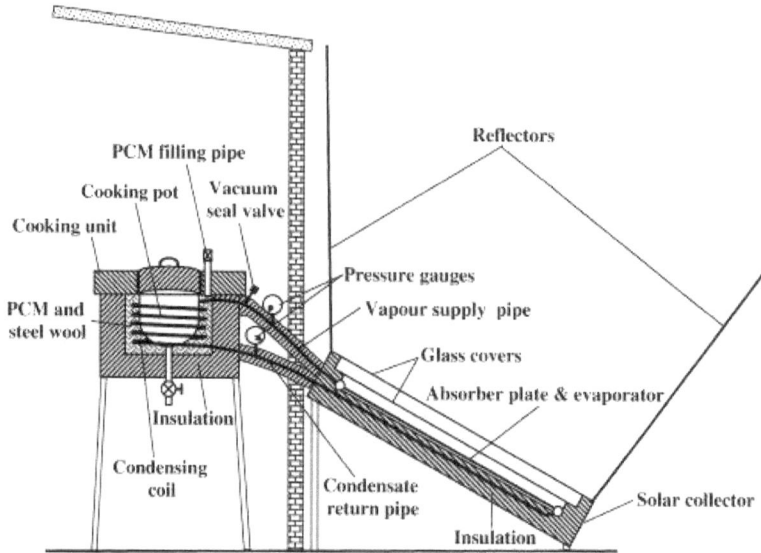

Fig. (15). Schematic diagram of modified solar cooker [34].

The double-layer glass cover of the flat plate collector was integrated into the indirect cooking unit by Singh *et al.* [35], as shown in Fig. (**16**). Coconut oil was used as a base fluid to transfer the heat from the collector to the cooking unit. Coconut oil was circulated in the pipe by the thermosyphon effect, and no pump was used in the system. Due to high thermal conductivity copper plate was used as an absorber plate. It was observed that the cooking unit achieved a maximum temperature of 150°C.

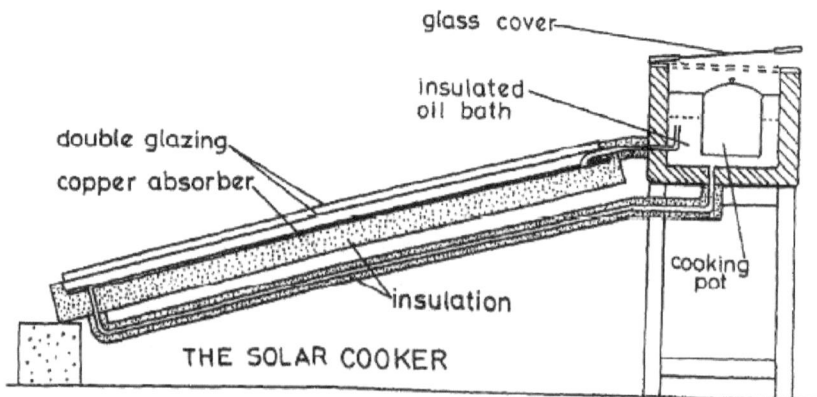

Fig. (16). Double-layer glass cover FPC integrated into cooking unit [35].

4. APPLICATION OF SOLAR COOKER

Gadhia [36] of India holds the credit for popularizing commercial cookers. He has developed and installed one of the biggest commercial solar cookers capable of serving over 50,000 people per day (as shown in Fig. **6**). The Scheffler-type concentrators of 2m to 16m diameters were used in this cooker. In Scheffler-type concentrators, the maximum temperature at the focal point reaches 1020°C. The application of solar cooker for different purposes is shown in Table **1.**

Fig. (6). Scheffler type concentrators [36].

Table 1. The efficiency and temperature of various solar cookers [2, 3, 37-39].

S. No.	Solar Cooker Type	Temperature Range (°C)	Efficiency (%)	Application
1	Box with one reflector	130–140	77.4	Cooking
2	Box with PCM	115	N/A	Cooking
3	Box Pyramid configuration	140	54	Cooking
4	Cookit	90–110	67.4	Cooking
5	Copenhagen Cooker	90–160	N/A	Cooking

(Table 1) cont.....

6	Jim's all season	90–110	N/A	Cooking
7	Parabolic	Over 250	31.5	Cooking & Fry
8	Fresnel Bing Gu type	Over 250	N/A	Cooking & Fry
9	Parabolic Trough Gosun	Over 250	N/A	Cooking & Fry
10	'Parvathi' Cooker	Over 250	N/A	Cooking & Fry

CONCLUSION

Despite being used in various applications, solar cookers offer cheap and clean technology to cook food. Solar cookers have become a game-changer technology to mitigate fuel costs, environmental problems, desertification and deforestation and offer a sustainable model for cooking food. Hence, it is being promoted by most non-profit organizations. This technology gives a practical alternative to the poor people living in underdeveloped countries of the tropical region, who otherwise need to arrange firewood before they can cook. Global deforestation in the past century has made it difficult and time-consuming to find firewood and other combustible material. According to a study, the cooking fire contributes to around 42% of the total soot in the south Asian atmosphere.

Solar cooking offers a free cooking environment because it does not require anyone to constantly watch over or cook, thus helping in saving man-hours for other tasks. The smoke caused by open fires can harm the lungs and adversely impact the respiratory system. In the light of the facts mentioned earlier, solar cookers seem to be our life and environment savior technology capable of dealing with deforestation, food shortages, and global warming. Drying food and purifying water are some of the new applications; ironing clothes is another innovative usage of this technology by South Africans. This literature review stresses that solar cookers offer versatile utilization with no limitation due to their cost effectiveness, being healthy for humans, sustainable to our environment and so on.

CONSENT FOR PUBLICATION

Not applicable.

CONFLICT OF INTEREST

The authors declare no conflict of interest, financial or otherwise.

ACKNOWLEDGEMENTS

Declared none.

REFERENCES

[1]   A. Saxena, V. Goel, and M. Karakilcik, "Solar food processing and cooking methodologies", *Appl. Sol. Energy,* pp. 251-294, 2018.

[2]   M. Aramesh, M. Ghalebani, A. Kasaeian, H. Zamani, G. Lorenzini, O. Mahian, and S. Wongwises, "A review of recent advances in solar cooking technology", *Renew. Energy,* vol. 140, pp. 419-435, 2019.
        http://dx.doi.org/10.1016/j.renene.2019.03.021

[3]   U.C. Arunachala, and A. Kundapur, "Cost-effective solar cookers: A global review", *Sol. Energy,* vol. 207, pp. 903-916, 2020.
        http://dx.doi.org/10.1016/j.solener.2020.07.026

[4]   E. Cuce, and P.M. Cuce, "A comprehensive review on solar cookers", *Appl. Energy,* vol. 102, pp. 1399-1421, 2013.
        http://dx.doi.org/10.1016/j.apenergy.2012.09.002

[5]   "Top solar cookers compared (infographic). Solar cookers, ovens, and stoves: Comparing the options", Available from: https://gosun.co/blogs/news/top-solar-cookers-compared

[6]   M.S. Abd-Elhady, A.N.A. Abd-Elkerim, S.A. Ahmed, M.A. Halim, and A. Abu-Oqal, "Study the thermal performance of solar cookers by using metallic wires and nanographene", *Renew. Energy,* vol. 153, pp. 108-116, 2020.
        http://dx.doi.org/10.1016/j.renene.2019.09.037

[7]   H.P. Garg, and R.S. Adhikari, "Renewable energy programme and vision in India", *Renew. Energy,* vol. 14, no. 1-4, pp. 473-478, 1998.
        http://dx.doi.org/10.1016/S0960-1481(98)00106-2

[8]   N.M. Nahar, "Design and development of a large size non-tracking solar cooker", *J. Eng. Sci. Technol,* vol. 4, no. 3, pp. 264-271, 2009.

[9]   R. Wimmer, C. Pokpong, M.J. Kang, and M. Ardeshir, "Analysis of user needs for solar cooking stove acceptance", *Sustain. Energy Build.: Res. Adv,* no. 3, pp. 43-51, 2012.

[10]   P.P. Otte, "Solar cooking in Mozambique—an investigation of end-user's needs for the design of solar cookers", *Energy Policy,* vol. 74, no. C, pp. 366-375, 2014.
        http://dx.doi.org/10.1016/j.enpol.2014.06.032

[11]   F. Riva, M.V. Rocco, F. Gardumi, G. Bonamini, and E. Colombo, "Design and performance evaluation of solar cookers for developing countries: The case of Mutoyi, Burundi", *Int. J. Energy Res.,* vol. 41, no. 14, pp. 2206-2220, 2017.
        http://dx.doi.org/10.1002/er.3783

[12]   J.M.F. Mendoza, A. Gallego-Schmid, X.C. Schmidt Rivera, J. Rieradevall, and A. Azapagic, "Sustainability assessment of home-made solar cookers for use in developed countries", *Sci. Total Environ.,* vol. 648, pp. 184-196, 2019.
        http://dx.doi.org/10.1016/j.scitotenv.2018.08.125 PMID: 30114589

[13] S.D. Pohekar, and M. Ramachandran, "Multi-criteria evaluation of cooking energy alternatives for promoting parabolic solar cooker in India", *Renew. Energy,* vol. 29, no. 9, pp. 1449-1460, 2004.
http://dx.doi.org/10.1016/j.renene.2003.12.017

[14] K. Ashok, "A review of solar cooker designs", *TIDE,* vol. 8, no. 1, pp. 1-37, 1998.

[15] N. El Moussaoui, S. Talbi, I. Atmane, K. Kassmi, K. Schwarzer, H. Chayeb, and N. Bachiri, "Feasibility of a new design of a Parabolic Trough Solar Thermal Cooker (PSTC)", *Sol. Energy,* vol. 201, pp. 866-871, 2020.
http://dx.doi.org/10.1016/j.solener.2020.03.079

[16] A.A.M. Omara, A.A.A. Abuelnuor, H.A. Mohammed, D. Habibi, and O. Younis, "Improving solar cooker performance using phase change materials: A comprehensive review", *Sol. Energy,* vol. 207, pp. 539-563, 2020.
http://dx.doi.org/10.1016/j.solener.2020.07.015

[17] E. Cuce, P.M. Cuce, C.J. Wood, and S.B. Riffat, "Toward aerogel based thermal superinsulation in buildings: A comprehensive review", *Renew. Sustain. Energy Rev.,* vol. 34, pp. 273-299, 2014.
http://dx.doi.org/10.1016/j.rser.2014.03.017

[18] P.M. Cuce, "Box type solar cookers with sensible thermal energy storage medium: A comparative experimental investigation and thermodynamic analysis", *Sol. Energy,* vol. 166, pp. 432-440, 2018.
http://dx.doi.org/10.1016/j.solener.2018.03.077

[19] A. Harmim, M. Boukar, and M. Amar, "Experimental study of a double exposure solar cooker with finned cooking vessel", *Sol. Energy,* vol. 82, no. 4, pp. 287-289, 2008.
http://dx.doi.org/10.1016/j.solener.2007.10.008

[20] A. Harmim, M. Merzouk, M. Boukar, and M. Amar, "Performance study of a box-type solar cooker employing an asymmetric compound parabolic concentrator", *Energy,* vol. 47, no. 1, pp. 471-480, 2012.
http://dx.doi.org/10.1016/j.energy.2012.09.037

[21] E. Cuce, "Improving thermal power of a cylindrical solar cooker via novel micro/nano porous absorbers: A thermodynamic analysis with experimental validation", *Sol. Energy,* vol. 176, pp. 211-219, 2018.
http://dx.doi.org/10.1016/j.solener.2018.10.040

[22] B. Kerr, and J. Scott, "Use of the solar panel cooker for medical pressure steam sterilization", *Solar Cookers and Food processing International Conference,* 2006 pp. 1-6 Granada, Spain

[23] P.K. Gupta, A. Misal, and S. Agrawal, "Development of low cost reflective panel solar cooker", *Mater. Today Proc.,* vol. 45, no. 2, pp. 3010-3013, 2021.
http://dx.doi.org/10.1016/j.matpr.2020.12.004

[24] S.M. Ebersviller, and J.J. Jetter, "Evaluation of performance of household solar cookers", *Sol. Energy,* vol. 208, pp. 166-172, 2020.
http://dx.doi.org/10.1016/j.solener.2020.07.056 PMID: 33012849

[25] E.J. Steven, "The Solar Funnel Cooker: How to Make and Use the BYU Solar Cooker/Cooler, Brigham Young University (BYU), Solar Cookers International Network", Available from: http://solarcooking.org/plans/funnel.htm

[26] I.M.M. Hassan, "Optical Evaluation of Funneled Panel Solar Cooker and Design Evolution, Middle East", *J. Appl. Sci.,* vol. 7, no. 4, pp. 992-1004, 2018.

[27] J. Chepkurui, and S. Biira, "Thermal performance evaluation of the funnel solar cooker of different funnel lengths implemented in Nagongera, Uganda", *Tanzan. J. Sci.,* vol. 46, no. 1, pp. 53-60, 2020.

[28] C. Andrés, A.R. García, E.A. Pagoaga, X. Ruivo, C.C. Lopez, and J. Max, "Raytracing optical analysis of a solar funnel cooker", *Third International Conference Consol-Food, Advances in Solar Thermal Food Processing,* 2020.

[29] M. Hosseinzadeh, A. Faezian, S.M. Mirzababaee, and H. Zamani, "Parametric analysis and optimization of a portable evacuated tube solar cooker", *Energy,* vol. 194, p. 116816, 2020.
http://dx.doi.org/10.1016/j.energy.2019.116816

[30] S.M.M. Ahmed, M.R. Al-Amin, S. Ahammed, F. Ahmed, A.M. Saleque, and M. Abdur Rahman, "Design, construction and testing of parabolic solar cooker for rural households and refugee camp", *Sol. Energy,* vol. 205, pp. 230-240, 2020.
http://dx.doi.org/10.1016/j.solener.2020.05.007

[31] M.S. Al-Soud, E. Abdallah, A. Akayleh, S. Abdallah, and E.S. Hrayshat, "A parabolic solar cooker with automatic two axes sun tracking system", *Appl. Energy,* vol. 87, no. 2, pp. 463-470, 2010.
http://dx.doi.org/10.1016/j.apenergy.2009.08.035

[32] A. Lecuona, J.I. Nogueira, R. Ventas, M-C. Rodríguez-Hidalgo, and M. Legrand, "Solar cooker of the portable parabolic type incorporating heat storage based on PCM", *Appl. Energy,* vol. 111, pp. 1136-1146, 2013.
http://dx.doi.org/10.1016/j.apenergy.2013.01.083

[33] M. Hosseinzadeh, R. Sadeghirad, H. Zamani, A. Kianifar, and S.M. Mirzababaee, "The performance improvement of an indirect solar cooker using multi-walled carbon nanotube-oil nanofluid: An experimental study with thermodynamic analysis", *Renew. Energy,* vol. 1481, no. 20, pp. 31644-X, 2020.
http://dx.doi.org/10.1016/j.renene.2020.10.078

[34] H.M.S. Hussein, H.H. El-Ghetany, and S.A. Nada, "Experimental investigation of novel indirect solar cooker with indoor PCM thermal storage and cooking unit", *Energy Convers. Manage.,* vol. 49, no. 8, pp. 2237-2246, 2008.
http://dx.doi.org/10.1016/j.enconman.2008.01.026

[35] I. Haraksingh, I.A. Mc Doom, and O.S.C. Headley, "A natural convection flat-plate collector solar cooker with short term storage", *Renew. Energy,* vol. 9, no. 1-4, pp. 729-732, 1996.
http://dx.doi.org/10.1016/0960-1481(96)88387-X

[36] D. Gadhia, "Commercial Scheffler cooker", *Solar Cooking Wiki,* 2020.

[37] Q.U. Islam, and F. Khozaei, "The Review of Studies on Scheffler Solar Reflectors", *International Conference on Innovation in Modern Science and Technology,* 2019 pp. 589-595.

[38] K. Ashok, "A Treatise on Solar Cookers", *International Alternate Energy Trust,* 2018.

[39] E.O.M. Akoy, and A.I.A. Ahmed, "Design, construction and performance evaluation of Solar cookers", *J. Agric. Sci. Eng.,* vol. 1, no. 2, pp. 75-82, 2015.

CHAPTER 9

Semi-Transparent Photovoltaic Thermal (SPVT) Modules and their Application

Arvind Tiwari[1*]

[1] *Bag Energy Research Society, Jawahar Nagar (Margupur)-221701, Ballia (UP), India*

Abstract: In this review, an attempt has been made for various applications of semi-transparent photovoltaic thermal (SPVT) modules used for PVT-CPC water air collector, drying, space heating/cooling of the building, greenhouse integration for agricultural production of vegetables, and power generation. It has been observed that semi-transparent photovoltaic thermal (SPVT) modules are more efficient and economical for many sectors and have more advantages than opaque photovoltaic thermal modules. The brief details of each case have been discussed. Furthermore, a greenhouse integrated semi-transparent photovoltaic thermal (GiSPVT) system has been elaborated for vegetable growth with different packing factors.

Keywords: Semi-transparent PV module, Solar energy, Thermal energy.

1. INTRODUCTION

It has now been established that solar-cell-based photovoltaic (PV) technology is best suited for the sustainability of the environment and climate [1]. A photovoltaic (PV) system produces electrical power by harnessing solar energy. Solar power generation plants come under renewable energy sources (RES) as they do not involve fossil fuels, such as coal, petroleum, and natural gas sources for power generation. Solar PV plants are classified broadly based on their location.

Furthermore, photovoltaic thermal modules are generally known as PVT, generating both DC electrical and thermal energy using solar energy. The history and development of PVT technology have been described by Tiwari and Dubey [2]. It is also important to mention that PVT technology is a self-sustained system and does not require grid power to operate solar thermal technology under the forced

*Corresponding author Arvind Tiwari: Bag Energy Research Society, Jawahar Nagar (Margupur)-221701, Ballia (UP), India; E-mail: dr.arvind.tiwari@gmail.com

Manoj Kumar Gaur, Brian Norton & Gopal Tiwari (Eds.)

mode of operation. Furthermore, the photovoltaic module has been categorized as opaque and semi-transparent transparent PV modules. It has been found that semi-transparent PV modules are more efficient in terms of electrical efficiency [3]. There are many applications of PVT technology which include:

➢**PVT-CPC water/ air collectors:** Tiwari *et al.* [4] have developed a general expression for an analytical thermal and electrical model for outlet fluid temperatures, the rate of thermal, electrical energy, overall thermal energy and exergy, and an instantaneous thermal efficiency. The developed thermal and electrical models are valid for conventional flat plate collectors, PVT, and CPC collectors. The PVT-CPC collector can be used for active solar distillation [5], swimming pool heating [6], biogas digester heating [7], and vapor absorption refrigeration (VAR) system [8].

➢**PVT solar dryers:** Based on the literature survey, it can be observed that a mixed-mode solar dryer is the most appropriate technology for vegetables/fruit crop drying to retain its color, quality, and market value. The semi-transparent integrated dryer is self-sustaining and most suitable [9, 10].

➢**PVT building:** Basically, there are three types of PV integration into a building as per requirement, namely (a) on the roof to provide only electrical power and reducing solar flux into the room below it for cooling purposes, (b) rooftop installation to provide electrical power and shadow over the roof for thermal cooling, and (c) integration on the roof as a greenhouse for use as an electrical power source as well as greenhouse roof for either floriculture, dryer or sunbath during winter [11]. In the first case, opaque PV modules have been used, while in the second and third cases, opaque and semi-transparent PV modules have been used.

➢**PVT greenhouse:** There are many categories of a greenhouse concerning shape, cost, uses, technologies, *etc.* [12]. It has been found that ridge and furrow type greenhouse is most economical from the commercial point of view. Based on a review of the research work, it has been found that only semi-transparent PV modules, which generate more electrical power, can also be used as a roof of a greenhouse for the photosynthesis process needed for plant growth inside a greenhouse.

2. COMPARISON BETWEEN THE PERFORMANCE OF OPAQUE AND SEMI-TRANSPARENT PV MODULES

In this section, we will discuss the energy balance of a single opaque and semi-transparent PV module with the following assumptions:

(i)　One-dimensional heat conduction.

(ii) The system is in a quasi-steady state condition.

(iii) The ohmic losses between solar cells of PV modules are negligible due to high electrical conductivity.

(iv) The heat capacity of transparent glass, tedlar, and ethyl vinyl acetate (EVA) is negligible.

2.1. Energy Balance for Opaque (Glass to Tedlar) (Fig. 1a)[13]

In this case, solar radiation, $I(t)$, after transmission from the glass cover $\tau_g I(t)$ is absorbed by a solar cell with an area A_m and packing factor, β_c is $\tau_g \alpha_c \beta_c I(t) A_m$. The remaining solar radiation, $\tau_g (1 - \beta_c) I(t)$, is absorbed by Tedlar (α_T) on the non-packing area of the PV module, which is $\tau_g \alpha_T (1 - \beta_c) I(t) A_m$. The temperature of solar cells increases when light is reflected on them. Therefore, there will be (a) an upward rate of overall heat loss $\left[U_{t,ca}(T_c - T_a) A_m \right]$ from the solar cell to ambient air through the top glass cover and (b) bottom rate of overall heat loss $\left[U_{b,ca}(T_c - T_a) A_m \right]$ from the solar cell to ambient air through tedlar in addition to electrical power generation as $\tau_g \eta_c \beta_c I(t) A_m$. The thermal circuit diagram corresponding to Fig. (**1a**) in terms of various heat loss and gain has been shown in Fig. (**1c**).

Following Fig. (**1a**), an energy balance equation for an opaque PV module can be expressed as follows:

$$\tau_g[\alpha_c\beta_c I(t) + (1 - \beta_c)\alpha_T I(t)]$$
$$= \left[U_{tc,a}(T_c - T_a) + U_{bc,a}(T_c - T_a) \right] + \tau_g \eta_c \beta_c I(t) \tag{1}$$

The above equation can be rearranged as

$$\tau_g[\alpha_c\beta_c I(t) + (1 - \beta_c)\alpha_T I(t)] = U_{Lm}(T_c - T_a) + \eta_m I(t) \tag{2}$$

where, $\quad U_{Lm} = (U_{tc,a} + U_{bc,a})$ and $\quad \eta_m = \tau_g \eta_c$

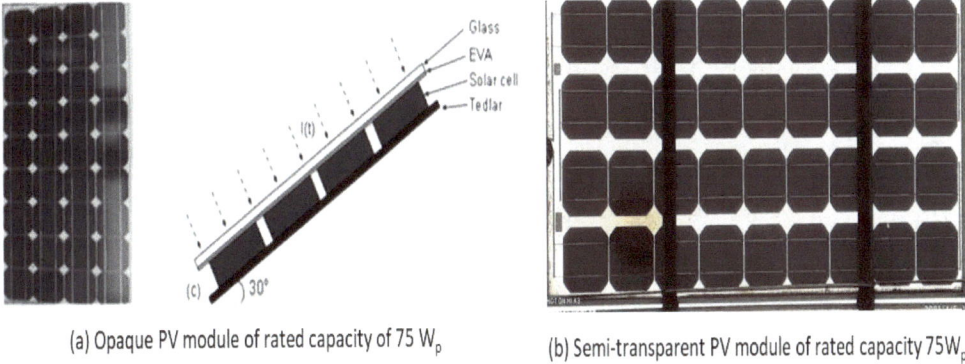

(a) Opaque PV module of rated capacity of 75 W$_p$

(b) Semi-transparent PV module of rated capacity 75W$_p$

Fig. (1). (a) Single opaque (glass-tedlar) and **(b)** semi-transparent (glass-glass) PV module.

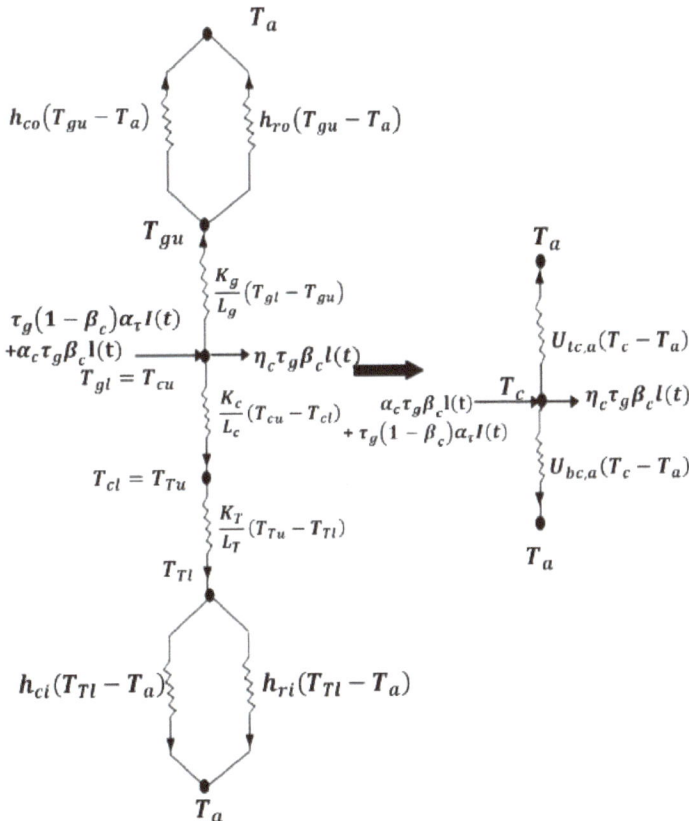

Fig. (1). (c). Thermal circuit diagram of opaque PV module shown in Fig. **(1a)**.

From Eq. (2), one can get

$$T_c - T_{ref} = (T_a - T_{ref}) + \frac{[\tau_g\{\alpha_c\beta_c + (1-\beta_c)\alpha_T - \eta_c\beta_c\}]I(t)}{U_{Lm}} \tag{3}$$

With the help of the following equation

$$\eta_c = \eta_{mo}\left[1 - \beta_{ref}[T_c - T_{ref}]\right]$$

Equation 3 can be written as:

$$\eta_c = \frac{\eta_{ref}\left[1 - \beta_{ref}\left\{(T_a - T_{ref}) + \frac{[\tau_g\{\alpha_c\beta_c + (1-\beta_c)\alpha_T\}]I(t)}{U_{Lm}}\right\}\right]}{\left[1 - \frac{\eta_{ref}\beta_{ref}\tau_g\beta_c}{U_{Lm}}I(t)\right]} \tag{4}$$

After knowing the electrical efficiency of solar cells from the above equation, one can calculate the electrical efficiency of PV modules as,

$$\eta_m = \tau_g\eta_c \tag{5}$$

The threshold intensity $I(t)_{th}$ can be obtained by having a denominator of Eq. (4) equal to zero, and it is obtained as:

$$I(t)_{th} = \frac{U_{Lm}}{\eta_{ref}\beta_{ref}\tau_g\beta_c} \tag{6}$$

2.2. Energy Balance for Semitransparent (Glass to Glass) PV Module (Fig. 1b)

In this case, solar radiation received on the non-packing area of the opaque PV module, (Fig. 1b), will be further transmitted by the glass as $\tau_g^2(1 - \beta_c)I(t)A_m$. For a semi-transparent PV module, the energy balance equation can be written as:

$$\alpha_c\tau_g\beta_cI(t) = [U_{tc,a}(T_c - T_a) + U_{bc,a}(T_c - T_a)] + \tau_g\eta_c\beta_cI(t) \tag{7}$$

Similarly, an expression for the electrical efficiency of a semi-transparent PV module can be obtained as:

$$\eta_c = \frac{\eta_{ref}\left[1 - \beta_{ref}\left\{(T_a - T_{ref}) + \frac{\alpha_c \tau_g \beta_c}{U_{Lm}} I(t)\right\}\right]}{\left[1 - \frac{\eta_{ref}\beta_{ref}\tau_g\beta_c}{U_{Lm}} I(t)\right]} \qquad (8)$$

After knowing the electrical efficiency of solar cells by Eq. (8), one can calculate the electrical efficiency of PV modules as:

$$\eta_m = \eta_c \tau_g \qquad (9)$$

Similarly, in this case also, the threshold intensity $I(t)_{th}$ is given as follows:

$$I(t)_{th} = \frac{U_{Lm}}{\eta_{ref}\beta_{ref}\tau_g\beta_c} \qquad (10)$$

The comparison of electrical efficiency of the solar cell of opaque (glass to tedlar) and semi-transparent (glass to glass) PV modules is given in Fig. (2). One can conclude that solar cell electrical efficiency in semi-transparent (glass to glass) PV modules is more due to its low operating temperature. This can also be seen from the energy balance equation of the semi-transparent PV module (Eq. 7).

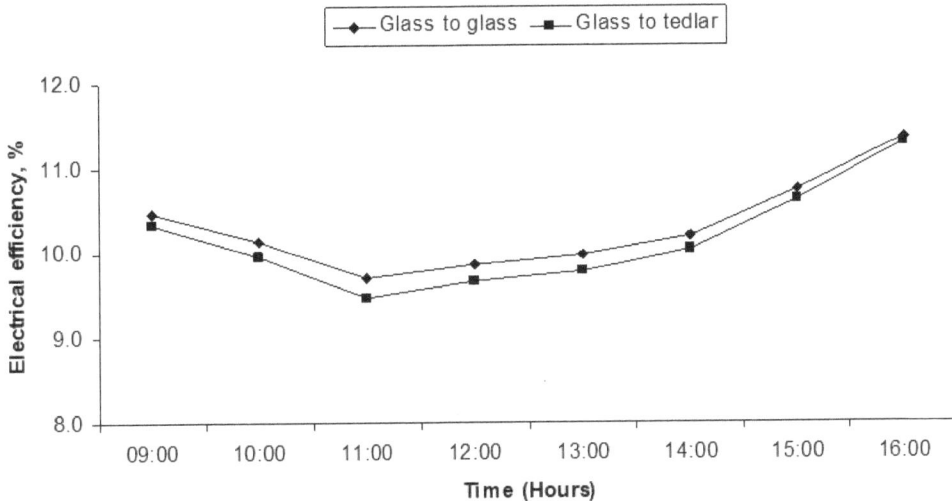

Fig. (2). Comparison of electrical efficiency of solar cell for an opaque and semi-transparent PV module [3].

Furthermore, this comparison between the performance of eight opaque and eight semi-transparent PV modules connected in series was also carried out, as shown in Fig. (**3a**). In this case, the electrical efficiency of semi-transparent PV modules was about 0.7% higher than opaque PV modules. We have also seen a crack in one of the semi-transparent PV modules, (Fig. **3b**), but there was no effect on the electrical efficiency of the semi-transparent PV module.

Based on the above finding, many authors have used semi-transparent PV modules in the solar dryer, building, and greenhouse as a roof to make the system self-sustained. In the next sections, we will discuss the applications of semi-transparent PV modules.

Fig. (3). (a). Eight opaque and eight semi-transparent PV modules connected in series rated power 600Wp.

Fig. (3). (b). Cracked semi-transparent PV module of 75Wp rated capacity.

3. SEMI-TRANSPARENT PVT DRYER

Fig. **(4).** shows the photograph of a single slope semi-transparent PVT dryer installed at SODHA BERS COMPLEX (SBC), Varanasi (UP). The semi-transparent roof of the dryer consists of eight PV modules of each 35Wp rating to provide electrical power to the fan fitted at the top of vertical height. The packing factor of the semi-transparent PV module is 42.5%, so the packing portion of the PV module provides shadow on crops and heats indirectly by convection from the back of the solar cell. So, there is direct and indirect heating of drying crops from the semi-transparent roof.

Fig. **(4).** Semi-transparent PV module roof of solar dryer.

The semi-transparent PVT dryer has been installed on one of the wind towers of SBC, so exiting hot air is also indirectly utilized for drying the crop above it. Hence, it can be considered a mixed-mode dryer to retain color, test, and market value. The fan transfers moist air from the inside dryer to the outside for fast drying in the forced mode of operation. In this dryer, there is four sliding wire mesh fitted inside the chamber so that the user can dry at least four vegetable/medicinal plants for later use. The importance of a semi-transparent PVT dryer is to use extra electrical energy for street lighting in the staircase if the dryer is not in use.

The design, fabrication, thermal modeling, energy, exergy analysis, and payback period with and without load have been carried out by Shyam *et al.* [14]. They have concluded that the energy payback times (EPBT) based on overall thermal energy are 3.74 years and 4.10 years with load and without load conditions, respectively. Based on exergy, the energy payback time (EPBT) is 12.40 years for both cases. A review on solar dryers, including semi-transparent PVT dryers, has been carried out by Singh *et al.* [15].

4. APPLICATION OF SEMI-TRANSPARENT PV MODULE IN BUILDING

This section will discuss the application of semi-transparent PV modules in two modes, namely (a) an integration into a building to create a greenhouse effect on the roof and (b) a rooftop. The building has been chosen as an energy-efficient passive building known as SODHA BERS COMPLEX (SBC), located in Varanasi, Tiwari *et al.* (2016). The SBC has been integrated with the following cooling concepts:

(i) Modified Trombe wall

(ii) Wind tower

(iii) Underground earth shelter

(iv) Day-lighting arrangement

(v) Height of each floor

(vi) Rain harvesting and

(vii) Cross-ventilation

(viii) Orientation

4.1. Building-Integrated Semi-Transparent Photovoltaic Thermal (BiSPVT) System (Fig. 5a)

There is a similarity between the semi-transparent PVT dryer and the BiSPVT system. The BiSPVT system has been installed on the roof instead of the wind tower. However, the BiSPVT system has more applications than a semi-transparent PVT dryer. There are about 96 semi-transparent PV modules with a rated capacity of 75Wp per module. There is three set/array of PV modules with a generation capacity of 2.5kW$_p$ each (32 PV module) in roof integration. (Fig. **5a**), [16] shows clean PV modules, and Fig. (**5b**) shows one with a dusty condition. Fig. (**5c**) shows 2.5 kW$_p$ for a rooftop for comparison of performance in terms of greenhouse room air, electrical efficiency, and electrical power. The building integrated semi-transparent photovoltaic thermal (BiSPVT) system can be used for sunbath, solar drying, floriculture, vegetation in the off-season, co-generation by using a micro-wind turbine, *etc.* The battery bank and charge controller cum inverter arrangement have been shown in Figs. (**5d** and **5e**), respectively. The detailed periodic modeling with [17] and without [18] water flow over the roof has also been carried out.

4.2. Semi-transparent Rooftop (Fig. 5c)

As mentioned in the previous section, there are 48 PV modules in the rooftop system, out of which 40 are semi-transparent, and 8 are opaque PV modules at the lower portion of the rooftop, as shown in Fig. (**3a**). The rooftop has a separate battery bank and charge controller, as shown in Figs. (**5d** and **5e**). On the rooftop, the upper and lower portion of the semi-transparent PV module is in contact with ambient air, so cooling of the PV module can occur with air movement.

However, the comparison of BiSPVT and rooftop with and without dust deposition has not been made. In the recent past, we have compared the performance in terms of temperature and electrical performance and observed the following:

> In both cases, as shown in Fig. (**5b** and **5c**), the transmissivity of the non-packing (double glazed) area is reduced up to 9 to 10% and hence cleaning, particularly in summer, must be done once a week. The wastewater during cleaning can be fed into rainwater harvesting available to SBC. Furthermore, the dust deposition reduced the greenhouse temperature for cooling and the cost of high-grade power.

> The cleaning of the semi-transparent PV module in both cases increases the electrical efficiency. However, an increase in the rooftop is more in comparison with BiSPVT due to the low operating temperature of the rooftop. It is about 0.25%.

> The room air temperature below BiSPVT is higher by 3.5°C than the room below the rooftop in winter, particularly from December to February. It is due to higher greenhouse room air temperature.

> The electrical efficiency of the rooftop PV module in winter and summer is more in comparison with the BiSPVT PV module.

> In off-grid power generation, an inclination of the PV module should be at (latitude 15°) [19].

> The BiSPVT and rooftop should be installed only for single or double-story buildings.

> The ventilation in the building through the air duct available at the upper portion of BiSPVT can be used for both heating and cooling.

(a) Cleaned outside view of building integrated semi-transparent photovoltaic thermal (BiSPVT) system of the rated capacity of 7.5kWp.

(b) Dusty outside View of building integrated semi-transparent photovoltaic thermal (BiSPVT) system of the rated capacity of 7.5kWp.

(c) Outside view of rooftop of SODHA BERS COMPLEX (SBC) of the rated capacity of 2.5 kWp.

(d) Battery bank.

(e) Four sets of charge controller/inverters for each floor with a rating capacity of 2.5kWp.

Fig. (5). Semi-transparent PV modules integrated into the roof of Sodha Bers Complex (SBC).

4.3. Greenhouse Integrated Semi-Transparent PV Thermal (GISPVT) System

This section will discuss vegetable crop yield and the installation of 30 kW$_p$ (equivalent to 116 kW$_p$) solar power generation over waste desert land, which has not been utilized for a long time. Furthermore, the present system will also protect vegetables from harsh, sudden changes in weather conditions in any season.

4.4. Cultivation of Vegetables

Three types of semi-transparent PV panels, namely (a) 80W$_p$, (b) 50W$_p$, and (c) 25 W$_p$ with different packing factors (PF), have been used. The dissimilar packing factor allows different solar radiation for each zone. All zones have been waterproofed with a panel cleaning arrangement. The entire greenhouse is divided into two sections one is the southside, and another is the northside, The southside is facing the South, as shown in Figs. (**6a** and **6b**). Southside is categorized as a semi-transparent PV panel's side. Northside is categorized as transparent window glass with an open roof window for hot air circulation from inside to outside. The westside view is shown in Fig. (**6c**) with an opening by a sliding glass window. The south roof semi-transparent PV module is divided into six zones, as given in Table **1**.

(**a**) Complete view of GiSPVT from the south end.

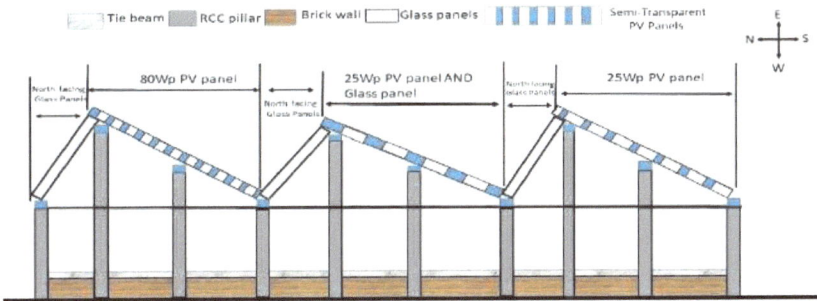

(**b**) Schematic side view of GiSPVT.

(**c**) Westside view of GiSPVT without glass windows.

Fig. (6). Ridge and furrow type uneven greenhouse integrated semi-transparent photovoltaic thermal system.

The effective floor area of GiSPVT is about 24.38m x 73.15m. The whole floor area is divided into six-zone as mentioned in Table 1. Each zone has a 24.38m x 24.38m effective area. For the cultivation of the vegetable crop in the desert area, root media has been prepared in a ratio of 40% (local soil), 40% (sand), and 20% manure [Farmyard manure (FYM)], as shown in Fig. (**7a**). Vegetable crops have been grown in the brick channel and pot due to unfertilized/desert land below GiSPVT to conserve water and manure.

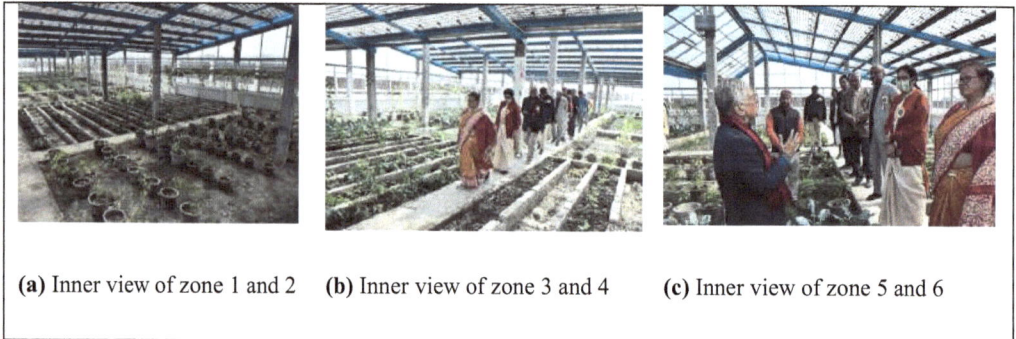

(a) Inner view of zone 1 and 2   (b) Inner view of zone 3 and 4   (c) Inner view of zone 5 and 6

Fig. (7). Inside view of the different zone of the GiSPVT system.

Table 1. Different zone of GiSPVT for the cultivation of vegetable crops.

Zone -1:	80 W_p module with a packing factor (PF) of 80%, (Fig. **8a**)
Zone -2:	50 W_p module with a packing factor (PF) of 40%, (Fig. **8a**)
Zone -3:	50 W_p module with packing factor (PF) of 40% and 6 mm window glass, (Fig. **8b**)
Zone -4:	25 W_p module with packing factor (PF) of 22% and 6 mm window glass, (Fig. **8b**)
Zone -5:	25 W_p module with a packing factor (PF) of 22%, (Fig. **8c**)
Zone -6:	25 W_p module with a packing factor (PF) of 22%, (Fig. **8c**)

Six vegetable crops, namely bottle gourd, french beans, tomato capsicum, cucumber, and broccoli which are summer crops, were selected for off-season cultivation between October 2020 and March 2021. The brick channel has been constructed in zone-1, zone-2 and zone 3, and zone 5. For comparison, the number of pots has also been placed in each zone, as shown in Fig. (**7**). The sowing of each vegetable seed has been done in the channel and in the pot for zone-1, zone-2, zone 3, and zone 5 below 2.5cm below the root media surface. After sowing, water was sprayed over the channel and the pot. The germination of some vegetables has been shown in Fig. (**8a**). After two weeks to three weeks, all plants were ready for transplantation, and it was done in both channel and pot, (Fig. **8b**), respectively. Further, Fig. (**8c**) shows a complete inside view of the growth of plants in channel and pot, respectively. The flowering of tomatoes and the yield of cucumber and capsicum in the plant are shown in Fig. (**9**).

The summarized results for the growth of plants are as follows:

(i) It has been observed that bottle gourd and cucumber need larger root media and space to propagate their many branches and high solar intensity for good photosynthesis, unlike other vegetable crops. Furthermore, the distance between the two plants should be at least 1m away. Hence the bottle gourd and cucumber give the best productivity in zone 3 and 5, respectively.

(ii) The French beans have the best productivity and early yield in the pot compared to the brick channel. It may be due to the high value of FYM in the brick channel for French beans. The yield of French beans was better in all zones. It may need low intensity, unlike bottle gourd and cucumber.

(iii) The tomatoes, capsicum, and broccoli can be grown properly in the pot, brick channel, and all zones. However, the final productivity of each one, as shown in Fig. (**10**), was best in zone 5 (Table **2**).

(a) Root media preparation in a brick channel with a ratio of 40% (local soil), 40% (sand) and 20% manure [Farmyard manure (FYM)].

(b) Transplantation of six vegetable crops in a brick channel as well as in a pot in zone 1 and 2, respectively.

(c) Photograph showing growth of six vegetable crops inside GiSPVT.

Fig. (8). View of prepared root media, sowing, germination, transplantation and growth of vegetable crops.

(iv) The bottle gourd, tomatoes, and capsicum have a long spell of harvesting in comparison with other vegetable crops.

(v) There is a variation of solar intensity between 108 W/m^2 to 318 W/m^2 from zone I to zone IV for clear sky conditions with the outside solar intensity of 640 W/m^2 in December. During the cloudy period in January, only diffuse radiation of level 10-40 W/m^2 was observed inside GiSPVT.

(vi) An average relative humidity during the day has been reported as 77-80% compared to outside, *i.e.*, 70%. It becomes about 95% during late-night times due to less respiration from the plant.

(vii) GiSPVT greenhouse room air temperature is always higher by 4-5^{0}C than outside ambient air temperature during harsh winter conditions with the completely cloudy condition between January 1st to February 2nd, 2021, due to emission of radiation from the floor. The calibrated thermometer in each zone has recorded the temperature inside the greenhouse. The average daily outside ambient air temperature has been recorded as 16°C.

(viii) Inside GiSPVT, there is a variation of about 1.5 to 2°C air temperature from zone 1 to 6. However, zone 6 has the maximum temperature.

(ix) The yield of bottle gourd, tomatoes, and capsicum extended to the first week of April 2021.

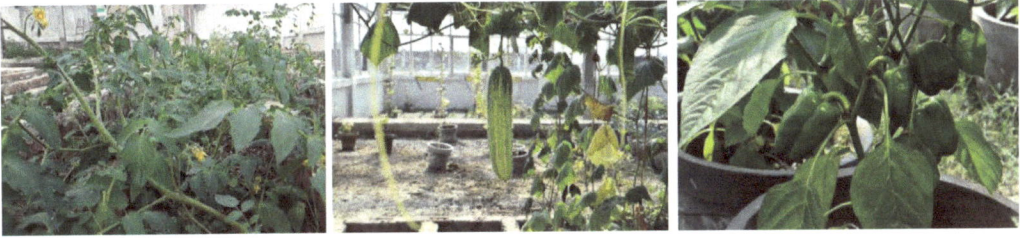

Fig. (9). Flowering and fruiting of some vegetable crops.

Fig. (10). The photograph of some of the vegetables after harvesting and weighing.

Following manual maintenance during the operation of GiSPVT were taken:

(i) The watering of the root media was done once in two weeks between November to January. It was then reduced due to high inside room temperature between February and April. This also helped to maintain cooling inside.

(ii) The roof vent and sliding doors were closed from November to January to avoid heat losses from inside to outside. The same were opened at high inside temperatures ($\geq 37°C$).

(iii) The dry cow dung cake was burned inside during healthy plant size to maintain CO_2 levels between 10 am to 5 pm. It also helped to increase greenhouse room air temperature.

(iv) Two floodlight arrangements were made in zone 1 and 2 to help for photosynthesis at night if required.

Table 2. Total production of vegetables till 15ᵗʰ March, 2021.

S. No.	Vegetable	Production (kg)				Total (kg)
		80 Wp	**40 Wp**	**25 Wp + Glass**	**25 Wp**	
1.	French Beans	1.34	3.07	1.78	5.34	**11.53**
2.	Tomato	5.46	16.5	20.01	20.13	**62.1**
3.	Capsicum	2.12	8.77	15.18	19.42	**45.49**
4.	Broccoli	0.37	4.21	5.45	6.6	**16.63**
5.	Bottle Gourd	0	0.96	0	15.17	**16.13**
6.	Cucumber	2.14	5.43	13.44	10.55	**31.56**

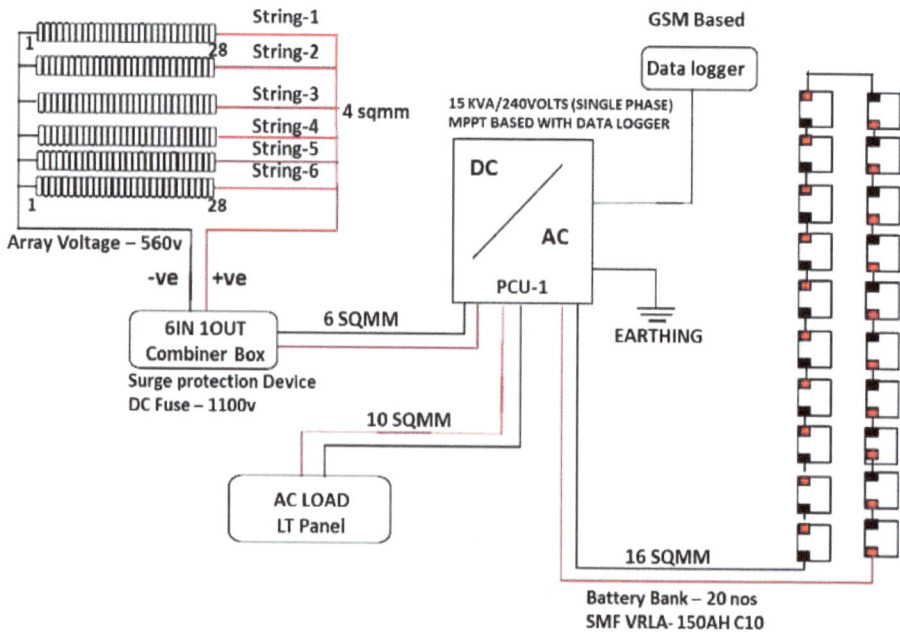

Fig. (11). The block diagram of the 10 kWp PV system.

4.5. Electrical Power from GiSPVT

Fig. (**11**) shows the electrical line diagram of the 10kWp semi-transparent PV roof for zone 1 installed on the south roof of GiSPVT. The rated capacity of 80 W_p of each module with a packing factor (PF) of 80%, (Fig. **8a**), has been considered to examine the effect of packing factors on the performance of each crop mentioned in section 5.1. The total production of vegetables from the GiSPVT up to 15th March, 2021 is shown in Table **2**. The electrical energy generated on a typical day in March was around 40kWhr. The electrical power available from GiSPVT has been used in SODHA ENERGY RESEARCH PARK (SERP-2021) along with nearby houses.

CONSENT FOR PUBLICATION

Not applicable.

CONFLICT OF INTEREST

The authors declare no conflict of interest, financial or otherwise.

ACKNOWLEDGEMENTS

The author is very grateful to the Department of Science and Technology, Government of India, for sponsoring the ongoing project entitled "Mission Innovation (MI) for Uneven GiSPVT for off-grid electrical power for the agricultural sector." The author would also like to thank all the project staff for helping in the project work.

REFERENCES

[1]   B. Agrawal, and G.N. Tiwari, *Developments in Environmental Durability for Photovoltaics.* Pira International Ltd.: UK, 2008.

[2]   G.N. Tiwari, and S. Dubey, *Fundamentals of Photovoltaic Modules and Their Applications.* Royal Society of Chemistry RSC: UK, 2010.

[3]   G.N. Tiwari, and R.K. Mishra, *Advanced Renewable Energy Sources.* RSC publishing: UK, 2012.

[4]   G.N. Tiwari, M. Meraj, M.E. Khan, R.K. Mishra, and V. Garg, "Improved Hottel-Whillier-Bliss equation for N-photovoltaic thermal-compound parabolic concentrator (N-PVT-CPC) collector", *Sol. Energy,* vol. 166, pp. 203-212, 2018.
       http://dx.doi.org/10.1016/j.solener.2018.02.058

[5]   G.N. Tiwari, A.K. Mishra, M. Meraj, A. Ahmad, and M.E. Khan, "Effect of shape of condensing cover on energy and exergy analysis of a PVT-CPC active solar distillation system", *Sol. Energy,* vol. 205, pp. 113-125, 2020.
       http://dx.doi.org/10.1016/j.solener.2020.04.084

[6]   A.K. Singh, G.N. Tiwari, R.G. Singh, and R.K. Singh, "Active heating of outdoor swimming pool water using different solar collector systems", *J. Sol. Energy Eng.,* vol. 142, no. 4, pp. 1-41, 2020.

[7]   A.K. Singh, R.G. Singh, and G.N. Tiwari, "Thermal and electrical performance evaluation of PVT-CPC integrated fixed dome biogas plant", *Renewable Energy Sources,* vol. 154, 2020.

[8]   G.N. Tiwari, M. Meraj, and M.E. Khan, "Exergy analysis of N-photovoltaic thermal-compound parabolic concentrator (N-PVT-CPC) collector for constant collection temperature for vapor absorption refrigeration (VAR) system", *Sol. Energy,* vol. 173, pp. 1032-1042, 2018.

http://dx.doi.org/10.1016/j.solener.2018.08.031

[9]   G.N. Tiwari, and P. Barnwal, *Fundamentals of Solar Dryers.* Anamaya Publisher: New Delhi, 2008.

[10] G.N. Tiwari, and A. Tiwari, *Handbook of Solar Energy.* Springer, 2016.

[11] B. Agrawal, and G.N. Tiwari, *Building Integrated Photovoltaic Thermal Systems.* Royal Society of Chemistry RSC: UK, 2010.

[12] G.N. Tiwari, *Greenhouse Technology for Controlled Environment, Alpha Science.* Narosa Publishing House: New Delhi, 2003.

[13] A. Tiwari, and M.S. Sodha, "Performance evaluation of hybrid PV/thermal water/air heating system: A parametric study", *Renew. Energy,* vol. 31, no. 15, pp. 2460-2474, 2006.

http://dx.doi.org/10.1016/j.renene.2005.12.002

[14] A-H.I.M. Shyam, I.M. Al-Helal, A.K. Singh, and G.N. Tiwari, "Performance evaluation of photovoltaic thermal greenhouse dryer and development of characteristic curve", *J. Renew. Sustain. Energy,* vol. 7, no. 3, p. 033109, 2015.

http://dx.doi.org/10.1063/1.4921408

[15] P. Singh, V. Shrivastava, and A. Kumar, "Recent developments in greenhouse solar drying: A review", *Renew. Sustain. Energy Rev.,* vol. 82, pp. 3250-3262, 2018.

http://dx.doi.org/10.1016/j.rser.2017.10.020

[16] G.K. Mishra, and G.N. Tiwari, "Performance evaluation of 7.2 kWp standalone building integrated semi-transparent photovoltaic thermal system", *Renew. Energy,* vol. 146, pp. 205-222, 2020.

http://dx.doi.org/10.1016/j.renene.2019.06.143

[17] G.K. Mishra, G.N. Tiwari, and T.S. Bhatti, "Effect of water flow on performance of building integrated semi-transparent photovoltaic thermal system: A comparative study", *Sol. Energy,* vol. 174, pp. 248-262, 2018.

http://dx.doi.org/10.1016/j.solener.2018.09.011

[18] A. Deo, G.K. Mishra, and G.N. Tiwari, "A thermal periodic theory and experimental validation of building integrated semi-transparent photovoltaic thermal (BiSPVT) system", *Sol. Energy,* vol. 155, pp. 1021-1032, 2017.

http://dx.doi.org/10.1016/j.solener.2017.07.013

[19] G.N. Tiwari, A. Deo, V. Singh, and A. Tiwari, "Energy efficient passive building: A case study of SODHA BERS complex", *Foundations and Trends® in Renewable Energy,* vol. 1, no. 3, pp. 109-183, 2016.

http://dx.doi.org/10.1561/2700000003

CHAPTER 10

Developments in Solar PV Cells, PV Panels, and PVT Systems

Deepak[1], Shubham Srivastava[1], Sampurna Panda[2], C. S. Malvi[1*]

[1]*Madhav Institute of Technology & Science, Gwalior, Madhya Pradesh, India*

[2]*Energy Poornima University, Jaipur, Rajasthan, India*

Abstract: With the advancement in technology and manufacturing techniques, various solar cell materials evolved, and their practical implementation led to modification in the design and installation of photovoltaic panels. Different solar cells are compared in this chapter considering their efficiency, performance, temperature coefficient, *etc.* Developments in PV panel and photovoltaic thermal (PVT) systems are outlined with their respective applications and advantages. It was found that the cost and efficiency of any solar cell are crucial parameters for deciding its implementation in PV panels. Additionally, the solar panel's temperature deflates its efficiency and lowers the thermal conversion. In order to overcome this problem, a PV system was incorporated with different thermal storage materials and cooling mediums, such as air, water, oil, fluids, *etc.*, lowering the temperature of solar panels and making them able to store the excess solar thermal energy to use it during the sun-off period. It was concluded that thin solar cells, such as perovskite and DSSC solar cells, are widely used where flexibility is important and thermal storage materials are utilized with nanoparticles for better thermal efficiency.

Keywords: Bifacial, Electricity, Power, PVT, Solar cell, Solar energy.

1. INTRODUCTION

Due to abrupt changes in the climatic condition of the world and the total consumption of conventional resources, it is high time to utilize renewable energy most effectively. Although actions have been taken globally in this field, its abundant nature and future potential are still concerning for researchers and scientists. Generally, solar energy utilization is directly linked with solar rooftops

*Corresponding author C. S. Malvi: Madhav Institute of Technology & Science, Gwalior, Madhya Pradesh, India; E-mail: csmalvi@mitsgwalior.in

Manoj Kumar Gaur, Brian Norton & Gopal Tiwari (Eds.)

generating electricity, but solar panels' efficiency is inversely proportional to their temperature. In 2009, the Jawaharlal Nehru National Solar Mission (JNNSM) was launched. The goal was to initiate 20 GW of grid-connected solar projects by 2022. The government raised the target to 100 GW by 2022 in May 2015. With solar energy projects in Tamil Nadu, Rajasthan, Gujarat, and Maharashtra, solar power has become India's fastest-growing industry and continues to produce electricity.

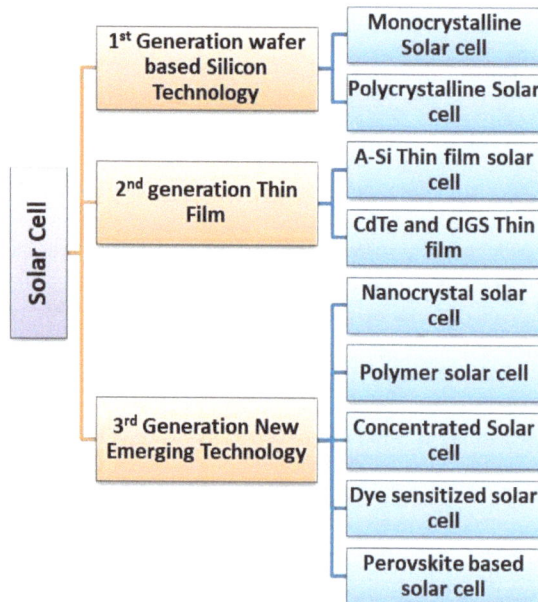

Fig. (1). PV Material Chart [2].

These states are also the top five states with India's highest wind electricity production. Solar energy prices have decreased from Rs. 17.90 per unit in 2010 to around Rs. 7 per unit in 2015. Moreover, solar power achieved grid parity in 2017-18 with technological advances and market rivalry. The Ministry of Energy Resources for Non-Convection, the Government of India, is attempting to increase its power capacity and meet the 100 GW mark by 2022 [1].

2. MATERIAL AND CLASSIFICATION OF SOLAR CELL

A description of the manufacturing materials of solar cells can be found in Fig. (1). Silicon is widely used to manufacture solar cells because it has high efficiency.

Owing to its high costs, however, most researchers are searching for better technologies that can reduce the cost of a solar cell.

Table 1. Advancement in photovoltaic technology with time.

Year	Development	Reference
1839	Antoine – Ceser Becquerel discovered voltage output by exposing solid electrodes to electrolytes. This impact has been called the PVE.	[2]
1876	The first development was the selenium cell invented by W.G., R.E. and Adams. Photovoltaic effects were observed in solid selenium.	[3]
1883	Charles Fritz developed the first true solar cell. He used a thin layer of gold for coating semi-conductor selenium, which had an efficiency of less than 1 %.	[4]
1904	Albert Einstein wrote the first article on the photoelectric effect	[3]
1927	A new cell is produced using copper and copper oxide type, which has less than 1 % efficiency.	[2]
1941	Russell developed the photovoltaic silicon cell	[2]
1954	In silicon photovoltaic cells, the performance of the bell labs has increased rapidly to 6% and 11%.	[4]
1958	Photovoltaic cells were used in space for the first time.	[3]

3. GENERATION OF PHOTOVOLTAIC CELL

Some roadmaps are generated from the research and development (R&D) strategies, which depend on the process and technology. In 2001, a general roadmap of photovoltaic (PV) was presented by Martin Green [5]. According to this PV roadmap, the production of crystalline silicon modules cannot be lower than 1 USD/Wp, which leads to the generation of other technologies. This roadmap categorized PV cell fabrication technologies and materials into three generations, as shown in Fig. (**2**). A wafer-based crystalline silicon cell represents the quite efficient first-generation but is not too encouraging because of higher prices.

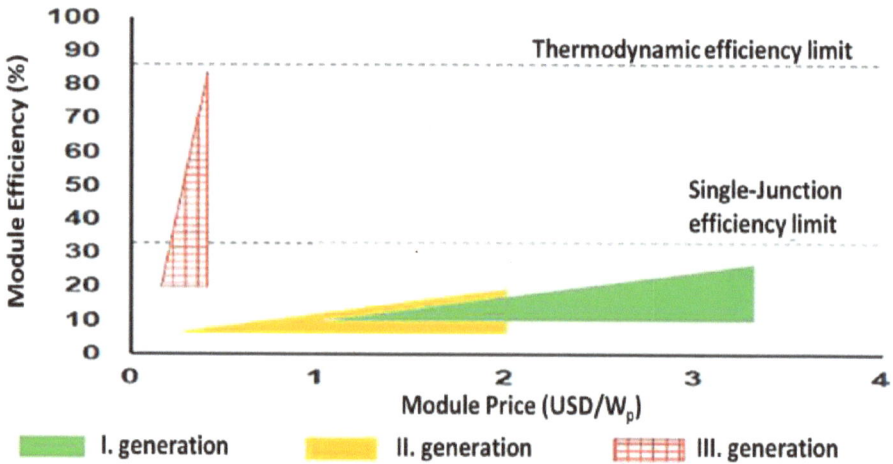

Fig. (2). Efficiency and price of PV module generation.

3.1. First-generation Solar Cells

Silicon wafers are used to create first-generation integrated circuits. Air conditioning technology is the oldest but one of the most commonly used systems. The generations of solar cells are further divided into two types [6]: monocrystalline and polycrystalline silicon solar cells.

3.1.1. Mono-crystalline Silicon Solar Cell (Mono-Si)

The Czochralski technique is applied for manufacturing monocrystalline solar cells [7]. During this Czochralski process, huge crystals are cut from large ingots. Producing high-quality and flawless single-crystal leads requires extremely accurate processing. However, the efficiency of these solar cells varies from 14% to 18%. The leading manufacturer of Sun-Power (2015) claims that its production efficiency is over 20% [8].

3.1.2. Polycrystalline Silicon Solar Cell (p-Si)

This cell is made by mixing different silicon crystals. Cooling a graphite fill process yields cost savings that make it possible. These solar cells are widely used nowadays. The main explanation for the poor efficiency of these cells is their low material quality [7]. Table **2** shows the temperature effect on polycrystalline cell efficiency.

3.2. Second Generation Solar Cells

Thin-film photovoltaic technologies are defined as follows: amorphous silicon (a-Si), micromorph silicon (a-Si/µc-Si), cadmium telluride (CdTe), copper indium diselenide (CIGS), *etc.* Second-generation solar cells are less costly than their predecessors. Solar cells can be manufactured by using silicon wafers. The thickness of a light-sensing film of previous solar cells is 350 µm. This kind of solar panel has a bandgap of 0.1365 eV. Researchers have made major strides in solar conversion efficiency [9].

3.2.1. Amorphous Solar Cell

The solar cell comprises a solar panel constructed of amorphous silicon. Polylactic acid can also be used using inexpensive methods; hence it can be readily prepared. These substrates provide great durability and adaptability during the development method [10]. This solar power module has a dark color to deflect the sun (Black colour conducting band). These tools typically fail or do not work as advertised. Fig. (**3**) shows the structure of amorphous silicon cells. In 2016, the researchers fabricated an amorphous silicon module with a maximum efficiency of 13.8% [11].

3.2.2. Cadmium Telluride (CdTe) Solar Cell

One of the major semi-conductor materials used in processing LEDs and LCDs is cadmium telluride (CdTe). The construction and repair costs of thin-film solar panels are minimal and favorable. Between the cadmium layers is the p-n relation. The procedure entails the treatment of solar cells based on CdTe [3]. Semi-conductors of forms P and T are currently used as substrates.

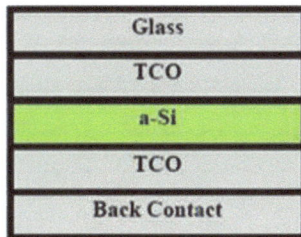

Fig. (3). Structure of Amorphous Si.

In multi-layered CdTe solar cells, different substrates are added. It suits well between 9.5 and 11 % in general, as cadmium is extremely toxic to humans and animals by ingestion. There will be many threats to the community. The Cd will further harm our environment and lives. The overall cell efficiency of CdTe solar cells is about 15% (Aramoto, Ferekides and Wu, 1993). At 295K, CdTe achieves the highest efficiency. According to the NREL, wind energy generated worldwide is estimated to be 17.3 % [12]. The structure of the CdTe cell is shown in Fig. (**4**).

3.2.3. Copper Indium Gallium Dimethyl Gallium Selenide (CIGS) Cells

There are four elements of this semi-conductor. The striking characteristics of these metals are well known. CIGS has reached productivity of around 10 and 12 %. Technologically focused [3] on CIGS solar cell is one of the most innovative high-density, thin-film technologies. It is prepared by sputtering, evaporation, and electroplating. Fig. (**5**) shows the diagram of the CIGS solar cell.

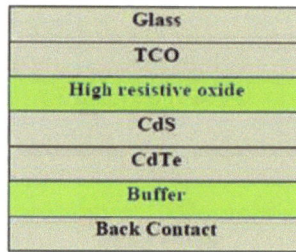

Fig. (4). Schematic CdTe solar cell.

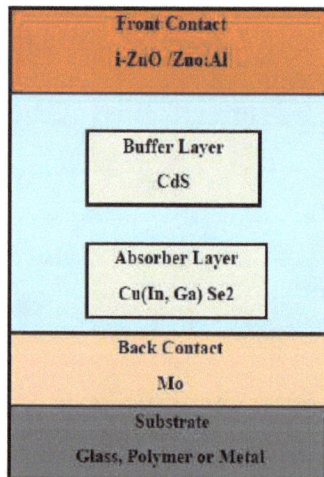

Fig. (5). Diagram of CIGS solar cell.

3.3. Third Generation Solar cells

Several research works on photovoltaic and organic photovoltaic cells have been conducted. The developed solar cells for third-generation are as follows:

- Polymer-based solar cells
- Nano crystal-based solar cells
- Concentrated solar cells

3.3.1. Nano crystal-based Solar Cells

Quantum Dots (QD) is another name for a nano solar cell. The band gap between QD cells is size-tunable. QD solar cells are mainly composed of transition group semi-conductors. Nanocrystals usually have a diameter of a few nanometres. Fig. (6) demonstrates the working principle of the quantum dot solar cells [13]. They are useful due to the presence of centrifugal force.

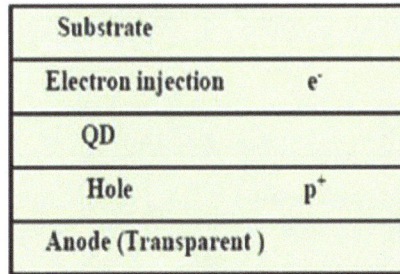

Substrate	
Electron injection	e^-
QD	
Hole	p^+
Anode (Transparent)	

Fig. (6). Diagram of Quantum Dot Cell [14].

3.3.2. Polymer-based Solar Cells

Solar cells are made of polymer. They are flexible due to their polymer substrate. These are the polymer solar cells constructed from successively linked thin functional layers having a ribbon coating on their polypropylene foils (PTFE, or PET). The PSC also exhibits photovoltaic effects, in which solar energy is converted into electric current. An analysis found its performance to be 3.0 % in computer-optimized polymer solar cells [14].

3.3.3. Concentrated Solar Cell (CSC)

CSC is the latest type of engineering advancement. Solar energy has the potential to focus on solar fuel. This approach allows large mirrors and lens concentration of solar energy in silicon cells [15]. In Fig. (7), solar energy produces enormous amounts of thermal energy. In doing so, solar energy is transformed into electricity. This can be divided into high, medium and low CSC. CPV technology has many benefits like this appliance has none of the natural and anthropogenic components of the earth. This table provides a comparison of strength and safety factors.

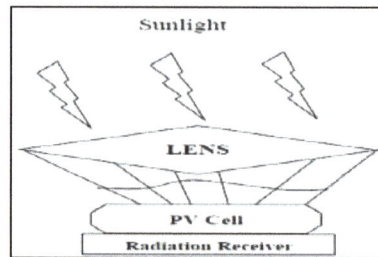

Fig. (7). Diagram of CSC [17].

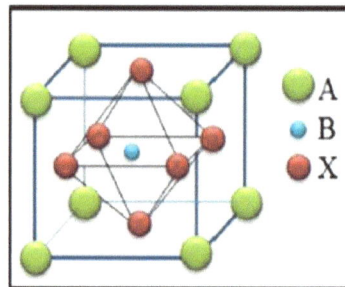

Fig. (8). Crystal Structure of Organometal Perovskite [18].

3.3.4. DSCC (Dye-sensitized solar cells)

DSSC is now promoted in the Swiss Federal Institute of Technology. Among the different electrodes of solar cells based on DSCC, dye molecules have been widely used. The main components of DSCC are a semi-conductor electrode, a material for electron transfer, a sensitizer and some redox reagent intermediaries (carbon or Pt) and a counter electrode (carbon or Pt). The traditional easy alternatives make DSCC more conceivable and affordable. Furthermore, they have the infrared coating of nanostructured titanium dioxide. DSCC greatly improves us because of more than 10% visible chromophores [16]. The titanium wave microstructure is developed by a two-stage electrochemical process appropriate for solar cell and photocatalysis substrate [17]. However, deterioration in the energy market and stability problems are interrelated, a key problem faced by such cells.

Table 2. Comparison of Different Technologies of Solar Cell.

Type of cell	Size	Cost	High-temperature performance	Temperature coefficient ($\times 10^{-3}$ K^{-1})	Efficiency (%)	Other Details	References
First-generation Solar Cells							
Mono-Si	Needs low volume for equivalent power	C-Si is more expensive and two times costlier	Does not work well at high temperatures (10 – 15%)	4-Mar	17 – 18	Widely used technology	[7, 8]
p-Si	Less volume is required in order to produce the same power	Most expensive and two times more expensive than thin-film solar cells silicon	Do not work well at high temperatures (Drops 20 %)	4.2 – 4.5	12 – 14	Cost-to-performance economic preference	[7, 9]
Second-generation Solar Cells							
A-Si solar cell	Provides a wide variety of design goods	50 % cheaper than silicone	Provides reliable ventilation in hot as well as cold conditions	1.1 – 2.6	8-Apr	Requires more time for installation	[3,10, 11]

CdTe solar cell	A broad range of products, including versatile, light, robust structures.	Offers 50% or more savings compared to crystalline silicon	Perform well under all weather conditions (0% drops)	2.5	--	Used in the manufacturing process	[7,19]
CIGS solar cell	Offering a wide range of high-quality items with lightweight and robust construction	50% less costly and cheaper than crystalline silicon technology	Provides reliable ventilation in hot as well as cold conditions	3.5 – 3.6	10 – 15	Some of the coal-fired power plants in China have high-efficiency ratings.	[3, 6]

3.3.5. Perovskite-based Solar Cell

It is an innovative study in the solar cell. A thin-film solar cell contains many useful properties compared to traditional silicon cells. It comprises a class of minerals called Perovskite compounds ABX3, where A denotes chlorine, bromine or fluorine. The cations have different sizes. The inflation rate has reached around 9.7 %. An analysis of Jacobsson found that profit had fallen by 25 %. Once the temperature rises from -80°C to 80°C [18], then it works very well and gives an efficiency of approximately 31%. However, one of the problems of Perovskite-style solar cells is the stability and longevity, and due to the deterioration of material in this cell, productivity gets degraded.

4. DEVELOPMENT IN PV TECHNOLOGY

Photovoltaic is the technology that converts solar energy into electrical energy with the help of the photoelectric effect. The team of researchers is continuously doing research that leads to a reduction in cost and improvement in the efficiency of photovoltaic panels. Currently, this requirement is only met by crystalline silicon modules. The efficiency of this module is about 16-22%. Fig. (**9**) shows the development of the crystalline silicon module with time [19]. The research trend is shifting from multi-crystalline to monocrystalline starting wafers, which have higher efficiency. Also, the concept of bifacial modules is used because it uses reflected radiation to increase output power [20]. The module service life depends upon the back-cover material and lamination technology. If the back-cover material is plastic, then the service life of the module is 25 years, while if the back cover is a glass sheet, then its life reaches up to 30-40 years [21]. The glass module life is higher because it has better mechanical stability, humidity and UV conditions and higher resistance to temperature. Also, the chances of micro cracks are reduced during its installation and operation. The main disadvantages of using glass as a back sheet are higher cost, fragile nature, and low specific power (kW/Kg).

The current cost of crystalline silicon modules is cheaper than other modules due to the wide availability of silicon wafers, ingots, and silicon feedstock. The photovoltaic systems have undergone unprecedented cost decline in electricity-conversion technologies, including silicon and thin-film solar cells. The next cost reduction could be possible using new technology, such as bifacial, N-type material, *etc.*

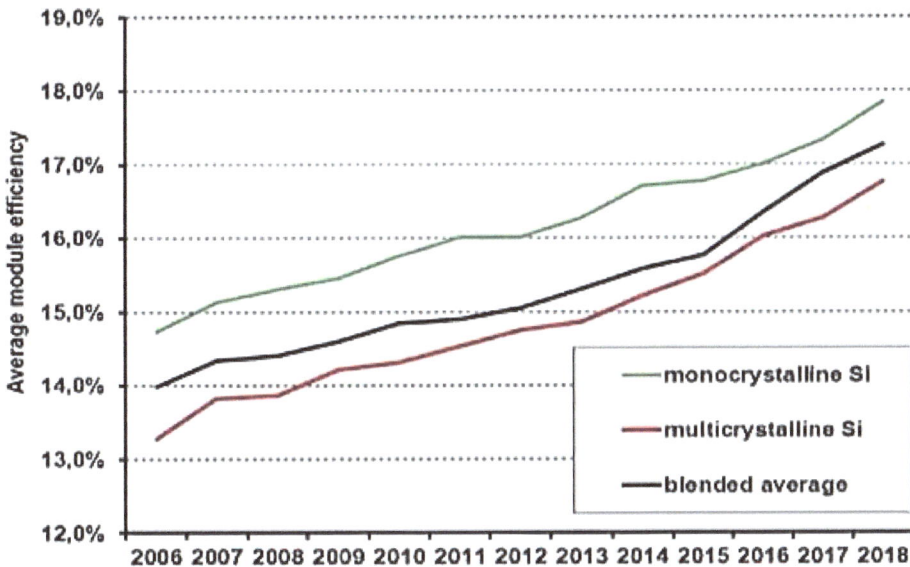

Fig. (9). Development in Crystalline Silicon Module with Time [19].

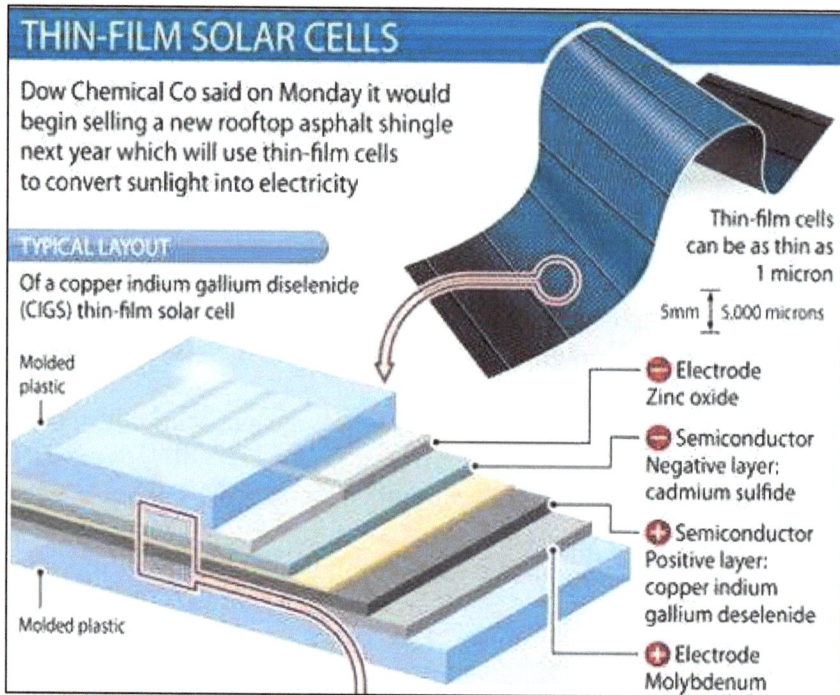

THIN-FILM SOLAR CELLS

Dow Chemical Co said on Monday it would begin selling a new rooftop asphalt shingle next year which will use thin-film cells to convert sunlight into electricity

TYPICAL LAYOUT

Of a copper indium gallium diselenide (CIGS) thin-film solar cell

Thin-film cells can be as thin as 1 micron

5mm | 5,000 microns

Molded plastic

Molded plastic

Electrode
Zinc oxide

Semiconductor
Negative layer:
cadmium sulfide

Semiconductor
Positive layer:
copper indium
gallium deselenide

Electrode
Molybdenum

Fig. (10). Flexible Solar Cell [23].

4.1. Thin-film Solar Module

It is also known as a Flexible Solar Module. It is best suited for curved surfaces and is used in building-integrated photovoltaics (BIPV). Due to its flexibility and lightweight, it requires less shipment cost, installation cost, structure cost, *etc.* The technology of flexible solar cells is not mature enough compared to the existing solar module. The major advantages of a flexible solar module include high efficiency, low cost, flexibility in nature, lightweight, and higher specific power. The commonly used substrate for a flexible solar cell is metallic foils [titanium (Ti) and stainless steel (SS)], plastic films [polyethylene terephthalate (PET), polyimide, polyethylene naphthalate (PEN)], paper substrate [bank notes, cellulose paper, plain white paper and security bond], and flexible glass substrates [22]. Thin-film solar cells are CIGS (Copper Indium Gallium Selenide), CdTe (Cadmium Telluride), and a-Si:H (Hydrogenated amorphous silicon). The efficiency of CIGS increases from 10–13% to 14–16% [20]. The flexible solar cell is shown in Fig. **(10)**.

4.2. Bifacial Solar Module

It is a new concept that can increase the efficiency of photovoltaic systems. It was first investigated during the 1960s but scientifically introduced by Luque *et al.* in literature [24]. It can receive solar radiation from the top and bottom of the module surface. The efficiency of a bifacial solar module is greater than a mono-facial module. Fig. (**11**) shows a bifacial solar cell. This technology may be expected to become more popular in the global market. According to the International Technology roadmap of photovoltaic [25], the bifacial solar cell technology shares will increase by more than 15% by 2024 in the world market. The bifacial module conversion efficiency is shown in Table **3** [26-28].

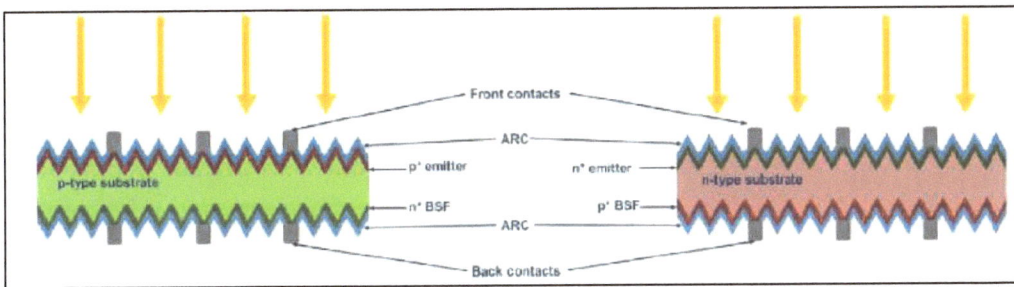

Fig. (11). Bifacial Solar cell.

Table 3: Efficiency of Different Bifacial Solar cell.

Description	Concentrator (Sun)	Efficiency (%)
Cz nPERT	1	19.4
(Rating: 263W, Cells: 60, Size: 156 x 156 mm2)	-	-
Cz n-type, copper plating Triex	1	19.98
(Rating: 390 W, Cells: 72, Size: 152.4 x 152.4 mm2)	-	-
Fz n-type Si rear-contact V-groove and inverted pyramids	1	20.66/10.54

(Table 3) cont.....

(Rating: 9.42W, Cells: 20, Size: 66.3 x 32.5 mm2)	-	-
Description	**Concentrator (Sun)**	**Efficiency (%)**
Cz nPERT (Rating: 263W, Cells: 60, Size: 156 x 156 mm²)	1	19.4
Cz n-type, copper plating Triex (Rating: 390 W, Cells: 72, Size: 152.4 x 152.4 mm²)	1	19.98
Fz n-type Si rear-contact V-groove and inverted pyramids (Rating: 9.42W, Cells: 20, Size: 66.3 x 32.5 mm²)	1	20.66/10.54

5. DEVELOPMENT IN PVT TECHNOLOGY

To utilize the available solar energy effectively, photovoltaic thermal (PVT) systems have become the prime interest of researchers working in the field of energy now a day, which generate electricity and heat simultaneously. This system is appropriate for such institutions and offices, with limited installation area and demand for both electricity and heat. Further obtained heat can be utilized either to heat the space or water [29].

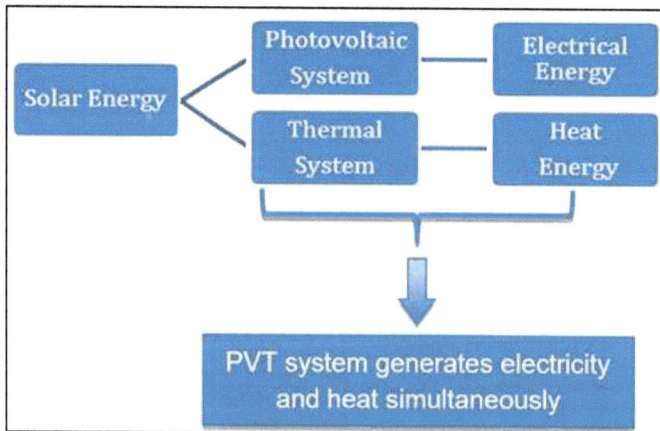

Fig. (12). Solar PVT System.

5.1. PVT System

Solar PV technology only converts a fraction of incident solar energy into electricity. The remaining energy is wasted into heat, which ultimately increases the module's temperature and decreases the system's overall efficiency. PVT system utilizes that waste heat for thermal application and provides a cooling medium to a solar module, thus increasing its efficiency. The schematic diagram of the PVT system is shown in Fig. (**12**).

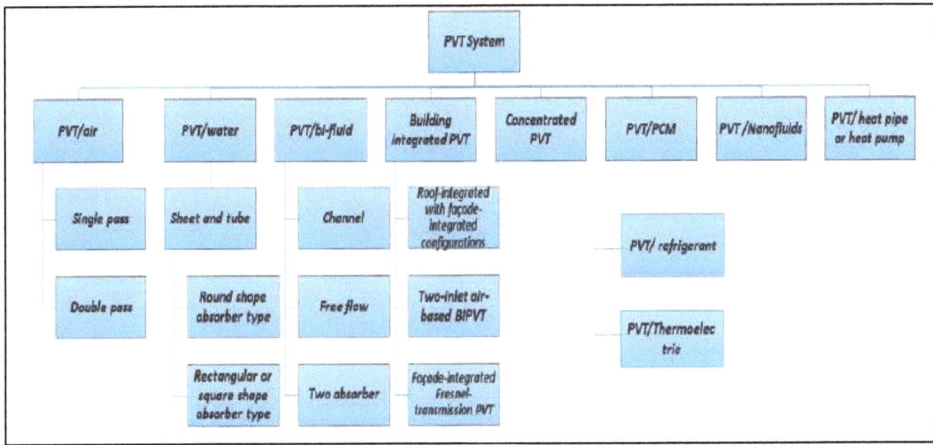

Fig. (13). Classification of a PVT system.

5.2. Classification of PVT Systems

Classification of the PVT system is generally based on the type of working fluid used, and it can be divided into PVT/air, PVT/water, and PVT/bi-fluid (air-water), as shown in Fig. (**13**). Another type of classification is based on the circulation of working fluids (natural and forced). However, its classification is based on multiple criteria; to classify it generally, we can have criteria of the absorber and working fluid used in it.

5.2.1. PVT/air

In this system, the air is circulated over a PV panel with a different configuration of an absorber, either inactive or in passive mode, to enhance the efficiency of the PV system. Many researchers developed the mathematical model and carried out simulations and experiments to optimize the PVT system's methods, material, and configuration [30-32]. Air has a low density and low heat capacity; these properties limit the application of air in a PVT system. One of the air-based PVT systems is shown in Fig. (**14**).

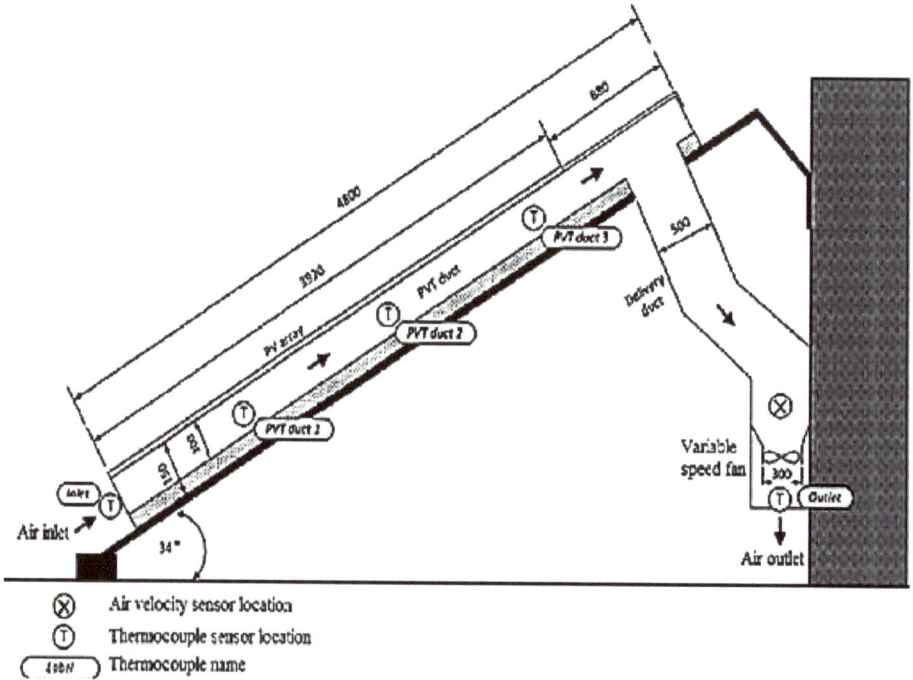

Fig. (14). Air-based PVT system [33].

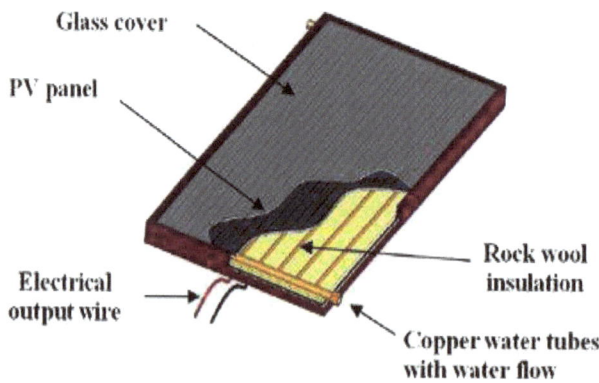

Fig. (15). PVT/water system [35].

5.2.2. PVT/water

To overcome the limitations of air-based PVT systems, water is used as a working fluid in conventional PVT systems because of its higher heat carrying capacity than air. However, an additional heat exchanger is needed for energy conversion, which ultimately increases the cost of the basic PVT system. We can find many industrial and domestic PVT/water systems applications, such as heating water and simultaneous electricity production [33]. PVT/water system generally uses a parallel-connected flat plate solar collector with a thermo-siphon water circulation mechanism for domestic application. In contrast, several series-connected flat plate collectors with solar-driven pumps for water circulation are used in industrial applications. Gond *et al.* used a phase change material in a flat plate collector [34]. A comparative study on different flow configurations in flat plate collectors is presented by Malvi *et al.* [35]. Fig. (**15**) shows one example of a water-based PVT system.

5.2.3. PVT/bi-fluid

In this PVT system, both air and water properties are utilized to maximize its performance. It is also called the PVT combo system and was first developed by Tripanagnostopoulos [36]. Fig. (**16**) shows the bi-fluid-based PVT system. Different configurations can be obtained by interchanging or modifying the air channel and water circulation tube location. This system shows greater efficiency and performance because of the higher temperature reduction of PV panels compared to PVT/air and PVT/water.

5.2.4. Building-integrated PVT (BIPVT)

Another interesting application of solar PV, solar thermal and PVT is found in buildings, where these are directly integrated into building segments such as façades and rooftops, as shown in Fig. (**17**). This building integration (BI) provides a better aesthetic view and building efficiency because a part of energy consumption is directly borne by solar energy.

5.2.5. Concentrated PVT

Concentrated PV (CPV) systems have become a fascinating field for researchers because of the low cost of reflectors compared to solar cells. Fig. (**18**) shows the concentrated PVT system. CPV systems operate at a higher temperature than conventional ones; the excess heat generated in CPV is utilized for thermal applications, called CPV thermal (CPVT) system. CPVT system can be classified as a low concentration and high concentration CPVT based on its application, such as line or point focused *etc.* or type of reflectors or concentrators [37]. Reflectors are used to reflect and concentrate the solar radiation over a high-efficiency solar cell; beneath this solar cell, fluid flows to carry away the excess heat generated by CPV.

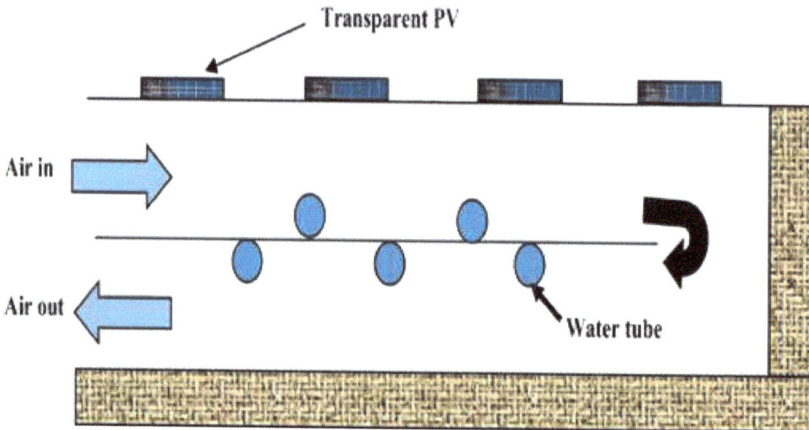

Fig. (16). PVT/bi-fluid system [37].

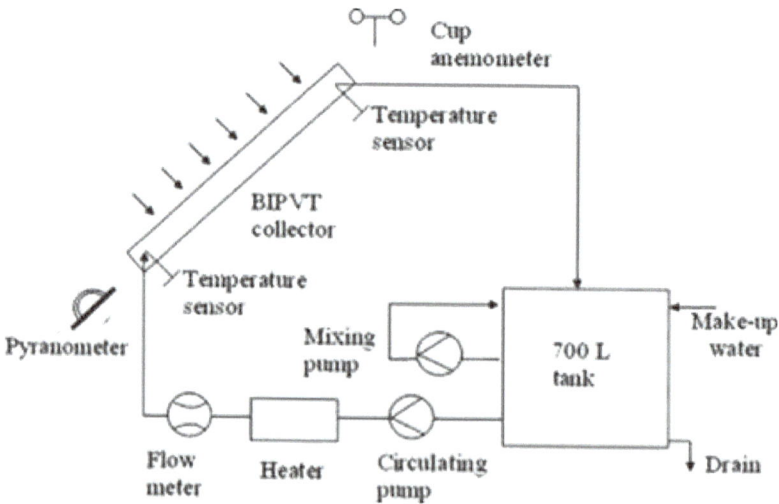

Fig. (17). Building-integrated PVT system [38].

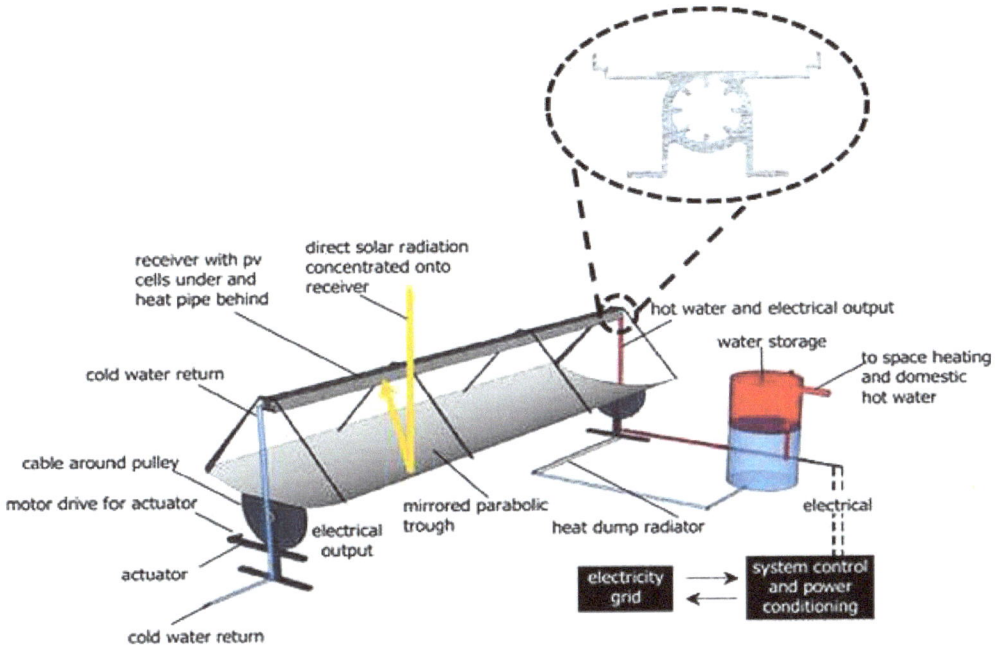

Fig. (18). CPVT system [39].

5.2.6. PVT/PCM

To manage the temperature and efficiency of the PV module, various research was conducted on utilizing the thermal energy storage media (Sensible and latent heat storage) in it. Initially, sand-rock minerals, cast iron, reinforced concrete, *etc.*, were used as sensible heat storage (SHS) media. However, these became obsolete because of small energy storage density and large volume requirements [38]. Contrary latent heat storage media (LHS) requires less storage volume and has the fascinating characteristic of storing the energy in the form of latent heat [39]. PCMs have been extensively used in many applications, such as cooling systems, automobiles, batteries, water storage, *etc.* Researchers and solar industry experts have recently incorporated PCMs into solar modules to improve the system's efficiency, as shown in Fig. (**19**). Beneath the PV module, PCMs are used; PCM carries away the heat of the panel and stores it in the form of latent heat and discharges that stored heat when required or during the off-sunshine hour.

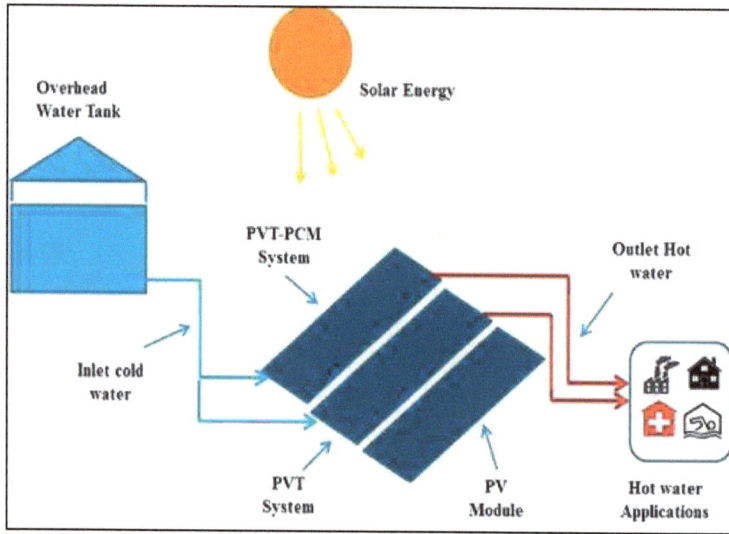

Fig. (19). PVT/PCM system.

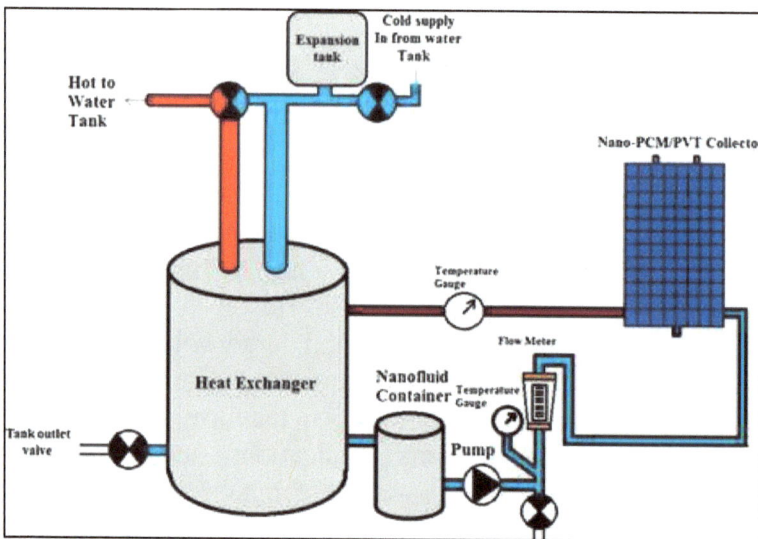

Fig. (20). PVT/ nanofluid system [44].

5.2.7. PVT/nanofluids

When nanoparticles are mixed with water, oil, ethylene glycol, and other base fluids, the thermal conductivity of the base fluids improves. Various methods are available to prepare nanofluid, and the properties of nanofluid depend on the base fluid type, nanoparticles type, and the ratio of their added mass. Various research was conducted on PVT/nanofluid systems that used water as a base fluid to cool the PV cell [40]. Koteswararao *et al.* [41] discussed the performance of PVT systems by utilizing Al_2O_3 and Cu nanoparticles in water or ethylene glycol. Al-Waeli *et al.* [42] analyzed the performance of PV/T systems by using three types of base fluids with nano-SiC. An example of a nanofluid-based PVT system is shown in Fig. (**20**).

5.2.8. PVT/thermoelectric

The working principle of the thermoelectric generator (TEG) is based on the See-beck effect and Peltier effects. TEGs are solid-state devices that have no moving parts, and there is no need for maintenance, but the device's cost is the only thing that bothers the research community. By varying the heat source type, which is solar cell type (such as monocrystalline, polycrystalline and thin-film *etc.*), and varying heat sink configuration and materials, we can obtain and analyze the performance of different PVT/thermoelectric systems (shown in Fig. **21**).

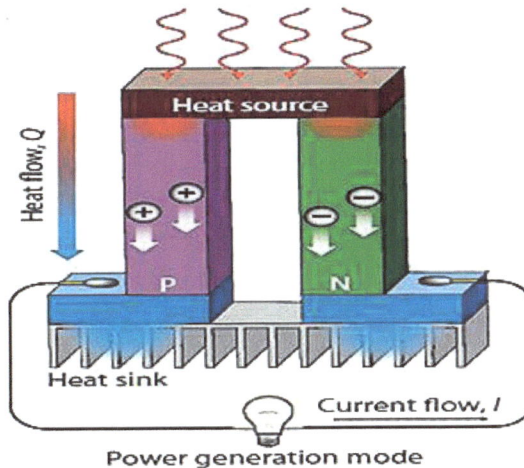

Fig. (21). Thermoelectric generator.

5.2.9. PVT/Heat Pipe

Heat pipes possess good thermal conductivity and good heat removal and transmission rate. The typical diagram of the heat pipe is shown in Fig. (**22**). In a

PVT/heat pipe system, one side of the heat pipe with several microchannels acts as an evaporative section, and the other side acts as a condensation section. The fluid was flowing in the condensation section condensed to discharge the heat absorbed from the PV panel.

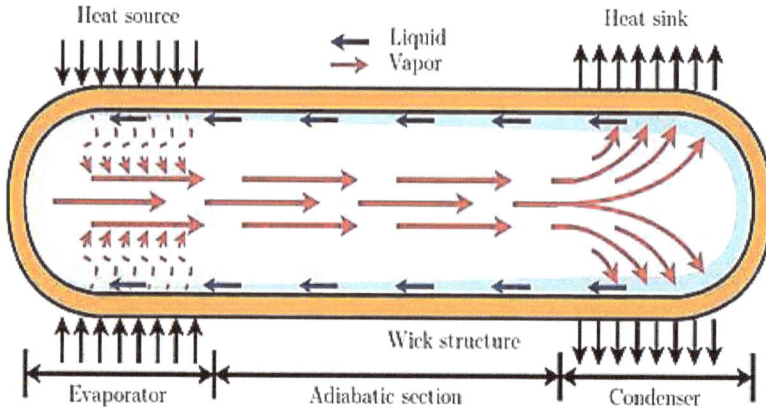

Fig. (22). Typical diagram of heat pipe [40].

5.2.10. PVT/Refrigerant

When working fluid is replaced with refrigerant in place of water or air in the PVT system, the modified system is proved efficient compared to the conventional PVT system. Although refrigerant fluid R134a is commonly used in this system, by varying the refrigerant type and configuration of the PVT system, the performance can be predicted for different PVT/refrigerant systems. Tsai developed the mathematical model of PVT/refrigerant-based system assisted with heat pump water heater and validated the results on the test rig [43].

6. RECENT TRENDS AND MODIFICATIONS IN PVT SYSTEM

It is obvious from the above classification of the PVT system that numerous studies were conducted to analyze the performance of solar PV systems. Researchers and scientists implemented new technologies (cells, materials, cooling medium, *etc.*) and modified the system configuration to improve its efficiency. The developments in solar cells and technologies lead the way in evaluating PVT systems. Although many studies are available on PVT systems, the commercialization of the product is still a very tough task because of the lack of agreement on a specific type of PV/T systems among researchers [38].

Table 4. Recent literature on PVT.

Authors	Year	PVT System Description	Results
Ahmadi *et al.* [44]	2021	PVT with PCM composite (PCM is filtered in PS-CNT foam)	PV cell temperature dropped to 6.8%, and the overall system performed well with 66.8–82.6% energy efficiencies.
Walshe *et al.* [45]	2021	PVT with organic luminescent down-shifting liquid beam splitters	63% optical efficiency and up to 20% improvement were achieved for the standalone PV system. The fluid showed a better performance than the base fluid used in the previous PVT system.
Rad *et al.* [46]	2020	PVT with PCM and aluminium shaving porous media	The exergy and electrical efficiency improved by 4.34% and 2.4 %, respectively. The PCM melting time was decreased by utilizing porous media.
Wahab *et al.* [47]	2020	Incorporation of nanofluid and PCM in PVT	The maximum thermal and electrical exergy efficiency obtained were 1.78% and 13.02% at 20 L and 40 L per minute, respectively, with a volume concentration of 0.1% nanofluid.
Kalkan *et al.* [48]	2019	PVT associated with fins	CFD ANSYS Fluent was used to analyze three parameters, namely channel height, air temperature, and PCM length. The overall efficiency was increased up to 19%.

(Table 4) cont.....

Kazemian *et al.* [49]	2018	PVT with PCM and ethylene glycol	PVT/PCM system performed well, and electrical efficiency was improved. The thermal exergy efficiency was enhanced when ethylene glycol with different concentrations was added to PVT/PCM.
Yamaguchi *et al.* [50]	2017	CO_2 based PVT	The efficiency of the hybrid PVT system increased by 2% compared to the conventional one.
Hjerrild *et al.* [51]	2016	PVT with nanofluid (silver-silica nano-discs and carbon nanotubes suspended in water)	It was concluded that the system proved efficient. However, an additional cost was associated while utilizing nanofluid.
Radwan *et al.* [52]	2016	Low concentrated PVT(LCPVT) with heat sink and nanofluid	Solar cell temperature was reduced by 38°C, and electrical efficiency was enhanced up to 19%.

CONCLUSION

It is evident from the research that silicon is used as a prime material for manufacturing solar cells. However, its high price and low energy conversion efficiency lead to the development of the thin-film solar cell. The efficiency and price of thin-film solar cells are better than the previous one but less than the Shockley-Queisser limit. Furthermore, newly developed technologies, such as Perovskite cell, DSSC *etc.*, are discussed for generating electricity using a solar cell. In addition to these technologies, a bifacial solar panel absorbs the incoming solar radiation from the top and bottom of the module surface, increasing the system's efficiency.

Moreover, the thermal management of solar panels also plays a significant role in energy conversion as the efficiency of a solar panel is inversely proportional to its temperature. In this regard, various PVT systems, such as PVT/air, PVT/water, PVT/ refrigerant *etc.*, have come into the picture, generating electricity and thermal energy simultaneously. Further addition of nanoparticles into its driving mechanism positively increases its efficiency.

CONSENT FOR PUBLICATION

Not applicable.

CONFLICT OF INTEREST

The authors declare no conflict of interest, financial or otherwise.

ACKNOWLEDGEMENTS

Declared none.

REFERENCES

[1]   R. Appa, M.J. Rama, R.L. Malleswara, N.K. Ram, S. Singh, and R.P. Ramakrishna, "Solar Energy In India - Present And Future", *Int. J. Eng. Sci. Invent.,* 2018.

[2]   "Energy Conversion :Development of solar cells", Available from: https://www.britannica.com/technology/solar-cell.

[3]   N. Rathore, N.L. Panwar, F. Yettou, and A. Gama, "A comprehensive review of different types of solar photovoltaic cells and their applications", *Int. J. Ambient Energy,* pp. 1-49, 2019.
      http://dx.doi.org/10.1080/01430750.2019.1592774

[4]   P. Hersch, and K. Zweibel, *Basic photovoltaic principles and Methods.* 1982.
      http://dx.doi.org/10.2172/5191389

[5]   M.A. Green, "Third generation photovoltaics: Ultra-high conversion efficiency at low cost", *Prog. Photovolt. Res. Appl.,* vol. 9, no. 2, pp. 123-135, 2001.
http://dx.doi.org/10.1002/pip.360

[6]   P.C. Choubey, A. Oudhia, R. Dewangan, and V.Y.T.P.G. Autonomous, "Autonomous, A review : Solar cell current scenario and future trends", *Recent Res. Sci. Technol.,* vol. 4, no. 8, 2012.

[7]   S. Sharma, K.K. Jain, and A. Sharma, "Solar Cells: In research and applications—A review", *Mater. Sci. Appl.,* vol. 6, no. 12, pp. 1145-1155, 2015.
      http://dx.doi.org/10.4236/msa.2015.612113

[8]   *Solar cell materials Course: Solid State II.* Department of Physics, University of Tennessee, 2008.

[9]   P. Kamkird, N. Ketjoy, W. Rakwichianm, and S. Sukchai, "Investigation on temperature coefficients of three types photovoltaic module technologies under Thailand operating condition", *Procedia Engineering,* vol. 32, pp. 376-383, 2012.

http://dx.doi.org/10.1016/j.proeng.2012.01.1282

[10] A.M. Barnett, J.A. Rand, R.B. Hall, J.C. Bisaillon, E.J. DelleDonne, B.W. Feyock, D.H. Ford, A.E. Ingram, M.G. Mauk, J.P. Yaskoff, and P.E. Sims, "High current, thin silicon-on-ceramic solar cell", *Sol. Energy Mater. Sol. Cells,* vol. 66, no. 1-4, pp. 45-50, 2001.

http://dx.doi.org/10.1016/S0927-0248(00)00157-4

[11] M. Kaur, and H. Singh, "A review: comparison of silicon solar cells and thin film solar cells", *Int. J. Core Eng. Manag.,* vol. 3, no. 2, pp. 15-23, 2016.

[12] A. Mohammad Bagher, "Types of Solar Cells and Application", *American Journal of Optics and Photonics,* vol. 3, no. 5, pp. 94-113, 2015.

http://dx.doi.org/10.11648/j.ajop.20150305.17

[13] H. Hoppe, and S.N. Sariciftci, "Polymer Solar Cells", In: *Photoresponsive Polymers II..* Springer Berlin Heidelberg: Berlin, Heidelberg, 2007, pp. 1-86.

http://dx.doi.org/10.1007/12_2007_121

[14] N. Venkata, S. Ganesh, and Y.V. Supriya, "Recent advancements and techniques in manufacture of solar cells : Organic solar cells", *Int. J. Electron. Comput. Sci. Eng.,* vol. 2, pp. 565-573, 2013.

[15] W.A. Badawy, "A review on solar cells from Si-single crystals to porous materials and Quantum dots", *J. Adv. Res.,* pp. 1-10, 2013.

http://dx.doi.org/10.1016/j.jare.2013.10.001 PMID: 25750746

[16] S. Suhaimi, M.M. Shahimin, Z.A. Alahmed, J. Chyský, and A.H. Reshak, "Materials for enhanced dye-sensitized solar cell performance: Electrochemical application", *Int. J. Electrochem. Sci.,* vol. 10, pp. 2859-2871, 2015.

[17] M. Poullou, C.S. Malvi, D.W. Dixon-Hardy, and R. Crook, "Titanium wave-like surface microstructure for multiple reflections in solar cell substrates prepared by an all-solution process", *Scr. Mater.,* vol. 62, no. 6, pp. 411-414, 2010.

http://dx.doi.org/10.1016/j.scriptamat.2009.12.003

[18] T.J. Jacobsson, W. Tress, J.P. Correa-Baena, T. Edvinsson, and A. Hagfeldt, "Room temperature as a goldilocks environment for CH3NH3PbI3 perovskite solar cells: The importance of temperature on device performance", *J. Phys. Chem. C,* vol. 120, no. 21, pp. 11382-11393, 2016.

http://dx.doi.org/10.1021/acs.jpcc.6b02858

[19] "Fraunhofer. Photovoltaics Report", Available from: https://www.ise.fraunhofer.de/ conte%0Ant/dam/ise /de/documents/publications/studies/Photovoltaics

[20] V. Benda, and L. Černá, "PV cells and modules – State of the art, limits and trends", *Heliyon,* vol. 6, no. 12, p. e05666, 2020.

http://dx.doi.org/10.1016/j.heliyon.2020.e05666 PMID: 33364478

[21] "Lithuania. Glass/glass PV Modules, c 2013-2020", Available from: http://www.viasolis.eu/page/glass-gl%0Aass-pv-modules

[22] J. Ramanujam, D.M. Bishop, T.K. Todorov, O. Gunawan, J. Rath, R. Nekovei, E. Artegiani, and A. Romeo, "Flexible CIGS, CdTe and a-Si:H based thin film solar cells: A review", *Prog. Mater. Sci.,* vol. 110, p. 100619, 2020.

http://dx.doi.org/10.1016/j.pmatsci.2019.100619

[23] "Flexible Solar cell. How Stuff Works.com, BP, Mia Sole, news reports", Available from: https://science.howstuffworks.com/environmental/energy/solar-cell.htm

[24] A. Luque, A. Cuevas, and J.M. Ruiz, "Double-sided n+-p-n+ solar cell for bifacial concentration", *Solar Cells,* vol. 2, no. 2, pp. 151-166, 1980.

http://dx.doi.org/10.1016/0379-6787(80)90007-1

[25] International Technology roadmap for Photovoltaics, Available from: https://resources.solarbusinesshub. com/solar-industry-reports/item/international-technology-roadmap-for-photovoltaic-results

[26] S.H. Yu, C.J. Huang, P.T. Hsieh, C.H. Chang, W.C. Mo, Z.W. Peng, C.W. Lai, and C.C. Li, "20.63 % nPERT cells and 20% PR gain bifacial module", *2014 IEEE 40th Photovoltaic Specialist Conference (PVSC)*, 2014. *IEEE*, Denver, CO, USA.
http://dx.doi.org/10.1109/PVSC.2014.6925347

[27] J.B. Heng, J. Fu, B. Kong, Y. Chae, W. Wang, Z. Xie, A. Reddy, K. Lam, C. Beitel, C. Liao, C. Erben, Z. Huang, and Z. Xu, ">23% high-efficiency tunnel oxide junction bifacial solar cell with electroplated Cu gridlines", *IEEE J. Photovolt.*, vol. 5, no. 1, pp. 82-86, 2015.
http://dx.doi.org/10.1109/JPHOTOV.2014.2360565

[28] Z. Zhou, P.J. Verlinden, R.A. Crane, R.M. Swanson, and R.A. Sinton, "21.9% efficient silicon bifacial solar cells", *Conference Record of the Twenty Sixth IEEE Photovoltaic Specialists Conference*, 1997. pp. 287-290.
http://dx.doi.org/10.1109/PVSC.1997.654085

[29] C.S. Malvi, D.W. Dixon-Hardy, and R. Crook, "Energy balance model of combined photovoltaic solar-thermal system incorporating phase change material", *Sol. Energy*, vol. 85, no. 7, pp. 1440-1446, 2011.
http://dx.doi.org/10.1016/j.solener.2011.03.027

[30] J. K. Tonui, and Y. Tripanagnostopoulos, "Air-cooled PV/T solar collectors with low cost performance improvements", *Solar Energy*, vol. 81, no. 4, pp. 498-511, 2006.
http://dx.doi.org/10.1016/j.solener.2006.08.002

[31] S. Agrawal, and G.N. Tiwari, "Exergoeconomic analysis of glazed hybrid photovoltaic thermal module air collector", *Sol. Energy*, vol. 86, no. 9, pp. 2826-2838, 2012.
http://dx.doi.org/10.1016/j.solener.2012.06.021

[32] F. Sarhaddi, S. Farahat, H. Ajam, A. Behzadmehr, and M. Mahdavi Adeli, "An improved thermal and electrical model for a solar photovoltaic thermal (PV/T) air collector", *Appl. Energy*, vol. 87, no. 7, pp. 2328-2339, 2010.
http://dx.doi.org/10.1016/j.apenergy.2010.01.001

[33] M. Erdil, "An experimental study on energy generation with a photovoltaic (PV)-solar thermal hybrid system", *Energy*, vol. 33, no. 8, pp. 1241-1245, 2008.
http://dx.doi.org/10.1016/j.energy.2008.03.005

[34] B.K. Gond, M.K. Gaur, and C.S. Malvi, "Manufacturing and performance analysis of solar flat plate collector with phase change material", *Int. J. Emerg. Technol. Adv. Eng.*, vol. 2, no. 3, pp. 456-459, 2012.

[35] C.S. Malvi, A. Gupta, M.K. Gaur, R. Crook, and D.W. Dixon-Hardy, "Experimental investigation of heat removal factor in solar flat plate collector for various flow configurations", *Int. J. Green Energy*, vol. 14, no. 4, pp. 442-448, 2017.
http://dx.doi.org/10.1080/15435075.2016.1268619

[36] Y. Tripanagnostopoulos, T. Nousia, M. Souliotis, and P. Yianoulis, "Hybrid photovoltaic/thermal solar systems", *Sol. Energy*, vol. 72, no. 3, pp. 217-234, 2002.
http://dx.doi.org/10.1016/S0038-092X(01)00096-2

[37] O.Z. Sharaf, and M.F. Orhan, "Concentrated photovoltaic thermal (CPVT) solar collector systems: Part II – Implemented systems, performance assessment, and future directions", *Renew. Sustain. Energy Rev.*, vol. 50, pp. 1566-1633, 2015.
http://dx.doi.org/10.1016/j.rser.2014.07.215

[38] A.H.A. Al-waeli, "A review of photovoltaic thermal systems: Achievements and applications", *International Journal of Energy Research,* vol. 45, no. 7, 2020.
http://dx.doi.org/10.1002/er.5872

[39] N. Agyenim, and P. Hewitt, "A review of materials, heat transfer and phase change problem formulation for latent heat thermal energy storage systems (LHTESS)", *Renewable and Sustainable Energy Reviews,* vol. 14, no. 2, pp. 615-628, 2010.
http://dx.doi.org/10.1016/j.rser.2009.10.015

[40] N. Karami, and M. Rahimi, "Heat transfer enhancement in a hybrid microchannel-photovoltaic cell using Boehmite nanofluid", *Int. Commun. Heat Mass Transf.,* vol. 55, pp. 45-52, 2014.
http://dx.doi.org/10.1016/j.icheatmasstransfer.2014.04.009

[41] V. Basam, "Experimental Analysis of solar panel efficiency with different modes of cooling", *International Journal of Engineering and Technology (IJET),* vol. 8, no. 3, pp. 1451-1456, 2016.

[42] A.H.A. Al-Waeli, K. Sopian, J.H. Yousif, H.A. Kazem, J. Boland, and M.T. Chaichan, "Artificial neural network modeling and analysis of photovoltaic/thermal system based on the experimental study", *Energy Convers. Manage.,* vol. 186, pp. 368-379, 2019.
http://dx.doi.org/10.1016/j.enconman.2019.02.066

[43] H.L. Tsai, "Modeling and validation of refrigerant-based PVT-assisted heat pump water heating (PVTA–HPWH) system", *Solar Energy,* vol. 122, pp. 36-47, 2015.
http://dx.doi.org/10.1016/j.solener.2015.08.024

[44] R. Ahmadi, F. Monadinia, and M. Maleki, "Passive/active photovoltaic-thermal (PVT) system implementing infiltrated phase change material (PCM) in PS-CNT foam", *Sol. Energy Mater. Sol. Cells,* vol. 222, p. 110942, 2021.
http://dx.doi.org/10.1016/j.solmat.2020.110942

[45] J. Walshe, P.M. Carron, S. McCormack, J. Doran, and G. Amarandei, "Organic luminescent down-shifting liquid beam splitters for hybrid photovoltaic-thermal (PVT) applications", *Sol. Energy Mater. Sol. Cells,* vol. 219, p. 110818, 2021.
http://dx.doi.org/10.1016/j.solmat.2020.110818

[46] V. Amin, A. Rad, S. Kasaeian, and F. Mousavi, "Empirical investigation of a photovoltaic-thermal system with phase change materials and aluminum shavings porous media", *Renew. Energy,* vol. 1481, no. 20, pp. 31887-5, 2020.
http://dx.doi.org/10.1016/j.renene.2020.11.135

[47] A. Wahab, M.A.Z. Khan, A. Hassan, and A. Hassan, "Impact of graphene nanofluid and phase change material on hybrid photovoltaic thermal system: Exergy analysis", *J. Clean. Prod.,* vol. 277, p. 123370, 2020.
http://dx.doi.org/10.1016/j.jclepro.2020.123370

[48] A. Kalkan, and M. Akif, "Numerical study on photovoltaic/thermal systems with extended surfaces", *International Journal of Energy Research,* vol. 43, no. C, 2019.
http://dx.doi.org/10.1002/er.4477

[49] A. Kazemian, M. Hosseinzadeh, M. Sardarabadi, and M. Passandideh-Fard, "Experimental study of using both ethylene glycol and phase change material as coolant in photovoltaic thermal systems (PVT) from energy, exergy and entropy generation viewpoints", *Energy,* vol. 162, no. 18, pp. 210-223, 2018.
http://dx.doi.org/10.1016/j.energy.2018.07.069

[50] H. Yamaguchi, T. Maki, C. Pumaneratkul, H. Yamasaki, Y. Iwamoto, and J. Arakawa, "Experimental study on the performance of CO2-based photovoltaic-thermal hybrid system", *Energy Procedia,* vol. 105, pp. 939-945, 2017.
http://dx.doi.org/10.1016/j.egypro.2017.03.422

[51] E. Hjerrild, S. Mesgari, F. Crisostomo, J. A. Scott, R. Amal, and R. A. Taylor, "Hybrid PV/T enhancement using selectively absorbing Ag–SiO2/carbon nanofluids", *Solar Energy Materials and Solar Cells,* vol. 147, pp. 281-287, 2016.
http://dx.doi.org/10.1016/j.solmat.2015.12.010

[52] A. Radwan, M. Ahmed, and S. Ookawara, "Performance enhancement of concentrated photovoltaic systems using a microchannel heat sink with nanofluids", *Energy Convers. Manage.,* vol. 119, pp. 289-303, 2016.
http://dx.doi.org/10.1016/j.enconman.2016.04.045

CHAPTER 11

Thermal Modeling of Greenhouse Solar Dryers

Amit Shrivastava[1*], M.K. Gaur[1], Pushpendra Singh[1]

[1]*Department of Mechanical Engineering, Madhav Institute of Technology and Science, Gwalior, India*

Abstract: In agricultural applications, the preservation of crops is essential. The most suitable method of preserving the crop is drying. The traditional methods used for drying include mechanical drying, where hydrocarbon fuel or burnable materials are used as a source of heat. The other method is open sun drying, where crops are placed in an open space. Both methods affect the quality and vital nutritious properties of the crops. The mechanical drying processes are costly, whereas the latter is dependent on the weather condition. To overcome these limitations, solar dryers are developed. Mathematical modeling is highly important for the perfect design and improvement of solar dryers. The performance of the drying system depends upon the different parameters that can be optimized using thermal modeling. This chapter covers the thermal modeling of greenhouse solar dryers for active and passive modes.

Keywords: Direct, Hybrid, Indirect, Solar dryer, Thermal modeling.

1. INTRODUCTION

The drying process used for crop preservation is one of the energy-intensive processes. Many societies and organizations are working to develop an effective and economical source of green energy to achieve optimum drying efficiency and minimum energy consumption in the drying processes [1]. Solar dryers can be grouped into direct, indirect, and mixed-mode dryers. In direct dryers, solar radiation, after transmitting through a transparent cover, strikes directly over the item to be dried. In contrast, in indirect dryers, the solar energy is utilized to heat the air externally in solar collectors, and then hot air is supplied to the opaque drying cabinet [2]. A mix of these two types is a mixed-type solar dryer [3]. The dryers operate in two modes, namely active and passive mode. In active dryers, fans or blowers are used to maintain the air circulation inside the dryer, while in passive

---

*Corresponding author **Amit Shrivastava:** Department of Mechanical Engineering, Madhav Institute of Technology and Science, Gwalior, India; E-mail: pushpendra852@gmail.com

Manoj Kumar Gaur, Brian Norton & Gopal Tiwari (Eds.)

dryers, the air movement takes place naturally due to the buoyancy effect. The general classification of the dryer is shown in Fig. (**1**).

Thermal modeling is done using the energy balance principle of the first law of thermodynamics. The total energy input to the system must be equal to the energy going out of the system. In other words, the energy gain is always equal to the energy lost by the system. Different researchers established the energy balance of the different parts of the dryers like room air, the floor of dryer, the surface of crop, the surface of covering material, PV module surface, *etc.* The main aim of energy balance or thermal modeling is to develop an expression that can predict the temperature of the different parts of a particular type of dryer. The steps that can be followed to carry out the thermal modeling of any developed dryer are encapsulated in this chapter.

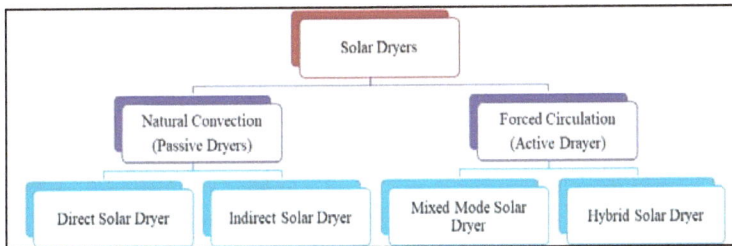

Fig. (1). General classification of solar dryers.

2. THERMAL MODELS

The thermal modeling helps in designing solar greenhouse dryers for a specific mass of crops and optimizing the design variables to maximize the performance in typical situations and for particular shapes and sizes. Using the thermal model simulation, we can investigate the effect of the design and shape of a solar dryer on the drying kinetics of the specific crop. The developed expression helps predict the required design parameters using the easily available parameters like solar radiation, ambient temperature, relative humidity of the air, the mass of crop to be dried, *etc.* This saves time as there is no need to perform the experiment to optimize the required parameter.

2.1. Moisture Ratio

The moisture ratio of a crop denotes a decrease in moisture content of the particular crop over time. The water mass inside the cops reduces with time inside the dryer. Mathematically, it is the ratio of moisture inside the crop at any instant to the initial moisture content of the crop. To predict the moisture ratio, various co-relations given by different researchers are provided in Table **1**. These expressions can predict the moisture inside the crop after a particular time interval.

Table 1. Mathematical model applied in the drying process.

S. No.	Author Model Name	Equation	Reference
1.	Henderson and Pabis	$M_r = a.\exp(-kt)$	[4]
2.	Prakash and Kumar	$M_r = at^3 + bt^2 + ct + d$	[5]
3.	Approximation of diffusion	$M_r = a.\exp(-kt) + (1-a)\exp(-gt)$	[6]
4.	Lewis	$M_r = \exp(-kt)$	[7]
5.	Modified Page equation-II	$M_r = \exp[-k(t/L^2)^n]$	[8]
6.	Logarithmic	$M_r = a.\exp(-kt) + c$	[9]
7.	Page	$M_r = \exp(-kt^n)$	[7]
8.	Modified Henderson and Pabis	$M_r = a.\exp(-kt) + b.\exp(-gt) + c\exp(-ht)$	[10]
9.	Verma *et al.*	$M_r = a.\exp(-kt) + b.\exp(gt) + c.\exp(-ht)$	[11]
10.	Thompson	$T = a.\ln(M_r) + b.[\ln(M_r)]^2$	[8]
11.	Two-term	$M_r = a.\exp(-k_0 t) + b.\exp(-k_1 t)$	[8]
12.	Wang and Singh	$M_r = 1 + at + bt^2$	[5]
13.	Simplified Fick's diffusion equation	$M_r = a.\exp[-c\,(t/L^n)]$	[12]
14.	Midilli and Kucuk	$M_r = \exp(-k.t^n) + bt$	[5]
15.	Page's Modified	$M_r = \exp[-(kt)^n]$	[6]
16.	Two-term exponential	$M_r = a.\exp(-k.t) + (1-a)\exp(-k.a.t)$	[13]

2.2. For Natural Convection Solar Dryer

The energy balance established by Jain and Tiwari [10] for the natural convection greenhouse dryer is given as follows:

2.2.1. Crop Surface

The transmitted solar radiation received by the crop surface is equal to the summation of energy stored inside the crop and the energy lost by the crop through convection and evaporation.

$$(1 - F_{nw})F_{co}\alpha_c\sum I_i A_i \tau_i$$
$$= M_c C_c \frac{dT_c}{dt} + h_{ca}(T_c - T_{ra})A_c + 0.016h_{ca}[P(T_c) - \gamma_{ra}P(T_{ra})]A_c \qquad (1)$$

2.2.2. Floor of dryer/ Bottom Surface

The amount of solar thermal energy received by the bottom surface of the solar dryer will be equal to the summation of conductive heat lost to the underground surface and convective and radiative loss from the ground surface to the room air.

$$(1 - F_{nw})(1 - F_{co})\alpha_{gr}\sum I_i A_i \tau_i = h_{g\infty}(T_{gr} - T_\infty)A_{gr} + h_{gra}(T_{gr} - T_{ra})(A_{gr} - A_c) \qquad (2)$$

2.2.3. Greenhouse Cabinet/ Room Air

The energy received by air inside the greenhouse is the summation of solar radiation absorbed by room air, convective and evaporative heat loss from the crop surface, and the convective and radiative loss from the ground surface. The energy lost from the room air is a summation of energy lost from room air to ambient through ventilation and the overall heat loss to the surrounding through walls, corners, improper sealing *etc.* The energy balance of the room air is calculated as:

$$(1 - F_n)(1 - F_{co})(1 - \alpha_{gr})\sum I_i A_i \tau_i + h_{ca}(T_c - T_{ra})A_c$$
$$+ 0.016h_{ca}[P(T_c) - \gamma_{ra}P(T_{ra})]A_c$$
$$+ h_{gr}(T_{gr} - T_{ra})(A_{gr} - A_c)$$
$$= C_d A_v \sqrt{2g\Delta H} \times \Delta P + \sum U_i A_i(T_{ra} - T_a) \qquad (3)$$

Heat loss through vent due to natural convection $= C_d A_v \sqrt{2g\Delta H} \times \Delta P$

Where, $\Delta H = \dfrac{\Delta P}{\rho_r g}$

$$\Delta P = P(T_{ra}) - \gamma_{ra}P(T_a)$$

The thermal models developed by Janjai *et al.* [14] for the passive greenhouse dryer are based on the following assumptions:

- No air stratification inside the dryer and airflow in a single direction.

- Thin-layer drying.

- Specific heat of the cover, crop, and air is constant.

The energy balance established by Janjai *et al.* [14] is given as follows:

2.2.4. Cover of Greenhouse

The thermal energy stored inside the cover is the algebraic summation of the radiative and convective heat transfer between the cover and crop surface, room air, and sky. Mathematically it is calculated as:

$$
\begin{aligned}
M_{co}C_{co}(dT_{co}/dt) \\
= A_{co} \\
\cdot [h_{cra}(T_{ra} - T_c) + h_{rcs}(T_s - T_{co}) + h_{coa}(T_{am} - T_c) \\
+ \alpha_c I_t] + A_c h_{rcc}(T_c - T_{co})
\end{aligned}
\tag{4}
$$

2.2.5. Air Inside the Greenhouse

The thermal energy stored in the room air is the algebraic sum of the convective heat transfer from crop and ground to room air, heat transfer through diffusion from crop surface to room air, heat lost through the vent, overall heat lost to ambient, and solar thermal energy received by room air. Mathematically, it is calculated as:

$$
\begin{aligned}
M_{ra}C_{ra}\left(\frac{dT_{ra}}{dt}\right) = A_c h_{ca}(T_c - T_{ra}) + A_{gr} h_{gra}\left(T_{gr} - T_{ra}\right) + D_c A_c C_{pv}\rho_c (T_c - T_{ra})\left(\frac{dM_c}{dt}\right) \\
+ [\rho_{ra}C_{ra}(V_{out}T_{out} - V_{in}T_{in})] + U_{co}A_{co}(T_a - T_{ra}) +
\end{aligned}
\tag{5}
$$

$$
[(1 - F_c)(1 - \alpha_{gr}) + (1 - \alpha_c)F_c]A_{co}I_t \tau_{co}
$$

2.2.6. Different Crop Surface

The net thermal energy stored inside the crop is the algebraic sum of convective heat loss from crop to room air, radiative heat loss from crop to cover, heat loss due to the diffusion of moisture, and the solar thermal energy absorbed by the crop surface. It is calculated as:

$$
\begin{aligned}
\left(M_c C_{pp} + C_{pl}M_p\right)(dT_c/dt) \\
= A_c h_{ca}(T_{ra} - T_c) + A_c h_{rcc}(T_{co} - T_c) \\
+ D_c A_c \rho_c L_c(dM_c/dt) + F_c \alpha_c A_c I_t \tau_{co}
\end{aligned}
\tag{6}
$$

2.2.7. Concrete Floor

The amount of energy stored in the concrete ground is the algebraic sum of convective heat exchange between ground to room air and the surface below the floor and the solar energy absorbed by the ground surface.

$$M_{gr}C_{gr}(dT_{gr}/dt)$$
$$= A_{gr}h_{gra}(T_{ra} - T_{gr}) + A_{gr}h_{g\infty}(T_\infty - T_f) \tag{7}$$
$$+ (1 - F_c)\alpha_{gr}A_{gr}I_t\tau_{co}$$

2.2.8. Heat Transfer Coefficients (HTCs)

The heat transfer coefficient shows the amount of heat transfer taking place through a particular part of the dryer from its 1 m^2 area and at a unit degree of temperature difference. The various heat transfer coefficients responsible for heat transfer in naturally ventilated dryers are as follows:

Radiative HTCs [14]:

From cover to the sky, $\quad h_{rcs} = \varepsilon_{co}\sigma(T_{co}^2 - T_s^2)(T_{co} - T_s)$ $\tag{8}$

From crop to cover, $\quad h_{rcc} = \varepsilon_{co}\sigma(T_c^2 - T_{co}^2)(T_c - T_{co})$ $\tag{9}$

Convective HTCs:

From cover to ambient air [14], $\quad h_w = 2.8 + 3.0V_w$ $\tag{10}$

From crop surface to room air [15], $\quad h_{ca} = \dfrac{K}{X}C(Gr \times Pr)^n$ $\tag{11}$

Evaporative HTCs:

From crop surface to room air [15],
$$h_{ev}$$
$$= 16.273$$
$$\times 10^{-3}h_{ca}\left(\frac{P(T_{ra}) - \gamma P(T_c)}{T_{ra} - T_c}\right) \tag{12}$$

2.3. For Forced Convection Solar Dryer

The energy balance established by Jain and Tiwari [10] for the forced convection greenhouse dryer is given as follows:

2.3.1. Crop Surface

$$(1 - F_{nw})F_{co}\alpha_c\Sigma I_i A_i \tau_i$$
$$= M_c C_c \frac{dT_c}{dt} + h_{ca}(T_c - T_{ra})A_c$$
$$+ 0.016h_{ca}[P(T_c) - \gamma_{ra}P(T_{ra})]A_c \tag{13}$$

2.3.2. Floor of dryer/ Bottom Surface

$$(1 - F_{nw})(1 - F_{co})\alpha_{gr}\Sigma I_i A_i \tau_i$$
$$= h_{g\infty}(T_{gr} - T_\infty)A_{gr} + h_{gra}(T_{gr} - T_{ra})(A_{gr} - A_c) \tag{14}$$

2.3.3. Greenhouse Cabinet/ Room air

In the case of force mode, the heat lost through the vent due to natural ventilation is replaced by energy lost due to air removed by a fan inside the dryer at a faster rate.

$$(1 - F_n)(1 - F_{co})(1 - \alpha_{gr})\Sigma I_i A_i \tau_i + h_{ca}(T_c - T_{ra})A_c$$
$$+ 0.016h_{ca}[P(T_c) - \gamma_{ra}P(T_{ra})]A_c$$
$$+ h_{gr}(T_{gr} - T_{ra})(A_{gr} - A_c) \tag{15}$$
$$= 0.33NV(T_{ra} - T_a) + \Sigma U_i A_i(T_{ra} - T_a)$$

The energy balance established by Janjai *et al.* [14] for a parabolic-shaped hybrid greenhouse is given as follows:

2.3.4. Cover of Greenhouse

$$M_{co}C_{co}(dT_{co}/dt) = A_{co} \cdot [h_{cra}(T_{ra} - T_c) + h_{rcs}(T_s - T_{co}) + h_{coa}(T_{am} - T_c) + \alpha_c I_t]$$
$$+ A_c h_{rcc}(T_c - T_{co}) \tag{16}$$

2.3.5. Air inside the Greenhouse

$$M_{ra}C_{ra}\left(\frac{dT_{ra}}{dt}\right)$$
$$= A_c h_{ca}(T_c - T_{ra}) + A_{gr}h_{gra}(T_{gr} - T_{ra})$$
$$+ D_c A_c C_{pv}\rho_c(T_c - T_{ra})(dM_c/dt) \tag{17}$$
$$+ [\rho_{ra}C_{ra}(V_{out}T_{out} - V_{in}T_{in})] + U_{co}A_{co}(T_a - T_{ra})$$
$$+ [(1 - F_c)(1 - \alpha_{gr}) + (1 - \alpha_c)F_c]A_{co}I_t\tau_{co}$$

2.3.6. Different Crop Surface

$$\left(M_c C_{pp} + C_{pl} M_p\right)\left(\frac{dT_c}{dt}\right) = A_c h_{ca}(T_{ra} - T_c) + A_c h_{rcc}(T_{co} - T_c) +$$
$$D_c A_c \rho_c \left[L_c + C_{pv}(T_c - T_{ra})\right]\left(\frac{dM_c}{dt}\right) + F_c \alpha_c A_c I_t \tau_{co} \tag{18}$$

2.3.7. Concrete Floor

$$M_{gr} C_{gr}\left(dT_{gr}/dt\right)$$
$$= A_{gr} h_{gra}(T_{ra} - T_{gr}) + A_{gr} h_{g\infty}(T_\infty - T_f)$$
$$+ (1 - F_c)\alpha_{gr} A_{gr} I_t \tau_{co} \tag{19}$$

3. Efficiency of the Drying System

Instantaneous thermal loss efficiency factor under passive and active mode is given by Eq. 20 and Eq. 21, respectively [16].

$$\eta_{i,nt} = 1 - \frac{\left[U_{gr}\left(\sum A_t - \sum A_{pv}\right) + U_{pv} \sum A_{pv}\right](T_{ra} - T_a)}{I_t A_{gr}} \tag{20}$$

$$\eta_{i,fr} = \frac{0.33NV(T_{ra} - T_a)}{I_t A_{gr}} 1 \tag{21}$$

U_{pv} is the overall heat transfer coefficient from greenhouse room air to surrounding through the PV module, which is calculated as:

$$U_L \sum A_t = U_{gr}\left(\sum A_t - \sum A_{pv}\right) + U_{pv} \sum A_{pv} \tag{22}$$

Where U_L is taken 6 W/m^2°C [16].

The electrical efficiency of the PV Panel attached to a dryer for supplying the electrical power to the fan is given as [16]:

$$\eta_{el} = \left(\frac{0.8 V_{ocv} I_{scc}}{A_{pv} I_{pv}}\right) \times 100 \tag{23}$$

V_{ocv} is the open-circuit voltage, I_{scc} is the short circuit current, and I_{pv} is the solar radiation intensity normal to the PV panel.

Daily drying efficiency was calculated by Nayak *et al.* using the relation given by Eq. 24 as [17],

$$\eta_d = \frac{M_{ev} \times LHV}{I_t A_d} \times 100 \tag{24}$$

Overall efficiency or system efficiency or energy efficiency indicates the performance of the entire drying system comprising solar collectors, drying chamber, and other energy sources. The overall efficiency of the hybrid dryer (η_{dr}) attached with auxiliary air heating devices is given as [18]:

$$\eta_{dr} = \frac{m_e \times LHV}{I_t A_d + P_{fn} + E_{as}} \times 100 \tag{25}$$

E_{as} is the energy supplied from additional energy sources.

Exergy efficiency is an important parameter to indicate the performance of solar dryers. It is the ratio of exergy available at the output (Ex_{out}) to the exergy input to the hybrid dryer (Ex_{in}) [19]. It is calculated by:

$$\eta_{Exe} = \frac{Ex_{out}}{Ex_{in}} = \frac{Ex_{evp} + Ex_{work}}{Ex_{in}} \tag{26}$$

Exergy input to the hybrid dryer is the summation of exergy by the sun, the exergy of PV module, exergy of solar collector, and exergy of any other auxiliary devices. The exergy at the output is the summation of exergy consumed in evaporating the moisture from crop surface (Ex_{evp}) and exergy of work (Ex_{work}).

Aritesty and Wulandani proposed the different relations to calculate the Dryer efficiency [20] as:

$$\eta_{dr} = \frac{m_c \times LHV + m_c C_c \Delta T_g}{Q_{bio} + Q_{so} + Q_{el}} \tag{27}$$

Q_{bio}, Q_{so}, and Q_{el} are bio, solar and electrical energy, respectively.

4. OTHER HEAT TRANSFER PARAMETERS

Total energy and exergy gain by the PV/T coupled dryer are given in Eq. 28 and Eq. 29, respectively [21].

$$\dot{Q}_{t,en} = \dot{m}_{ra}C_{ra}(T_{ra} - T_a) + \frac{\eta_{pv}A_{pv}I_t}{0.38} \tag{28}$$

$$\dot{Q}_{t,ex} = \dot{Q}_{u,th,ex} + \eta_{pv}A_{pv}I_t \tag{29}$$

$\dot{Q}_{u,th,ex}$ is the thermal exergy gain.

Solar energy going inside the dryer is calculated [22] as:

$$E_{i,d} = A_d \int_0^t I_n(t)dt \tag{30}$$

Energy going outside the dryer and flat plate collector attached to the dryer is given as [22]:

$$E_{o,d} = M_C \times LHV1 \tag{31}$$

$$E_{o,coll} = \int_0^t \dot{m}_t \times C_a(T_{oc} - T_{ic})dt \tag{32}$$

$\dot{m}_t$ is the mass flow rate at time t, $T_{oc,}$ and T_{ic} are outlet and inlet temperatures of solar collector, respectively.

The daily thermal output of the dryer can be determined as [17]:

$$\dot{Q}_{th} = M_c \times Latent\ heat\ of\ evaporation \tag{33}$$

The heat required to evaporate the moisture content in the drying product [23] is calculated as:

$$Q = M_c \times LHV \times \frac{\eta_{dr}}{\eta_{cl}} \tag{34}$$

η_{dr} and η_{cl} are dryer and collector efficiency.

The amount of useful energy inside the drying cabinet [18] is calculated as:

$$Q_u = \dot{m}_{ra}C_{ra}(T_{oc} - T_{ic}) \tag{35}$$

Energy taken by hot air to the ambient can be calculated using the relation described previously [24].

The convective heat transfer coefficient of air is calculated to determine the heat transfer inside the dryer using Eq. 36 [25].

$$h_{ca} = 0.884 \times \left[(T_{gr} - T_{ra}) + \frac{[P(T_{gr}) - Rh \cdot P(T_{ra})](T_{ra} + 273)}{268900 - P(T_{gr})} \right]^{1/3} \tag{36}$$

The heat utilization factor of the dryer is calculated by Chauhan and Kumar to indicate the performance of the dryer [25] as:

$$HUF = \frac{T_{gr} - T_{ra}}{T_{gr} - T_a} \tag{37}$$

CONCLUSION

The main focus of this chapter is to make aware of the existing thermal models for active and passive solar dryers. The suitability, efficiency, and capacity of producing dried products of high quality can be decided based on thermal models of solar dryers. Thermal modeling plays a crucial role in designing the optimized solar dryer by including operating and design parameters. It was found that the active or forced solar dryer gives higher performance for the crops with high water content; on the other hand, for low moisture content crops, the passive or natural solar dryers are found suitable. For a particular mass of crops, food, and vegetables, an optimum design will be very helpful in improving the rate of drying.

NOMENCLATURE

A	Area, m²	τ	Transmissivity
C	Constant	ρ	Density of air, kg/m³
c	Specific heat, J/kgK	γ	Relative humidity of air, %
C_d	Coefficient of diffusion	*Suffix*	
F	Fraction of radiation	a	Ambient air
Gr	Grashoff Number	c	Crop surface
g	Acceleration due to gravity, 9.81 m/s²	co	Cover of dryer
h	Heat transfer coefficient, W/m²K	ca	Crop surface to room air
I_t	Solar radiation incident to the dryer, W/m²	coa	Cover to ambient air

K	Thermal Conductivity, W/mK	*cra*	Cover to room air
LHV	Latent heat of vaporization, J/kg	*d*	dryer
M	Mass, kg	*ev*	Evaporative
M_r	Moisture ratio	*gr*	ground
N	Number of air exchange	*gra*	Ground to room air
n	Constant	*g∞*	Ground to below the ground surface
$P(T)$	Partial pressure at a particular temperature T, N/m^2	*nw*	North wall
Pr	Prandtl Number	*pv*	PV module
T	Temperature, °C	*ra*	Room air
U	Overall heat transfer coefficient, W/m^{2}K	*rcs*	Radiative cover to sky
V	Volume of the dryer, m^3	*rcc*	Radiative cover to crop
V_w	Wind velocity, 0 – 5 m/s	*sa*	Surface area of the dryer
X	Characteristic length, m	*t*	Tray
α	Absorptivity	*v*	Vent
σ	Boltzmann's constant, W/m^{2}K^4	–	–

CONSENT FOR PUBLICATION

Not applicable.

CONFLICT OF INTEREST

The authors declare no conflict of interest, financial or otherwise.

ACKNOWLEDGEMENTS

Declared none.

REFERENCES

[1] G.N. Tiwari, *Greenhouse technology for controlled environment.* Narosa publishing house: New Delhi, 2003.

[2] A.A. El-Sebaii, and S.M. Shalaby, "Solar drying of agricultural products: A review", *Renew. Sustain. Energy Rev.,* vol. 16, no. 1, pp. 37-43, 2012.
http://dx.doi.org/10.1016/j.rser.2011.07.134

[3] E. Tarigan, and P. Tekasakul, "A small scale solar agricultural dryer with biomass burner and heat storage back-up heater", *Proceedings of ISES World Congress 2007,* 2008. pp. 1956-1959 Berlin, Heidelberg.

http://dx.doi.org/10.1007/978-3-540-75997-3_398

[4] O. Prakash, and A. Kumar, "Solar greenhouse drying: A review", *Renew. Sustain. Energy Rev.,* vol. 29, pp. 905-910, 2014.

http://dx.doi.org/10.1016/j.rser.2013.08.084

[5] S. Janjai, "A greenhouse type solar dryer for small-scale dried food industries: Development and dissemination", *Int. J. Energy Environ.,* vol. 3, pp. 383-398, 2012.

[6] C. Ratti, and A.S. Mujumdar, "Solar drying of foods: Modeling and numerical simulation", *Sol. Energy,* vol. 60, no. 3-4, pp. 151-157, 1997.

http://dx.doi.org/10.1016/S0038-092X(97)00002-9

[7] H.P. Garg, and R. Kumar, "Studies on semi-cylindrical solar tunnel dryers: Thermal performance of collector", *Appl. Therm. Eng.,* vol. 20, no. 2, pp. 115-131, 2000.

http://dx.doi.org/10.1016/S1359-4311(99)00017-4

[8] D. Jain, "Modeling the performance of greenhouse with packed bed thermal storage on crop drying application", *J. Food Eng.,* vol. 71, no. 2, pp. 170-178, 2005.

http://dx.doi.org/10.1016/j.jfoodeng.2004.10.031

[9] R.K. Goyal, and G.N. Tiwari, "Parametric study of a reverse flat plate absorber cabinet dryer: A new concept", *Sol. Energy,* vol. 60, no. 1, pp. 41-48, 1997.

http://dx.doi.org/10.1016/S0038-092X(96)00144-2

[10] D. Jain, and G.N. Tiwari, "Effect of greenhouse on crop drying under natural and forced convection II. Thermal modeling and experimental validation", *Energy Convers. Manage.,* vol. 45, no. 17, pp. 2777-2793, 2004.

http://dx.doi.org/10.1016/j.enconman.2003.12.011

[11] A. Fudholi, K. Sopian, B. Bakhtyar, M. Gabbasa, M.Y. Othman, and M.H. Ruslan, "Review of solar drying systems with air based solar collectors in Malaysia", *Renew. Sustain. Energy Rev.,* vol. 51, no. C, pp. 1191-1204, 2015.

http://dx.doi.org/10.1016/j.rser.2015.07.026

[12] D.K. Rabha, P. Muthukumar, and C. Somayaji, "Experimental investigation of thin layer drying kinetics of ghost chilli pepper (Capsicum Chinense Jacq.) dried in a forced convection solar tunnel dryer", *Renew. Energy,* vol. 105, pp. 583-589, 2017.

http://dx.doi.org/10.1016/j.renene.2016.12.091

[13] L. Bennamoun, and A. Belhamri, "Mathematical description of heat and mass transfer during deep bed drying: Effect of product shrinkage on bed porosity", *Appl. Therm. Eng.,* vol. 28, no. 17-18, pp. 2236-2244, 2008.

http://dx.doi.org/10.1016/j.applthermaleng.2008.01.001

[14] S. Janjai, P. Intawee, J. Kaewkiew, C. Sritus, and V. Khamvongsa, "A large-scale solar greenhouse dryer using polycarbonate cover: Modeling and testing in a tropical environment of Lao People's Democratic Republic", *Renew. Energy,* vol. 36, no. 3, pp. 1053-1062, 2011.

http://dx.doi.org/10.1016/j.renene.2010.09.008

[15] M. Kumar, S.K. Sansaniwal, and P. Khatak, "Progress in solar dryers for drying various commodities", *Renew. Sustain. Energy Rev.,* vol. 55, no. C, pp. 346-360, 2016.

http://dx.doi.org/10.1016/j.rser.2015.10.158

[16] P. Barnwal, and A. Tiwari, "Design, construction and testing of hybrid photovoltaic integrated greenhouse dryer", *Int. J. Agric. Res.,* vol. 3, no. 2, pp. 110-120, 2008.

http://dx.doi.org/10.3923/ijar.2008.110.120

[17] S. Nayak, A. Kumar, J. Mishra, and G.N. Tiwari, "Drying and testing of mint (Mentha piperita) by a hybrid photovoltaic-thermal (PVT)-based greenhouse dryer", *Dry. Technol.,* vol. 29, no. 9, pp. 1002-1009, 2011.

http://dx.doi.org/10.1080/07373937.2010.547265

[18] S. Deeto, S. Thepa, V. Monyakul, and R. Songprakorp, "The experimental new hybrid solar dryer and hot water storage system of thin layer coffee bean dehumidification", *Renew. Energy,* vol. 115, pp. 954-968, 2018.

http://dx.doi.org/10.1016/j.renene.2017.09.009

[19] A. Lingayat, V.P. Chandramohan, and V.R.K. Raju, "Energy and exergy analysis on drying of banana using indirect type natural convection solar dryer", *Heat Transf. Eng.,* vol. 41, no. 6-7, pp. 551-561, 2020.

http://dx.doi.org/10.1080/01457632.2018.1546804

[20] E. Aritesty, and D. Wulandani, "Performance of the rack type-greenhouse effect solar dryer for wild ginger (Curcuma xanthorizza roxb.) drying", *Energy Procedia,* vol. 47, pp. 94-100, 2014.

http://dx.doi.org/10.1016/j.egypro.2014.01.201

[21] S. Tiwari, G.N. Tiwari, and I.M. Al-Helal, "Performance analysis of photovoltaic–thermal (PVT) mixed mode greenhouse solar dryer", *Sol. Energy,* vol. 133, pp. 421-428, 2016.

http://dx.doi.org/10.1016/j.solener.2016.04.033

[22] M.A. Eltawil, M.M. Azam, and A.O. Alghannam, "Solar PV powered mixed-mode tunnel dryer for drying potato chips", *Renew. Energy,* vol. 116, pp. 594-605, 2018.

http://dx.doi.org/10.1016/j.renene.2017.10.007

[23] P. Mehta, S. Samaddar, P. Patel, B. Markam, and S. Maiti, "Design and performance analysis of a mixed mode tent-type solar dryer for fish-drying in coastal areas", *Sol. Energy,* vol. 170, pp. 671-681, 2018.

http://dx.doi.org/10.1016/j.solener.2018.05.095

[24] R.T.A. Hamdani, T.A. Rizal, and Z. Muhammad, "Fabrication and testing of hybrid solar-biomass dryer for drying fish", *Case Stud. Therm. Eng.,* vol. 12, pp. 489-496, 2018.

http://dx.doi.org/10.1016/j.csite.2018.06.008

[25] P.S. Chauhan, and A. Kumar, "Heat transfer analysis of north wall insulated greenhouse dryer under natural convection mode", *Energy,* vol. 118, pp. 1264-74, 2017.

http://dx.doi.org/10.1016/j.energy.2016.11.006

CHAPTER 12

Case Study on Thermal and Drying Performance Index of Hybrid Solar Dryer with Evacuated Collector

Gaurav Saxena[1*], M.K. Gaur[1]

[1]*Department of Mechanical Engineering, Madhav Institute of Technology and Science, Gwalior, India*

Abstract: Solar drying is one of the oldest and most popular food preservation methods that involve moisture removal by a complex heat and mass transfer phenomenon. The process of the drying system is dependent on a number of operating parameters. In the present chapter determination of thermal and drying performance parameters is discussed. A hybrid solar drying system with the integration of an evacuated water tube solar water heater is installed and tested for drying hygroscopic leaf crops. The drying performance of the hybrid system is evaluated in terms of mass reduction and its derived influence on moisture content and drying rate. The derived parameters are compared with the corresponding evaluations under open sun drying. The rise in greenhouse environment temperature and crop surface temperature at hourly intervals as compared to the ambient condition were used as parameters for the thermal performance of dryer. The average values of SMER were 60% lesser than that of the simple PVT-hybrid system (without ETSC), but the drying performance parameters of mass reduction, drying rate and mass shrinkage ratio provide favourable results. The drying time was reduced by 3.5 and 2.5 hours, respectively, for the present sample size of two crops as compared to the open sun drying.

Keywords: Hybrid, Solar dryer, Evacuated tube, Solar collector, Drying.

1. INTRODUCTION

Using the power of the Sun for the preservation of foodstuff and other products from agriculture has been in practice for centuries. The recent solar drying methods have good efficiency, provide hygiene and are capable of preventing the crops from undue damage. Researchers have proposed several designs of solar dryers with arrangements for favourable drying conditions. Performance evaluation of these

[*]**Corresponding author Gaurav Saxena:** Department of Mechanical Engineering, Madhav Institute of Technology and Science, Gwalior, India; E-mail: gmanojkumar@rediffmail.com

Manoj Kumar Gaur, Brian Norton & Gopal Tiwari (Eds.)

modified drying systems is important in terms of comparative analysis with conventional designs. The selection of a solar dryer for a particular food product is determined by quality requirements, product characteristics and economic factors. A systematic classification of available solar food dryers, based on the design of system components and the mode of utilization of solar energy, is presented in Fig. (1).

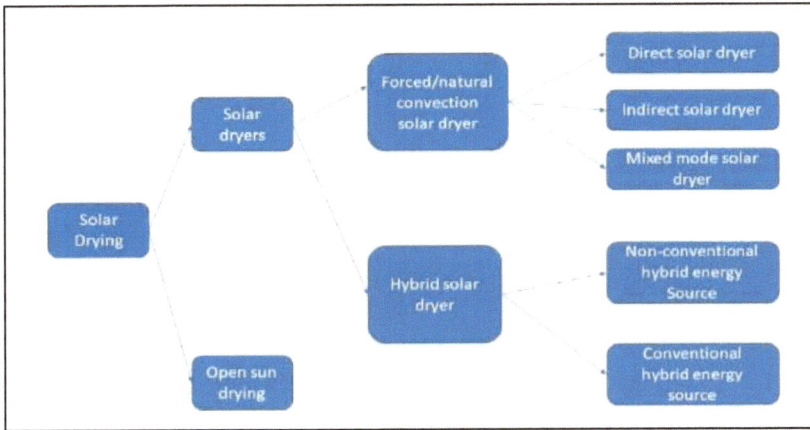

Fig. (1). Broad Classification of solar drying systems.

Flat plate collector type solar dryers are widely used for generating heated air for drying the products as compared to evacuated tube collectors. Recently some researchers have used evacuated tube heating as assistance for solar dryers and achieved favourable results.

Solar vacuum tubes were used by Mahesh *et al.* to increase the temperature of the supplied ambient air by heating. Testing of setup was carried out for drying various samples of vegetables and fruits. The result showed that conventional drying takes almost 150% more time as compared to the designed vacuum tubes type dryer [1]. Daghigh and Shaifeian used a double function heat pipe type vacuum tube drying. system and produced hot air at 45.5 °C. Outcomes of the analysis suggest that the heat provided by the evacuated tube was enough to replace the usage of auxiliary electric heating after certain hours of sunshine [2]. Ubale *et al.* tested the performance outcomes of vacuum tube solar collector for drying grape with forced convection heat transfer mode. The gross efficiency of the designed collector was found to be 24.3% as compared to 16-22% for flat plate collectors [3]. Thermal performance through experimental investigation was determined by Singh *et al.* for an evacuated tube-based solar dryer having shell and tube heat exchanger as main components integrated with drying space. The result showed 35.4 °C as the maximum rise in hot air temperature as compared to ambient air and 55% as the maximum evaluated efficiency of the setup [4]. Singh *et al.* developed a batch-type dryer using an evacuated tube for banana chips drying in a closed area. The influence of time of drying on energy, exergy, economic and other parameters were evaluated. The result showed comparatively better performance as compared to simple heat pump-based dryers [5]. Malakar *et al.* used heat pipe for development of solar dryer and performed experimentation for drying 10kg of garlic cloves 69% to 8% moisture content (wb). Using airflow velocities of 1, 2, and 3 m/s, the thermal performance was determined at no load and full load conditions. The highest temperature, drying rate and exergy efficiency achieved were 86.7°C, 1.56 kg H_2O/kg dry solid/h and 56.59% at 2 m/s airflow velocity [6].

2. MATERIAL AND METHOD

The complete assembly of hybrid greenhouse dryer (HSD) with evacuated water tube collector, as shown in Fig. (**2**) is mounted on the roof of Madhav Institute of Technology and Science, Gwalior, India (26.2183° N, 78.1828° E). The setup consists of frame type drying platform placed inside the greenhouse with two layers of floor area 200x185 cm each. Each layer of drying platform has 17 arrays of U-type copper tubes, each having length of 196cm for a series flow of heated water. Steel wire mesh with holes of 1.2x1.2 mm and wire diameter 0.3mm is placed over the copper tubes; this wire mesh gains heat by direct contact with copper tubes and transfers it to the drying product placed over it, besides this the greenhouse environment air also receives this additionally secondary heat (primary source is greenhouse heating). The fresh air enters from a 15 mm height wire mesh passage at the bottom of the greenhouse. The desired greenhouse temperature is maintained by controlling the rate of flow of heated water inside the copper tubes through a flow regulating valve [7].

Fig. (2). Actual view of installed PVT-hybrid greenhouse dryer.

3. PERFORMANCE ANALYSIS

The performance of the mentioned concept is evaluated by the comparison of the following two indices with the open sun drying:

3.1. Thermal Performance

The thermal performance of the dryer is expressed in terms of the following energy parameters: The relative rise in air temperature of the greenhouse environment and temperature of the crop surface, speciifc rate of moisture extraction, and energy☐ analysis.

3.2. Drying Performance

Performance of the present dryer, assisted by an auxiliary solar system is represented in terms of reduction in mass and its effect on moisture content and rate of drying and mass shrinkage ratio.

4. THERMAL PERFORMANCE

The following energy parameters are used for the analysis of thermal performance:

(a) The relative rise in air temperature of the greenhouse environment and temperature of the crop surface,

(b) Speciifc rate of moisture extraction, and Energy analysis☐

4.1. The Relative Rise in the Greenhouse Environment and Crop Surface Temperature

These two elementary parameters are used to express the thermal performance of the present setup. The reason for the selection of these parameters for representing the thermal performance of the system is because the former parameter directly affects the thermal efficiency whereas the later one is directly proportional to the drying efficiency of the system; this is justified from the following mathematical relations:

The thermal efficiency of any type of solar collector is defined as the ratio of useful gain in heat energy to the incident solar radiation on the surface of the collector. Hence, an analytical expression for thermal efficiency is expressed as Eq. 1 [8]:

$$\eta_{th} = \frac{\dot{m}_a C_a \, (T_{ot} - T_a)}{A_{SC} I + \dot{m} c_{Pw}(t_{wi} - t_{wo})} \times 100\% \tag{1}$$

This suggests that the rise in greenhouse environment temperature is directly influenced by the thermal efficiency of the collector.

System drying efficiency is expressed as Eq. 2 [9]:

$$\eta_d = \frac{\text{Rate of energy utilized to evaporate moisture } (J/sec)}{\text{Rate of heat received by the dryer}(J/sec)} \tag{2}$$

As such, an energy balance at the crop surface for the received and lost heat can be expressed in terms of the mathematical model as Eq. 3:

$$(1 - F_{nw})F_c\alpha_c \sum I_i A_i \tau_i + UA_c(T_w - T_c)$$
$$= m_c C_c \frac{dT_c}{dt} \tag{3}$$
$$+ \{h_c(T_c - T_r) + 0.016 h_c[P(T_c) - \gamma_r P(T_r)]\}A_c$$

Here,

$(1 - F_{nw})F_c\alpha_c \sum I_i A_i \tau_i$ = Fraction of solar radiation absorbed by the crop per second

$UA_c(T_w - T_c)$ = Rate of heat gained by the crop through direct

contact with the wire mesh

$M_c C_c \dfrac{dT_c}{dt}$ = Rate of heat absorbed by the crop

$h_c(T_c - T_r)A_c$ = Rate of heat lost from crop surface to greenhouse air

$0.016 h_c[P(T_c) - \gamma_r P(T_r)]A_c$ = Rate of heat utilization for evaporation of moisture

The direct influence of the increase in conductive transfer of heat can be observed in terms of useful gain in the amount of heat utilized for evaporation of moisture.

Following assumptions are made for comparative evaluation of drying efficiency:

- Increase in the heat utilization rate to evaporate moisture is the main influencing parameter.
- Considering the term $0.016h_c A_c$ as nearly constant and to reduce complexity to solve the exponential relationship of partial vapour pressure with temperature, a linear relationship of partial vapour pressure is considered as:

$$P(T) = AT + B \tag{4}$$

Eq. 4 suggests that the rise in crop surface temperature directly influences partial vapour pressure and the amount of heat utilized for evaporation of moisture. This shows that system drying efficiency will also be directly influenced by the crop surface temperature and can be considered as a parameter for the comparative performance evaluation with other drying methods.

The experimental observations of influencing ambient conditions and greenhouse environment are recorded by the data logger of the weather monitoring station after each regular interval of 30 minutes for one complete week, and average/mean values at a corresponding interval of time are presented in Table **1**.

Table 1. Recorded mean value of observations of influencing parameters during experimentation.

Time of the Day (Hr)	Local Weather Parameters				Environment Temperature (Inside the Green House)		
	Pyranometer Reading (W/m²)	R. Humidity (%)	Ambient Air Temp (Deg °C)	Wind Speed (m/s)	Temp. of Thermocouple (T2)	Temp. of Thermocouple (T3)	Temp. of Thermocouple (T4)
First Week							
11:00	611.64	47.89	23.97	1.467	31.34	34.16	41.42
11:30	639.73	44.54	25.08	0.200	34.76	43.42	58.35
12:00	646.66	39.57	25.90	0.333	35.80	44.83	59.34
12:30	647.93	36.92	26.36	0.067	35.84	44.67	58.00
01:00	637.44	33.86	26.94	1.867	36.86	44.79	57.63
01:30	609.34	33.75	27.35	1.000	36.40	45.68	58.87
02:00	541.75	32.83	27.78	0.533	36.98	45.31	59.23
02:30	479.05	30.32	27.35	1.000	37.21	43.82	56.76
03:00	382.25	31.58	27.25	1.733	36.60	43.46	54.93
03:30	185.63	31.21	27.16	1.333	36.45	43.00	54.21
04:00	133.50	36.77	27.00	0.000	37.19	37.66	44.40

(Table 1) cont......

	Second Week						
11:00	605.31	55.92	24.34	0.667	32.26	37.185	46.31
11:30	624.97	52.13	25.01	0.733	33.48	39.34	51.24
12:00	639.03	50.38	25.51	0.533	34.22	42.24	51.31
12:30	666.51	47.45	25.67	0.467	34.36	41.47	52.72
01:00	635.44	46.43	25.80	1.133	34.55	40.92	51.58
01:30	561.74	43.95	25.91	0.600	34.17	41.81	51.83
02:00	575.20	36.80	26.16	0.667	34.79	42.13	52.48
02:30	471.61	38.68	26.58	0.467	34.85	40.64	50.00
03:00	385.15	37.71	26.55	1.000	34.54	40.50	48.77
03:30	186.39	36.48	26.27	0.267	34.41	39.48	46.98
04:00	148.66	37.51	26.03	0.867	33.66	38.99	46.10

The drying parameters for crop-I and crop-II are presented in terms of observations of mass variation, Moisture content M (gram of water per gram of dry matter), Dimensionless moisture Ratio (Mt/Mo) and Drying Rate in gms/sec for both greenhouse and open sun drying in Tables **2** and **3** for both the crops I and II respectively.

Table 2. Drying performance parameter for Crop-I (Coriander).

Time of the Day (hrs)	Open Sun Drying				Solar Hybrid Green House Drying			
	Mean Sample Temp. (°C)	Moisture Content, M (Gram)	Dimension-less Moisture Ratio (Mt/Mo)	Mean Drying Rate* gm/min	Mean Sample Temp. (°C)	Moisture Content, M (Gram)	Dimension-less Moisture Ratio (Mt/Mo)	Mean Drying Rate* gm/min
Day-1								
11:00	33.3	11.5	1.00	0.4333	35.7	11.50	1.00	0.6000
11:30	37	8.25	0.72	0.3000	40.5	7.00	0.61	0.4333
12:00	36.5	6	0.52	0.1333	42	3.75	0.33	0.1333
12:30	35.3	5	0.43	0.1000	47.6	2.75	0.24	0.0667
01:00	38.5	4.25	0.37	0.1333	41.5	2.25	0.20	0.0333
01:30	39.6	3.25	0.28	0.1000	45.6	2.00	0.17	0.1667
02:00	34.6	2.5	0.22	0.0333	42.6	0.75	0.07	0.0667
02:30	37.1	2.25	0.20	0.0333	40.8	0.25	0.02	0.0333
03:00	32.8	2	0.17	0.0000	41.3	0.00	0.00	0.0000
03:30	32.3	2	0.17	0.0333	41.4	0.00	0.00	0.0000
04:00	31.2	1.75	0.15	....	40.9	0.00	0.00	....
Day-2								

(Table 2) cont.....

11:00	34.1	1.75	0.15	0.06667	41	0.00	0.00	0.00
11:30	40.2	1.25	0.11	0.1	42	0.00	0.00	0.00
12:00	38	0.5	0.04	0	42.2	0.00	0.00	0.00
12:30	33.2	0.5	0.04	0.06667	43	0.00	0.00	0.00
13:00	32.1	0	0.04	--------	42.7	0.00	0.00	-------

Table 3. Drying Performance Parameter for Crop-II (Fenugreek).

Time of the Day (hrs)	Open Sun Drying				Solar Hybrid Greenhouse Drying			
	Mean Sample Temp. (°C)	Moisture Content, M (Gram)	Dimensionless Moisture ratio (Mt/Mo)	Mean Drying Rate gm/min	Mean Sample Temp. (°C)	Moisture Content, M (Gram)	Dimensionless Moisture Ratio (Mt/Mo)	Mean Drying Rate* gm/min
Day-1								
11:00	31.2	13.29	1.00	1.033333	33.1	13.29	1.00	1.3333
11:30	33.3	8.86	0.67	0.533333	39	7.57	0.57	0.6667
12:00	36.9	6.57	0.49	0.233333	39.2	4.71	0.35	0.2667
12:30	37	5.57	0.42	0.266667	40.3	3.57	0.27	0.2000
01:00	37.8	4.43	0.33	0.233333	38.9	2.71	0.20	0.1333
01:30	37	3.43	0.26	0.133333	38.7	2.14	0.16	0.2333
02:00	34.4	2.86	0.21	0.1	39.9	1.14	0.09	0.1333
02:30	35.6	2.43	0.18	0.133333	42.8	0.57	0.04	0.0667
03:00	33.1	1.86	0.14	0.033333	42.9	0.29	0.02	0.0000
03:30	31.7	1.71	0.13	0.033333	40.5	0.29	0.02	0.0333
04:00	30.8	1.57	0.12	....	41.3	0.14	0.01	....
Day-2								
11:00	35	1.57	0.12	0.133333	45	0.14	0.01	0.033333
11:30	38.6	1.00	0.08	0.066667	39	0.00	0.00	0
12:00	39.9	0.71	0.05	0.066667	42	0.00	0.00	0
12:30	38.3	0.43	0.03	--------	43	0.00	0.00	-------

* The value of drying rates corresponding to each 30 min interval is constant.

4.2. Greenhouse Environment Temperature

The result shows a variable level of temperature differences achieved at different drying platforms, suitable for drying all types of hygroscopic crops. Fig. (**3**) shows that under the mode of active convection for rapid removal of moisture, a steady temperature state is maintained above the two drying platforms using a valve-controlled auxiliary supply of solar-heated water (referred by the author). The maximum relative temperature rise achieved was 9.92 °C and 18.9 °C above the lower and upper drying platform, respectively and highest relative rise in temperature of 33.4 °C was achieved at the upper zone of the greenhouse. Although the effect of rapid air change loads due to four DC fans causes a transient temperature state in the upper zone of the greenhouse.

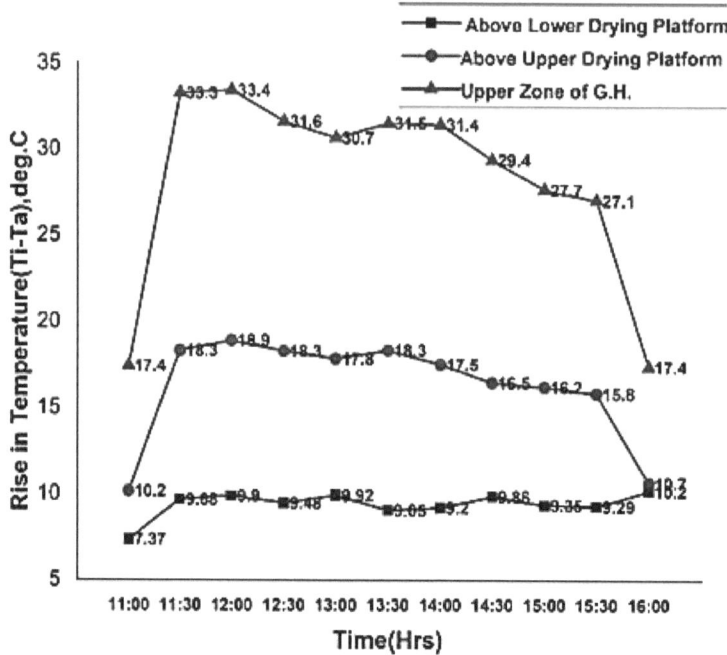

Fig. (3). Rise in greenhouse environment temperature [10].

4.3. Surface Temperature of the Crop

Increase in mean crop surface temperature as compared to the open drying under the sun and in terms of temperature differences are presented at every half-hourly interval for both the crops in Fig. (4). The maximum relative increase in crop surface temperature achieved was 12.3 and 10.5°C at the solar intensity of 647.94 and 133.51 (W/m^2) and dimensionless moisture ratios of 0.43 and 0.12 for coriander and fenugreek crops, respectively. The result shows that as the dried product gets converted into the partial solid state due to loss of bound and unbound moisture, a thermal equilibrium state is achieved for both the crops and temperature states for the complete solid to become nearly the same. This is due to cease in two processes *viz.* cease in the internal moisture transfer to the crop surface and later its evaporation by the usage of energy and further cease in the transfer of energy (mostly thermal) from the ambient environment required for the moisture evaporation from the surface of product/crop.

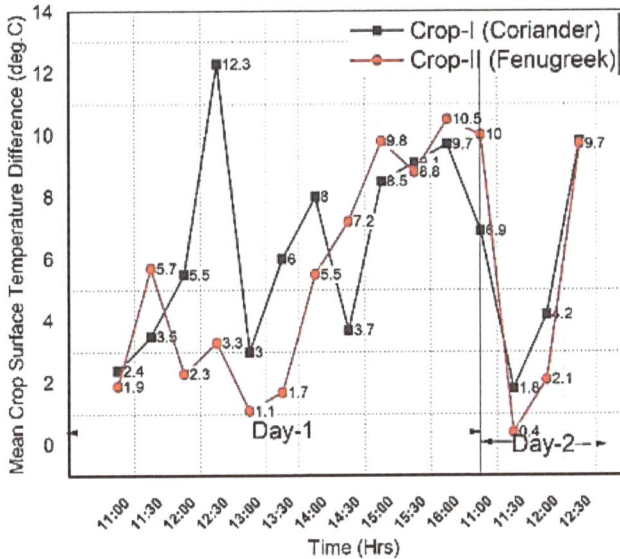

Fig. (4). Increase in mean surface temperature as compared to the open sun drying [10].

4.4. Specific Rate of Moisture Extraction

It is also termed as specific moisture extraction rate (SMER) and is defined as the requirement of energy for drying one kg of water from the product; SMER is calculated using Eq. 5 as [11],

$$SMER = \frac{W}{P} \tag{5}$$

Here, SMER is in kg/kWh, W is the mass of evaporated water from the product (kg) in the fixed interval of time and P (kWh) is total energy input to the dryer during the same duration.

Due to the 30-minute interval of data logger readings, the mean value of input energy is assumed constant for the half-hourly time interval. The variation of SMER for both the crops under consideration is shown in Figs. (**5a** and **b**). Since the heat supplied by the ETSC is used for direct conductive heat transfer to the crop, which may be supplied by other sources also, hence the comparative variation in SMER is shown with and without consideration of heat supplied by the integrated sustainable heat source of ETSC (under open sun drying).

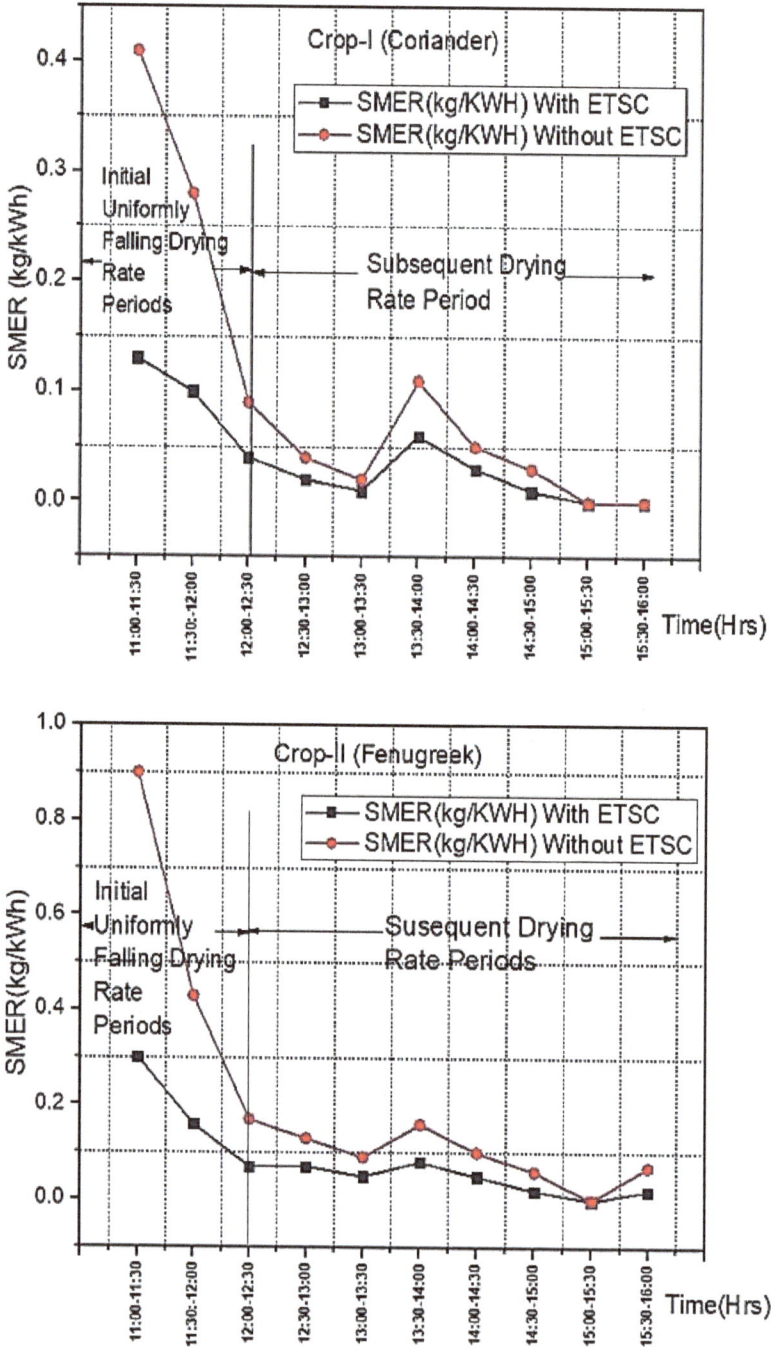

Fig. (5). (a) SMER for Coriander **(b)** SMER for Fenugreek.

4.5. Energy Analysis

The energy analysis includes the evaluation of system thermal efficiency, which is basically expressed as the ratio of the energy utilized for the evaporation of moisture to the energy provided to the drying unit. Mainly the supplied energy to the hybrid solar dryer is the incident solar isolation and energy inputs from other integrated subsystems. The thermal efficiency of the solar drying system provides the overall effectiveness of the system. Drying eiffciency of the system can be obtained using Eq. 6 [12].

$$\eta_d = \frac{W_m h_{fg}}{(A.I + P_f + P_p + P_h).t} \tag{6}$$

The drying efficiency of the system as presented in Figs. (**6a** & **b**) for both the crops, shows a similar trend of variation as obtained in the continuous falling rate period. The efficiency reduces initially with uniform rate and then variably.

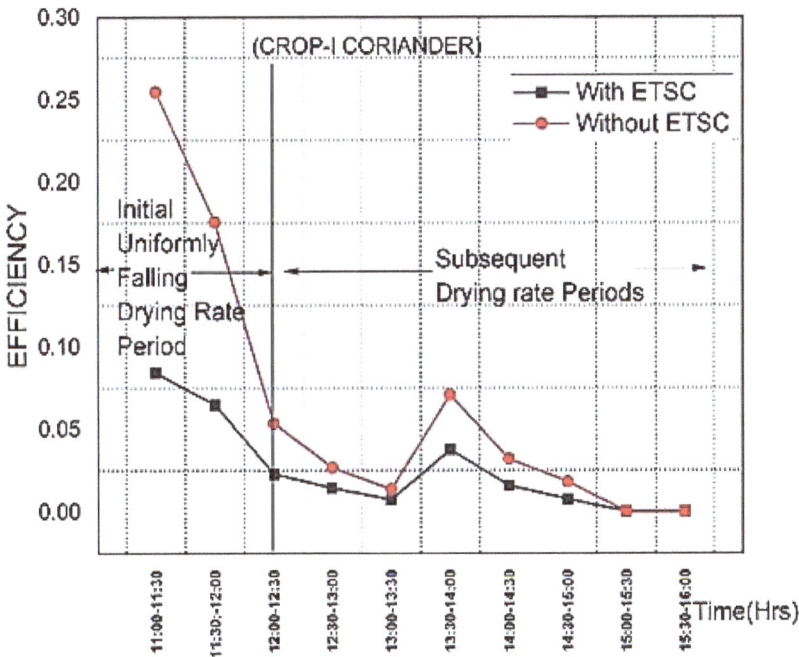

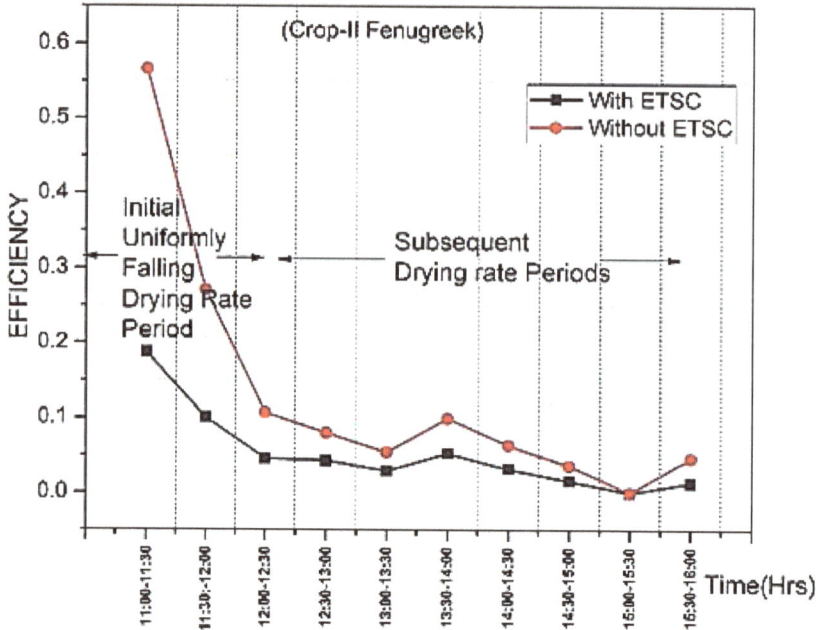

Fig. (6). (a) Drying efficiency for Coriander **(b)** Drying efficiency for Fenugreek.

5. DRYING PERFORMANCE

The performance of the present solar-assisted drying system is represented in terms of reduction in mass and its effect on moisture content as well as drying rate. For the research purpose under engineering applications, the dry basis moisture content is preferred since the weight change associated with each percentage point of moisture reduction on a dry basis becomes constant [13]. It is defined as the ratio of the weight of moisture present in produce or sample to the weight of the bone-dry material of the produce and is given in Eq. 7 as,

$$M_{db} = \frac{W_t - W_d}{W_d} = \frac{W_m}{W_d} \tag{7}$$

The drying rate is evaluated using Eq. 8 as,

$$\dot{m} = {dm}/{dt} = \frac{M_t - M_{t+\Delta t}}{\Delta t} \tag{8}$$

The values are determined for each consecutive time intervals Δt of 30 minutes.

Two Evacuated tube solar collectors of 200 LPD each are used to supply the heated water inside the copper tubes of the drying platform. The active flow is maintained by a 12V/18 W DC pump and through a regulating valve.

Dryer performance parameters are expressed in terms of the results of the observation recorded through experimentation and calculations are used to present the performance of the hybrid system in terms of these parameters:

(a) Mass reduction
(b) Drying rate and
(c) Mass shrinkage ratio

5.1. Mass Reduction Rate

The drying performance of any hybrid solar drying system is determined in terms of mass reduction and its effect on the moisture content as well as drying rate. The reduction in mean sample mass or rate of reduction in bound and unbound mixture from the four hygroscopic crops under consideration are shown in Figs. (**7a** and **b**). The drying curve for change in mass shows the complete time of drying for coriander and fenugreek reduced by 3.5 and 2.5 hours respectively, for the present sample size.

For both crops, no constant drying rate period is obtained. The complete drying process is observed in the falling rate period. The characteristics curves of drying rate for day-1 were found sufficient to provide the initial phase of uniformly falling drying rate followed by subsequent variable-I and variable-II, falling drying rate periods. The intermittent transient variations of the slope are considered as a single drying rate period for the presentation. The results show that the moisture release rate for the green vegetable is dependent on solar intensity, but for the present PVT-hybrid system, the change in slope is steadier due to backup supply from an auxiliary heat source.

5.2. Drying Rate

The drying rate periods are presented by the variation curves of moisture content (dry basis) and drying rate (g/min). The drying rate for each 30 minutes interval is assumed constant during calculation as well as presentation. For hygroscopic materials under consideration, the drying rates periods show a change in the slope of the curve as shown in Figs. (**8a** and **b**).

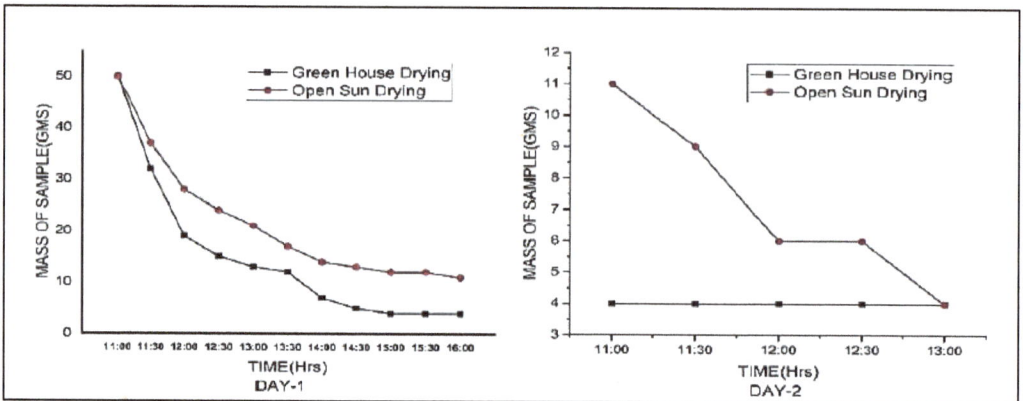

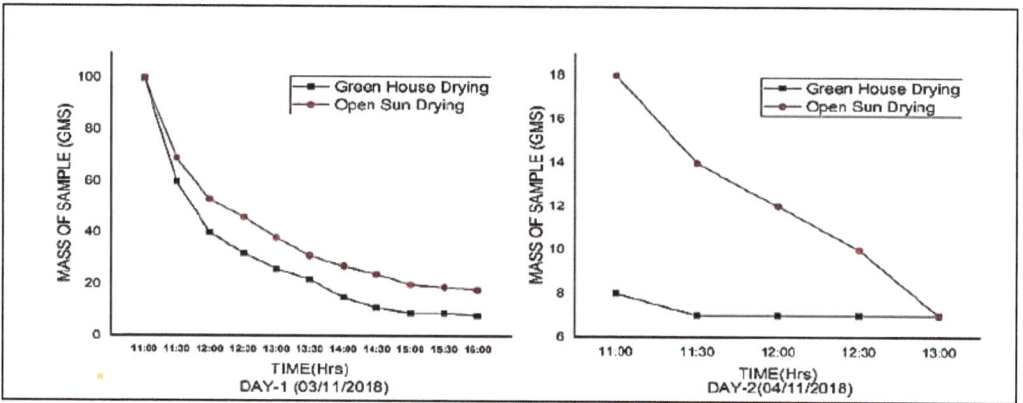

Fig. (7). (a) Variation in weight with time for Coriander **(b)** Variation in weight with time for Fenugreek.

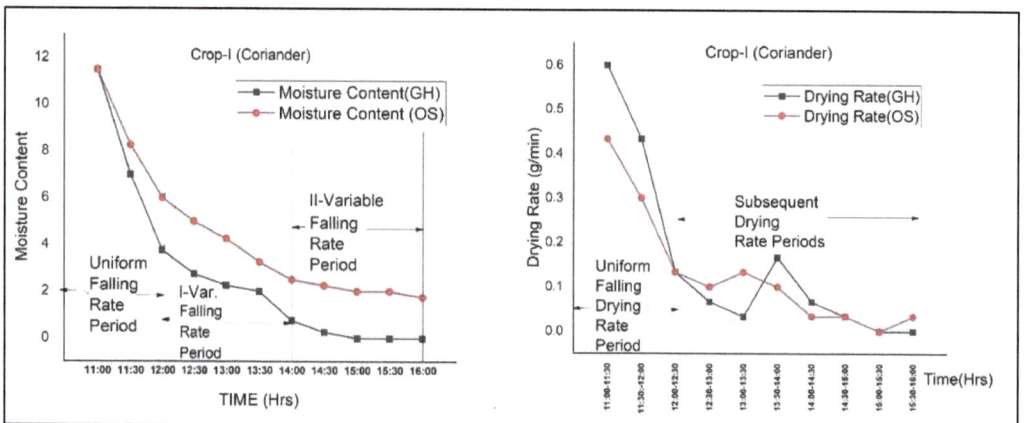

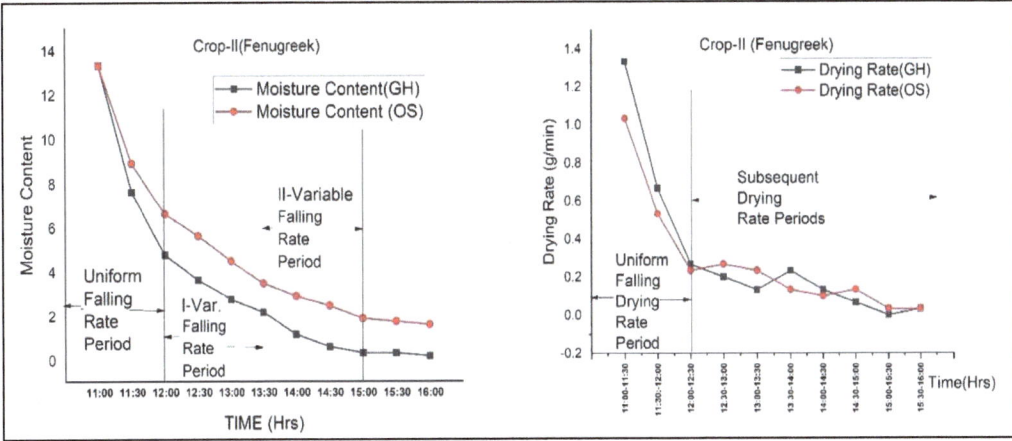

Fig. (8). (a) Drying rate periods for Coriander **(b)** Drying rate periods for Fenugreek.

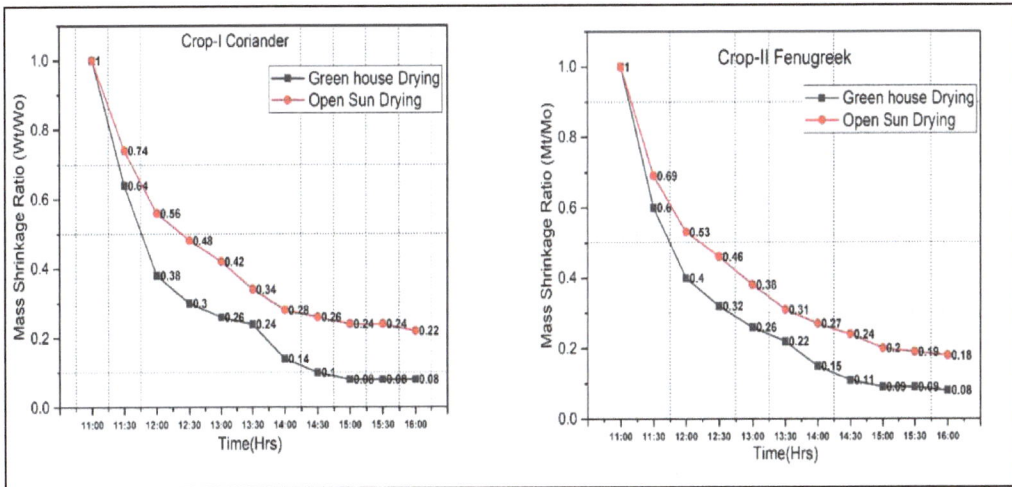

Fig. (9). (a) Mass shrinkage ratio for Coriander **(b)** Mass shrinkage ratio for Fenugreek.

5.3. Mass Shrinkage Ratio

Weight loss due to drying causes some structural changes in the dried product; the most important appeared variation in the structure of the crop is the mass shrinkage ratio expressed as Eq. 9 [14]:

$$SR = \frac{W_t}{W_o} \tag{9}$$

Here, W_t is the weight of the sample at time t and W_o is the initial weight of the sample.

The results of mass shrinkage ratio $\left(\frac{W_t}{W_o}\right)$ are shown in Fig. (**9a** and **b**). The results show that the shrinkage ratio for the crop dried in the solar greenhouse system is always greater than the open sun drying. The complete mass shrinkage ratio curve follows the trend of falling rate period for all the crops.

CONCLUSION

The performance of the installation is determined in terms of thermal and drying performance. Considering the results of the drying performance of the present PVT-hybrid vacuum tube assisted PVT-hybrid greenhouse dryer, the following conclusions can be drawn:

Results of Thermal Performance Analysis

- The maximum relative temperature rise achieved was 9.92 °C and 18.9 °C above the lower and upper drying platform respectively and a maximum relative rise in temperature of 33.4 °C was achieved at the upper zone of the greenhouse.
- The maximum relative increase in crop surface temperature (difference in greenhouse and open sun drying) achieved was 12.3 and 10.5 at the solar intensity of 647.94 and 133.51 (W/m^2) and dimensionless moisture ratios of 0.43 and 0.12 for coriander and fenugreek crops respectively.
- The SMER (kg/kWh) variation curves showed a uniformly reducing trend under drying for the installation with and without ETSC energy. The results showed an average of 60 and 61% lesser values of SMER with ETSC for the 02 crops respectively. This was due to the additional usage of vacuum tube heating for conductive heat transfer.
- The efficiency of the system was lesser with the integration of ETSC heating due to additional usage of ETSC heating. The comparison showed an average efficiency of 3.9 and 4.7 for the system with ETSC and 9.1 and 12.1 for the system without ETSC (simple PVT GH).

Results of Drying Performance Analysis

• The mass reduction curve shows favourable results for the present installation. The complete time of drying for the 02 crops was reduced by 3.5 and 2.5 hours respectively for the present sample size as compared to the open sun drying.
• The drying rate curves for moisture content (dry basis) and drying rate (g/min) showed a falling rate. For the 02 crops *viz.* coriander and fenugreek, "no constant drying rate period" is observed.

• Mass shrinkage ratio $\left(\frac{W_t}{W_o}\right)$ the curve showed favourable results for the installation when compared to the results of open sun drying. The mass shrinkage ratio curve, which is an important parameter for drying showed that for the same duration of drying, the present system showed 14 and 10% more drying as compared to the open sun drying.

The mentioned performance indices provide favourable results for comparison of the present new design of a hybrid solar dryer. Similarly, researchers may determine these parameters along with the design-specific parameters for the determination of dryer performance.

NOMENCLATURE

M_{db}	Moisture content dry basis, gm of water/gm dry matter	A	Area, m^2
W_t	Weight at time t, gm	t	Time, s
W_d	Weight of bone dry product, gm	m	Mass, kg
$\dot{m}$	Drying rate, gm/min	T	Temperature, °C
M_t	Moisture content of the product at t(dry basis)	R	Correlation coefficient
M_0	Initial moisture content at $t = 0$ (dry basis)	R^2	Coefficient of determination
P_f	Power consumed the DC fans, W	α	Absorptivity

P_P	Power consumed the DC pump, W	τ	Transmissivity
P_h	Input energy from the solar heated water, W	η_d	Drying efficiency
m_t	Mass of the sample at time t	η_{th}	Thermal efficiency
m_0	Initial mass of the sample at $t = 0$, gm		***Abbreviations***
m_d	Mass of dry matter of produce, gm	*SME R*	Specific moisture extraction rate
$\dot{m}_a$	Air mass flow rate, kg/sec	*SR*	Shrinkage ratio
C	Specific heat, J/kg°C		
h_{fg}	Latent heat of evaporation of moisture, J/kg°C		***Subscript***
A_{SC}	Solar collector area, m^2	i	i^{th} Observation
I	Solar radiation, W/m^2	ot	Outlet air
F	Fraction of solar radiation	a	Ambient air
U	Overall heat transfer coefficient, W/ m^2°C	nw	North wall
T_w	Temperature of hot water inside the drying platform, °C	c	Crop or crop surface
h	Convective heat transfer coefficient, W/m^2°C	w_i	Water inlet
T_r	Temperature of greenhouse air, °C	w_o	Water outlet
γ_r	Relative humidity of greenhouse air	r	Room

CONSENT FOR PUBLICATION

Not applicable.

CONFLICT OF INTEREST

The authors declare no conflict of interest, financial or otherwise.

ACKNOWLEDGEMENTS

Declared none.

REFERENCES

[1]   A. Mahesh, C.E. Sooriamoorthi, and A.K. Kumaraguru, "Performance study of solar vacuum tubes type dryer", *J. Renew. Sustain. Energy,* vol. 4, no. 6, p. 063121, 2012.
      http://dx.doi.org/10.1063/1.4767934

[2]   R. Daghigh, and A. Shafieian, "Energy-exergy analysis of a multipurpose evacuated tube heat pipe solar water heating-drying system", *Exp. Therm. Fluid Sci.,* vol. 78, pp. 266-277, 2016.
      http://dx.doi.org/10.1016/j.expthermflusci.2016.06.010

[3]   A.B. Ubale, D. Pangavhane, and A. Auti, "Performance analysis of forced convection evacuated tube solar collector used for grape dryer", *J. Eng. Sci. Technol.,* vol. 12, no. 1, pp. 42-53, 2017.

[4]   P. Singh, S. Vyas, and A. Yadav, "Experimental comparison of open sun drying and solar drying based on evacuated tube collector", *Int. J. Sustain. Energy,* vol. 38, no. 4, pp. 348-367, 2019.
      http://dx.doi.org/10.1080/14786451.2018.1505726

[5]   A. Singh, J. Sarkar, and R.R. Sahoo, "Experimental energy, exergy, economic and exergoeconomic analyses of batch-type solar-assisted heat pump dryer", *Renew. Energy,* vol. 156, pp. 1107-1116, 2020.
      http://dx.doi.org/10.1016/j.renene.2020.04.100

[6]   S. Malakar, V.K. Arora, and P.K. Nema, "Design and performance evaluation of an evacuated tube solar dryer for drying garlic clove", *Renew. Energy,* vol. 168, pp. 568-580, 2021.
      http://dx.doi.org/10.1016/j.renene.2020.12.068

[7]   M.K. Gaur, R.K. Pandit, G. Saxena, A. Kushwah, and P. Saxena, *"Method and apparatus for controlling the temperature of solar dryer"*, 201921001878 A

[8]   B.M.A. Amer, K. Gottschalk, and M.A. Hossain, "Integrated hybrid solar drying system and its drying kinetics of chamomile", *Renew. Energy,* vol. 121, pp. 539-547, 2018.
      http://dx.doi.org/10.1016/j.renene.2018.01.055

[9]   A. Fudholi, M.Y. Othman, M.H. Ruslan, and K. Sopian, "Drying of Malaysian Capsicum annuum L. (Red Chili) Dried by Open and Solar Drying", *Int. J. Photoenergy,* vol. 2013, pp. 1-9, 2013.
      http://dx.doi.org/10.1155/2013/167895

[10] G. Saxena, and M.K. Gaur, "Performance evaluation and drying kinetics for solar drying of hygroscopic crops in vacuum tube assisted hybrid dryer", *J. Sol. Energy Eng.,* vol. 142, no. 5, p. 051009, 2020.
      http://dx.doi.org/10.1115/1.4046465

[11] V. Shanmugam, and E. Natarajan, "Experimental investigation of forced convection and desiccant integrated solar dryer", *Renew. Energy,* vol. 31, no. 8, pp. 1239-1251, 2006.

http://dx.doi.org/10.1016/j.renene.2005.05.019

[12] A. Fudholi, K. Sopian, M.H. Yazdi, M.H. Ruslan, M. Gabbasa, and H.A. Kazem, "Performance analysis of solar drying system for red chili", *Sol. Energy,* vol. 99, pp. 47-54, 2014.

http://dx.doi.org/10.1016/j.solener.2013.10.019

[13] W.K. Lewis, "The rate of drying of solid materials", *J. Ind. Eng. Chem.,* vol. 13, no. 5, pp. 427-432, 1921.

http://dx.doi.org/10.1021/ie50137a021

[14] B.K. Bala, M.R.A. Mondol, B.K. Biswas, B.L. Das Chowdury, and S. Janjai, "Solar drying of pineapple using solar tunnel drier", *Renew. Energy,* vol. 28, no. 2, pp. 183-190, 2003.

http://dx.doi.org/10.1016/S0960-1481(02)00034-4

Thermal Analysis of Photovoltaic Thermal (PVT) Air Heater Employing Thermoelectric Module (TEM)

Neha Dimri[1]*

[1]*Laboratorie LOCIE, Université Savoie Mont Blanc (USMB), France*

Abstract: This chapter provides the description and analysis of a photovoltaic thermal (PVT) air heater, including a thermoelectric module (TEM). A PVT air heater offers several advantages over a PVT water heater. The problems such as corrosion and freezing do not exist when air is used as a working fluid. Also, the system design is less complex, incurs lower operation costs and can be easily integrated into buildings. Furthermore, it is not a cause of any major concern in case of air leaks from the duct. A PVT air heater poses some drawbacks as well, such as uneven cooling of PV panels and lower overall efficiency compared to a PVT water heater resulting from lower specific heat capacity of air. Nevertheless, the choice of the type of working fluid is subject to a variety of factors like efficiency, cost including capital investment, installation, operation and maintenance costs and the particular application.

Keywords: Bifacial, Electricity, Power, PVT, Solar cell, Solar energy.

1. CLASSIFICATION AND WORKING PRINCIPLE

A conventional photovoltaic thermal (PVT) air heater consists of an air duct placed underneath (or above) a photovoltaic module for the flow of air [1]. Huen and Daoud [2] reported a comprehensive literature review on thermoelectric modules coupled with solar technologies. A PVT air heater, therefore, produces both heated air at the outlet of the duct and electricity from the solar cells. The air flowing through the duct removes heat from the solar cells, which in turn reduces the solar cell temperature and improves the electrical efficiency of the PV module [3]. If a thermoelectric module (TEM) is incorporated in the design of the PVT air heater,

---

*Corresponding author Neha Dimri: Laboratorie LOCIE, Université Savoie Mont Blanc (USMB), France;
E-mail: nehadimri.91@gmail.com

Manoj Kumar Gaur, Brian Norton & Gopal Tiwari (Eds.)

it contributes to the total electrical production and the resulting new design is termed a PVT-TEM air heater. Fig. (**1**) illustrates the cross-sectional view of a PVT-TEM air heater, where the thermoelectric module (TEM) is placed below the PV module. A PV module consists of solar cells encapsulated between a top layer and a bottom layer. Typically, glass is used as the top covering layer to protect the solar cells against dirt and other environmental issues since it allows the solar radiation (*i.e.* short wavelength radiation) to transmit through to be absorbed by the solar cells. The solar cells absorb the solar radiation incident over the packing area of the PV module, transmit through the top glass cover, and produce electricity. The bottom layer of PV module could be either transparent (glass), making the design semi-transparent, or opaque (tedlar). TE module is essentially a combination of p-type and n-type semiconductors, connected thermally in parallel while electrically in series, and encapsulated between thermally conducting plates. Bismuth and antimony alloys are the most commonly used materials for TEM owing to their properties of low thermal conductivity and high electrical conductivity [4]. The working principles of semi-transparent and opaque PVT-TEM air heaters are discussed next.

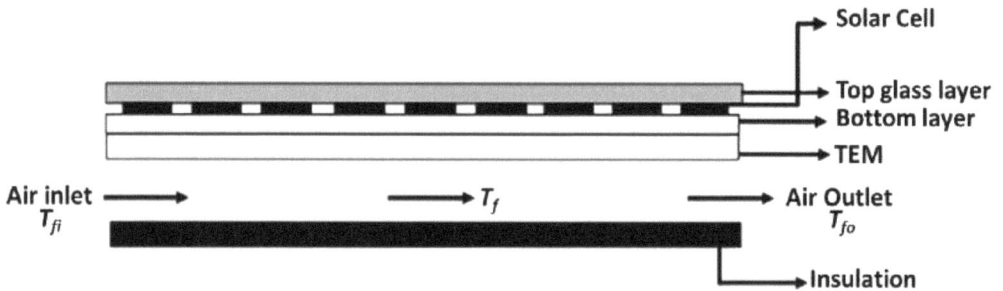

Fig. (1). A schematic representation of PVT-TEM air heater [5–7].

1.1. Semi-transparent PVT-TEM Air Heater

When glass is used as the bottom layer (Fig. **1**), the PVT-TEM air heater is known as a semi-transparent PVT-TEM air heater. Therefore, the top-side of TEM receives and absorbs the solar radiation entering through the non-packing area of the PV module $((1 - \beta_{sc})A_m)$, *i.e.* direct gain. Also, the top side of TEM receives thermal energy from the bottom of the solar cells, through indirect gain, reducing the solar cell temperature. The temperature of the top-side of TEM increases as a result of both direct and indirect gain. Moreover, heat is extracted owing to the air flowing in the duct from the bottom-side of TEM. This increases the temperature of the air at the outlet of the duct, and the heated air is utilized for space heating applications. The flow of air helps to maintain a temperature gradient across the TEM and hence, the TEM produces electricity through the Seebeck effect [8].

1.2. Opaque PVT-TEM Air Heater

Opaque PVT-TEM air heater has tedlar, *i.e.* opaque in nature, as the bottom layer (Fig. **1**). Therefore, tedlar absorbs solar radiation through the non-packing area of the PV module (direct gain). Further, thermal energy flows from underneath the solar cells to tedlar (indirect gain). This causes an increase in the temperature of the top side of TEM, which obtains thermal energy from the tedlar. Furthermore, the air flowing in the duct reduces the bottom-side temperature of TEM, leading to a temperature difference and generation of electricity. The heated air at the outlet presents a thermal energy gain from the air heater. The schematic representation depicted in Fig. (**1**) can be used in forced mode, wherein the electricity produced by the PV module and TEM is used to drive the pump for the circulation of air through the duct to make the system self-sustained.

In the next section, thermal models of semi-transparent and opaque PVT-TEM air heaters based on the energy balance between different components will be presented and discussed. The assumptions considered while writing the energy balances are stated below:

a) The PVT-TEM air heater is assumed to be in quasi-steady state.

b) Absence of temperature difference across the thickness of air column, solar cell, glass cover and tedlar.

c) The heat capacity of glass cover, tedlar and solar cells is neglected.

d) Ohmic losses in solar cells are negligible.

e) One-dimensional heat flow has been assumed.

f) The temperature of the top of tedlar (in the case of an opaque PVT-TEM air heater) is the same as the solar cell temperature.

g) Temperature of the top-side of the TEM module is the same as the temperature of the bottom layer (glass/tedlar).

h) Laminar flow is assumed for air flowing in the duct.

2. ENERGY BALANCE OF PVT-TEM AIR HEATER

This section presents the thermal models of semi-transparent (with glass bottom layer) and opaque (with tedlar bottom layer) PVT-TEM air heaters as shown in Fig. (1).

2.1. Thermal Analysis of Semi-transparent PVT-TEM Air Heater

Firstly, a mathematical model of a semi-transparent PVT-TEM air heater based on energy balances is presented. Fig. (2) shows the thermal circuit diagram of the semi-transparent PVT-TEM air heater, depicting the thermal resistances of each component of the air heater. Following (Fig. 2), the energy balance equations, considering an elemental length dx along the air column, are given as [5].

Semi-transparent PV Module

$$\tau_g \alpha_{sc} \beta_{sc} I(t) b dx = [U_{t,c-a}(T_{sc} - T_a) + h_g(T_{sc} - T_{tem,top})] b dx + \eta_{sc} \tau_g \beta_{sc} I(t) b dx \tag{1}$$

Bottom Glass Layer

$$h_g(T_{sc} - T_{tem,top}) b dx = U_{tec}(T_{tem,top} - T_{tem,bottom}) b dx \tag{2}$$

where, U_{tem} is the coefficient of overall heat transfer from the bottom glass cover to the bottom-side of TEM (Fig. 2) and is evaluated using:

$$U_{tem} = \left[R_c + \frac{L_{tem}}{K_{tem}}\right]^{-1} \tag{3}$$

As can be seen from Eq. (3), heat transfer across the TEM is governed by the thickness (L_{tem}) and thermal conductivity (K_{tem}) of TEM material and also by the thermal resistance (R_c) at thermoelectric leg-electrode contact interfaces. The thermal contact resistance (R_c) varies typically in the range of 1×10^{-6} to 5×10^{-4} m^2 K/W [9].

Thermoelectric Module (TEM)

The direct and indirect gains on the top side of TEM result in the production of electrical energy and residual thermal energy is transferred to the air flowing in the duct placed under the TEM.

$$\tau_g^2 \alpha_{tem}(1 - \beta_{sc})I(t)bdx + U_{tem}(T_{tem,top} - T_{tem,bottom})bdx$$
$$= h_{tf}(T_{tem,bottom} - T_f)bdx + \eta_{tem}U_{tem}(T_{tem,top} - T_{tem,bottom})bdx \qquad (4)$$

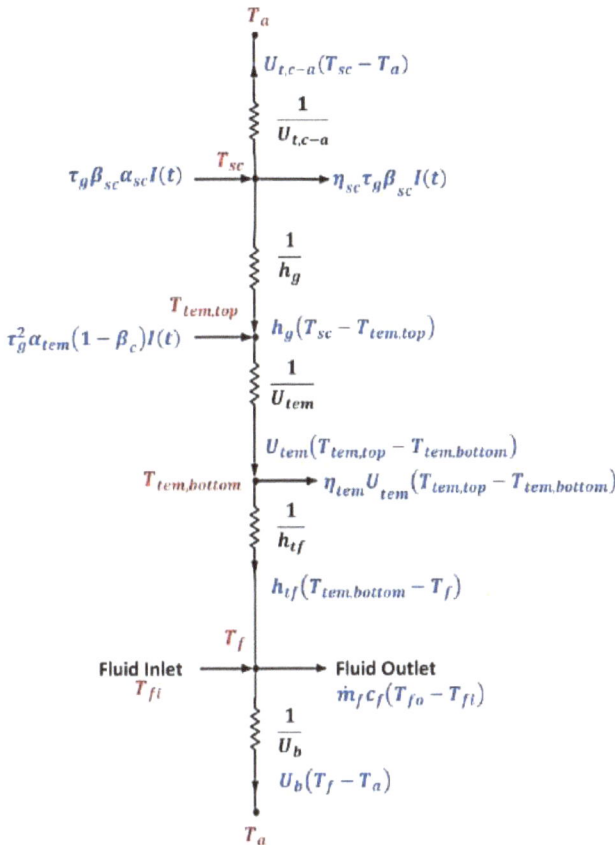

Fig. (2). Thermal resistances and the flow of thermal energy for semi-transparent PVT-TEM air heater [5].

η_{tem} represents the thermal to electrical energy conversion efficiency of TEM based on Seebeck effect. Depending on the thermoelectric material used the value of η_{tem} typically falls in the range of 5-8%. Some of the commonly used materials for the thermoelectric module (TEM) are bismuth telluride (Bi_2Te_3) and lead telluride (PbTe), among others [10].

Air Flowing in the Duct Placed Under TEM

$$h_{tf}\left(T_{tem,bottom} - T_f\right)bdx = \dot{m}_f c_f \frac{dT_f}{dx}dx + U_b\left(T_f - T_a\right)bdx \qquad (5)$$

The expressions for T_{sc} (solar cell temperature), $T_{tem,top}$ (TEM top-side temperature which is the same as bottom glass cover temperature), $T_{tem,bottom}$ (TEM bottom-side temperature) and T_f (fluid, *i.e.* air, temperature), calculated from Eq. 1 to Eq.4, are given below [5],

$$T_{sc} = \frac{(\alpha\tau)_{eff}I(t) + U_{t,c-a}T_a + h_g T_{tem,top}}{U_{t,c-a} + h_g} \qquad (6)$$

$$T_{tem,top} = \frac{h_{p1}(\alpha\tau)_{eff}I(t) + U_{tem,top-a}T_a + U_{tem}T_{tem,bottom}}{U_{tem,top-a} + U_{tem}} \qquad (7)$$

$$T_{tem,bottom} = \frac{(\alpha\tau)'_{eff}I(t) + (1 - \eta_{tem})U_{tem,bottom-a}T_a + h_{tf}T_f}{(1 - \eta_{tem})U_{tem,bottom-a} + h_{tf}} \qquad (8)$$

$$T_f = \left[\frac{h_{p3}(\alpha\tau)'_{eff}I(t)}{(U_{fa} + U_b)b} + T_a\right]\left[1 - exp\left(\frac{-(U_{fa} + U_b)bx}{\dot{m}_f c_f}\right)\right]$$
$$+ T_{fi}exp\left(\frac{-(U_{fa} + U_b)bx}{\dot{m}_f c_f}\right) \qquad (9)$$

The outlet fluid (air) temperature (T_{fo}) is attained using the boundary condition, $T_f|_{x=L} = T_{fo}$. The average fluid (air) temperature ($\underline{T}_f$) in the duct of PVT-TEM air heater, obtained by integrating Eq. (9), is given as [5]:

$$T_f = \left[\frac{h_{p3}(\alpha\tau)'_{eff}I(t)}{(U_{fa} + U_b)b} + T_a \right] \left[1 - \frac{\left\{ 1 - exp\left(\frac{-(U_{fa} + U_b)A_m}{\dot{m}_f c_f} \right) \right\}}{\frac{(U_{fa} + U_b)A_m}{\dot{m}_f c_f}} \right]$$

$$+ T_{fi} \frac{\left[1 - exp\left(\frac{-(U_{fa} + U_b)A_m}{\dot{m}_f c_f} \right) \right]}{\frac{(U_{fa} + U_b)A_m}{\dot{m}_f c_f}}$$

(10)

The assumed terms in Eq. (10) are listed in Appendix C.

T_f from Eq. 10 can be substituted in Eq. 6 to Eq. 8 to evaluate T_{sc} (average solar cell temperature), $T_{tem,top}$ (average TEM top-side temperature) and $T_{tem,bottom}$ (average TEM bottom-side temperature), respectively.

The analytical expression for solar cell efficiency (η_{sc}) is given as [5]:

$$\eta_{sc} = \eta_o \frac{\left[1 - \beta_o \{ XI(t) + YT_a + ZT_{fi} - T_o \} \right]}{[1 - \eta_o \beta_o I(t)W]}$$

(11)

The expressions derived for *X, Y, Z* and *W* are provided in Appendix C.

2.2. Thermal Analysis of Opaque PVT-TEM Air Heater

Similar to the thermal model of a semi-transparent PVT-TEM air heater, the energy balance equations for each component can be written to derive an analytical model of an opaque PVT-TEM air heater. The thermal resistances, flows of energy and temperatures at different points of the opaque air heater are illustrated in Fig. (**3**). Fig. (**3**) takes into consideration the packing factor of TEM, *i.e.* β_{tem} [6]. β_{tem}, analogous to β_{sc} represents the fraction of the total collector area covered with TEM. The energy balance for each component of opaque PVT-TEM air heater with β_{tem} as one (total area covered with TEM, as shown in Fig. (**1**), for an elemental area '*bdx*' along the air column, are given as follows [6, 7].

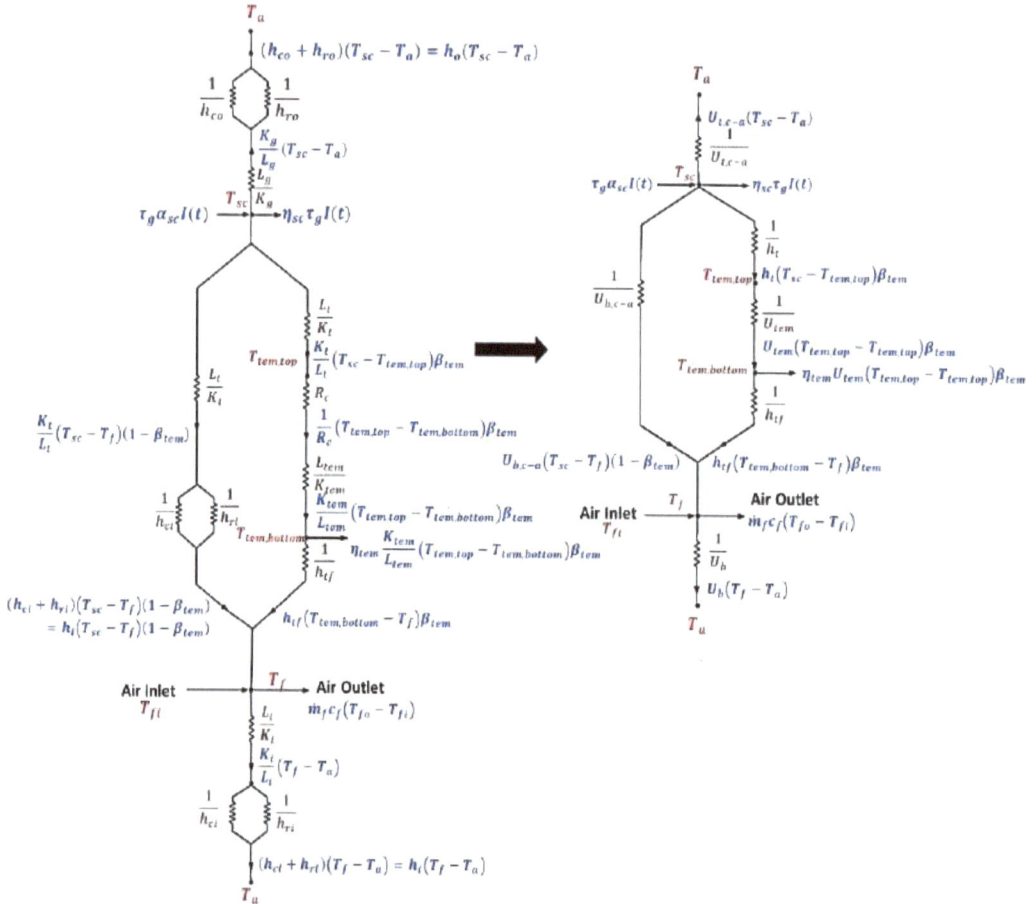

Fig. (3). Thermal resistances and thermal energy flow for opaque PVT-TEM air heater [6].

Opaque PV Module

In an opaque PV module, tedlar receives both direct gain from the solar radiation entering through a non-packing area of the PV module ($\tau_g \alpha_t (1 - \beta_{sc})I(t)$) and indirect gain from the bottom of solar cells ($h_t(T_{sc} - T_{tem,top})bdx$), as already explained in the working principle (Section 2).

$$[\tau_g \alpha_{sc} \beta_{sc} I(t) + \tau_g \alpha_t (1 - \beta_{sc}) I(t)] b dx$$
$$= U_{t,c-a}(T_{sc} - T_a) b dx + h_t(T_{sc} - T_{tem,top}) b dx \tag{12}$$
$$+ \eta_{sc} \tau_g \beta_{sc} I(t) b dx$$

Tedlar

The temperature of the top-side of TEM ($T_{tem,top}$) is equal to the temperature of tedlar (since temperature gradient across the thickness of tedlar is assumed to be negligible), as already stated under assumptions (Section 2). Therefore, $T_{tem,top}$ denotes both the temperature of tedlar and the temperature of top-side of TEM.

$$h_t(T_{sc} - T_{tem,top}) b dx = U_{tem}(T_{tem,top} - T_{tem,bottom}) b dx \tag{13}$$

TEM

TEM does not receive any direct gain from solar radiation in the case of an opaque PVT-TEM air heater, unlike the semi-transparent counterpart.

$$U_{tem}(T_{tem,top} - T_{tem,bottom}) b dx$$
$$= h_{tf}(T_{tem,bottom} - T_f) b dx \tag{14}$$
$$+ \eta_{tem} U_{tem}(T_{tem,top} - T_{tem,bottom}) b dx$$

Air Flowing Below TEM

The energy balance followed by the air flowing through the duct under the TEM is the same for opaque and semi-transparent PVT-TEM air heaters, *i.e.* given by Eq. 5.

Following a similar methodology adopted for semi-transparent PVT-TEM air heater, the expressions for different temperatures for opaque air heater are given by Dimri *et al.* [6,7].

2.3. Energy Gains from PVT-TEM Air Heater

As stated before, a PVT-TEM air heater generates electricity through both PV modules (by photovoltaic effect) and TEM (by Seebeck effect). Thus, the electrical energy produced by the PVT-TEM air heater is computed by adding the respective electrical gains from PV and from TEM, as per the following relation [5,6]:

$$E_{el} = E_{PV} + E_{TEM}$$
$$= \eta_m I(t) A_m$$
$$+ \eta_{tem} U_{tem} \left(T_{tem,top} - T_{tem,bottom} \right) \beta_{tem} A_m \tag{15}$$

This electrical energy is primarily used to drive the DC pump for the circulation of air through the duct. The remaining electrical energy, if any, could be used for running small electrical appliances or could be sold to the national grid to enhance the profitability of the system.

The electrical efficiency of PVT-TEM air heater is, therefore, given by [5,6]:

$$\eta_{el} = \frac{E_{el}}{I(t) A_m} = \eta_m + \frac{\left[\eta_{tem} U_{tem} \left(T_{tem,top} - T_{tem,bottom} \right) \beta_{tem} \right]}{I(t)} \tag{16}$$

The rate of thermal energy carried by the hot air produced by the air heater can be calculated as per the expression stated below [5,6]:

$$\dot{Q}_{th} = \dot{m}_f c_f \left(T_{fo} - T_{fi} \right) \tag{17}$$

Thus, the thermal efficiency (η_{th}) of the air heater is evaluated using [5,6]:

$$\eta_{th} = \frac{\dot{Q}_{th}}{I(t) A_m} = \frac{\dot{m}_f c_f \left(T_{fo} - T_{fi} \right)}{I(t) A_m} \tag{18}$$

The produced hot air can be used for room or space heating, drying applications (such as drying of crops) and air pre-heating processes. Thus, a PVT-TEM air heater has a vast majority of applications in the household, agricultural as well as industrial sectors. Depending on the type of thermal application desired, the air heater will be required to provide hot air within a specific temperature range. The temperature of hot air at the outlet of the duct can be controlled by adjusting the mass flow rate of air. As the mass flow rate of air increases, the time available for heat transfer from the bottom of TEM to the flowing air reduces and hence, the temperature of air at the outlet decreases (Fig. **4**). Thus, a lower mass flow rate allows the air flowing in the duct to reach a higher temperature at the outlet. Moreover, several PVT-TEM air heaters can be connected in series and in parallel to attain a higher temperature and higher quantity of hot air, respectively. Additionally, Fig. (**4**) shows that the thermal efficiency increases with an increase in mass flow rate, which is in concurrence with the expression given by Eq. 18.

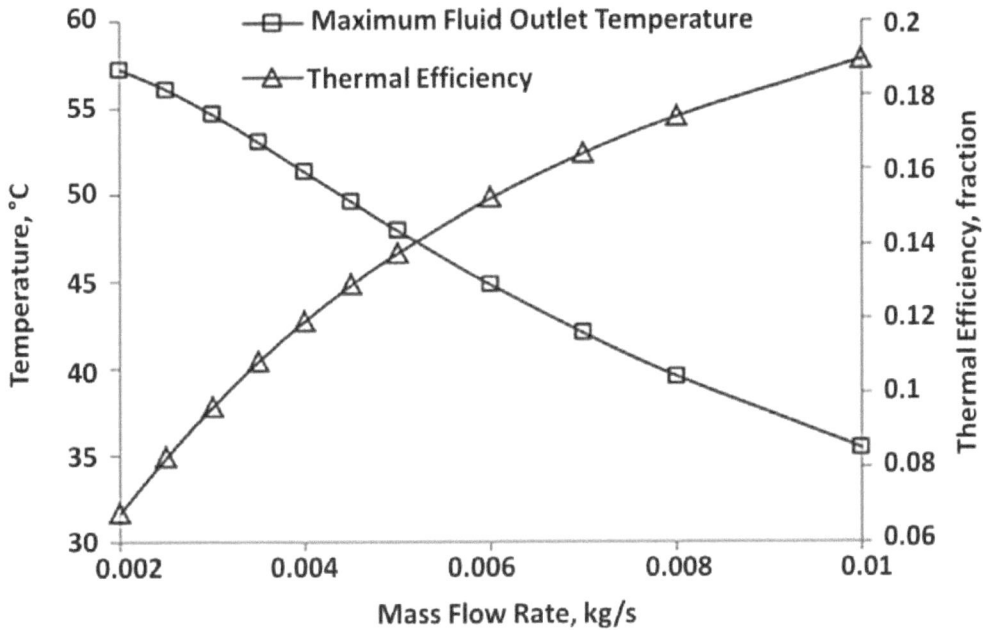

Fig. (4). Maximum air temperature at the outlet and thermal efficiency of semi-transparent PVT-TEM air heater as a function of mass flow rate [5].

It is a well-established notion that the electrical performance of solar cells lowers with a rise in solar cell temperature [3]. As expected, the solar cell temperature achieves the highest value around noon, *i.e.* during peak sunshine hour with the highest incident global solar radiation. Consequently, the electrical efficiency of a PVT-TEM air heater attains a minimum value during the same hour of the day, as shown in Figs. (**5a** and **b**). However it should be noted that the total electrical efficiency improves as a result of including TEM with a PV module. Moreover, upon adding an air duct to the combination of PV and TEM, the electrical efficiency increases further [5].

Also, the temperature of hot air flowing in the duct (T_f) is noticeably higher, as a result of heat transfers in accordance with the thermal model presented earlier, when compared to the ambient air temperature (T_a). This heated air adds to the thermal energy gain of the PVT-TEM air heater. The thermal energy and electrical energy gains from the PVT-TEM air heaters have been shown in Figs. (**6a** and **6b**).

Due to a higher magnitude of solar radiation available during the peak sunshine hours, the temperature difference between the air at outlet and inlet, and therefore thermal energy, is higher at peak sunshine hour. A similar result can be observed for electrical energy produced by PVT-TEM air heater as both the global solar radiation and the temperature gradient across TEM are higher at peak sunshine hour (Eq. 15).

It is worthwhile to point out that the energy gains, both thermal and electrical, in the absence of a storage unit are produced only during the day due to the non-availability of solar radiation at night. Also, the magnitudes of energy gains are higher when the solar radiation is higher, which may or may not be synchronous with the nature of energy demands. Thus, the importance of storage technologies must be stressed for any solar energy-based system.

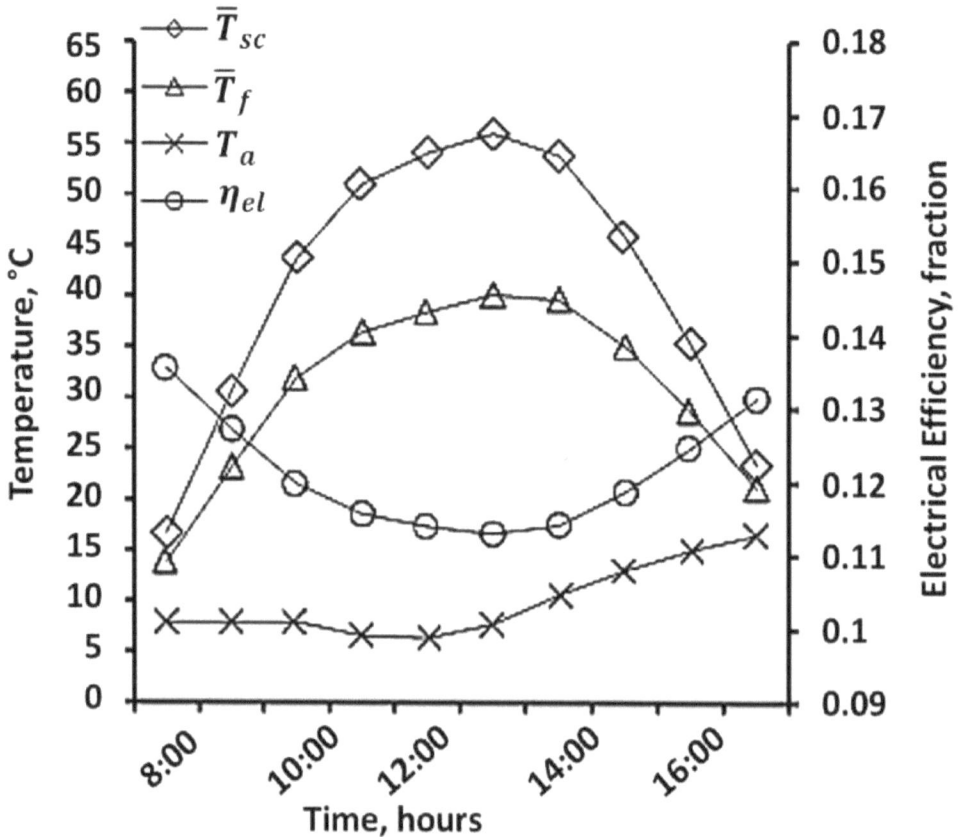

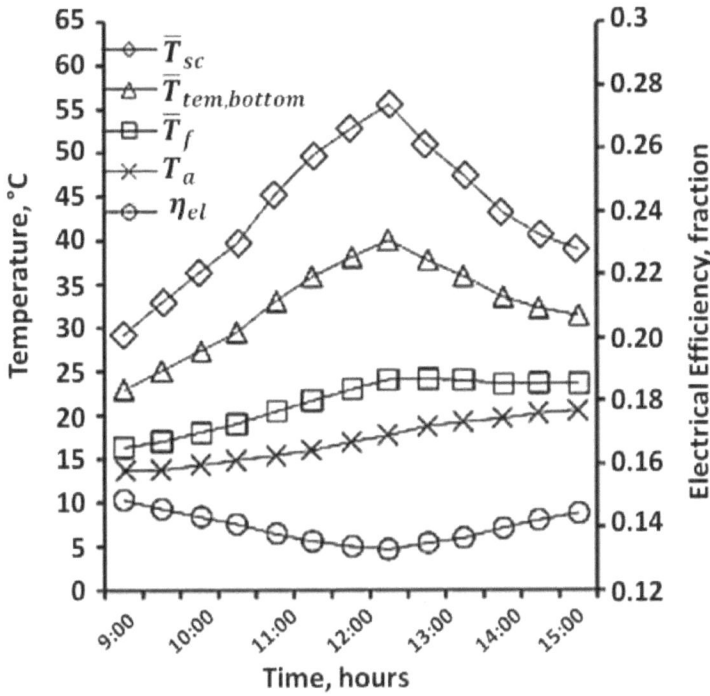

Fig. (5). Average temperatures of the components and total electrical efficiency of the PVT-TEM air heater. **(a)** semi-transparent PVT-TEM air heater [5] **(b)** opaque PVT-TEM air heater [6].

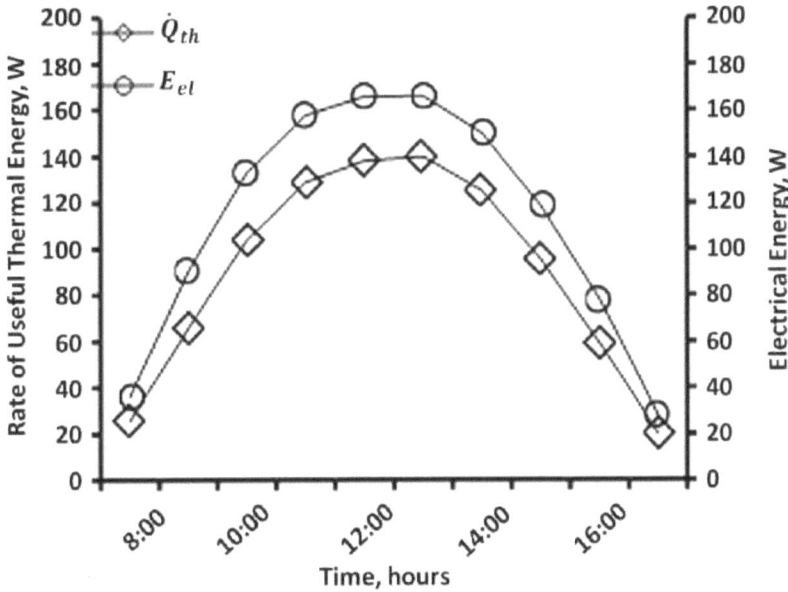

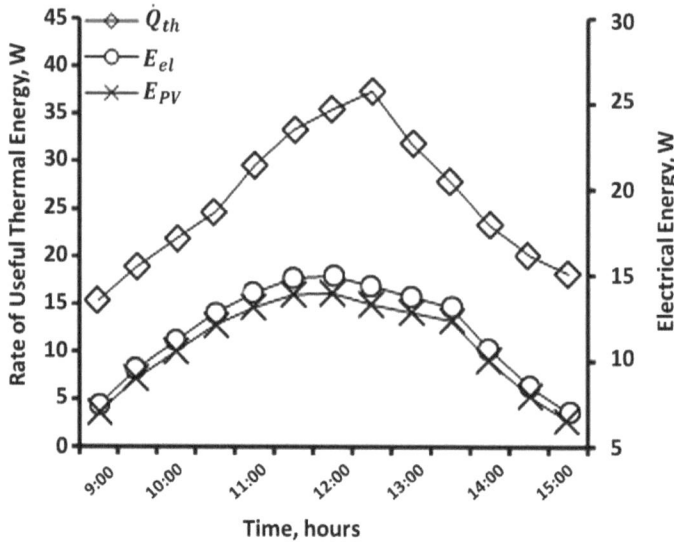

Fig. (6). Thermal energy and electrical energy produced by the PVT-TEM air heater. **(a)** semi-transparent PVT-TEM air heater [5] **(b)** opaque PVT-TEM air heater [6].

3. EXERGY ANALYSIS OF PVT-TEM AIR HEATER

Many studies suggest that it is important to complement the energy analysis (given in Section 3) with an exergy analysis in order to perform a thorough performance evaluation of a system [11–13]. Energy analysis, described in the previous section, is based on the first law of thermodynamics or the law of conservation of energy. The first law of thermodynamics cannot determine or quantify the system irreversibilities and therefore, in order to perform a more detailed analysis the second law of thermodynamics (based on entropy) needs to be considered. The term exergy arises from the second law of thermodynamics. Exergy denotes the maximum theoretical work as the system reaches equilibrium with respect to its reference environment. The measure of exergy helps in differentiating between the qualities of thermal energy (low-grade energy) and electrical energy (high-grade energy). Therefore, an exergy analysis aids in a detailed analysis of a system both in terms of the quantity as well as the quality of the produced energy.

In the case of a PVT-TEM air heater, both electrical energy and thermal energy are produced. The exergy associated with electrical energy (*i.e.* high-grade energy) equals the summation of the electrical work produced by different components of the system [14]. While the exergy associated with heat (thermal energy) is dependent on the temperature at which heat transfer takes place and the dead state temperature (*i.e.*, reference temperature).

The exergy associated with solar radiation, which is the input exergy ($\dot{E}_{xi}$) in case of the considered air heater, is calculated as per the relation [15,16]:

$$\dot{E}_{xi} = I(t) \left[1 - \frac{4}{3}\left(\frac{T_a}{T_s}\right) + \left(\frac{T_a}{T_s}\right)^4 \right] \tag{19}$$

Where, T_a and T_s are the ambient air temperature (in K) and the temperature of the Sun (usually taken as approximately 6000 K).

Exergy associated with thermal energy ($\dot{E}_{xth}$) is given by [17,18]:

$$\dot{E}_{xth} = \dot{m}_f c_f \left(T_{fo} - T_{fi}\right) - (T_a + 273) ln \frac{(T_{fo} + 273)}{(T_{fi} + 273)} \tag{20}$$

Subsequently, an overall exergy gain ($\dot{E}_{xo}$) of the PVT-TEM air heater can be obtained.

$$\dot{E}_{xo} = E_{el} + \dot{E}_{xth} \tag{21}$$

Moreover, overall exergy efficiency can be computed from:

$$\varepsilon = \eta_{el} + \frac{\dot{E}_{xth}}{\dot{E}_{xi} A_m} \tag{22}$$

Following Eqs.19-22, an evaluation of the PVT-TEM air heater from an exergy perspective can be made.

4. COMPARISON OF SEMI-TRANSPARENT AND OPAQUE PVT-TEM AIR HEATERS

In this section, the semi-transparent and opaque PVT-TEM air heaters shall be compared from both the energy and exergy perspectives. The thermal energy produced by a PVT-TEM air heater is higher when glass (semi-transparent air heater) is used as the bottom layer (Fig. **1**), instead of tedlar (opaque air heater), as shown in Fig. (**7**). This can be explained by the higher transmittance of solar radiation through the glass bottom layer (over a non-packing area) and thus, the higher magnitudes of heat exchanges between the following components and finally to the air flowing in the duct in case of the semi-transparent PVT-TEM air heater. On the contrary, the electricity produced by the PVT-TEM air heater is lower with glass as the bottom layer (*i.e.* semi-transparent air heater). This results from the higher temperature of the top-side of TEM, and consequently higher temperature of solar cells in a semi-transparent air heater. And since the electrical efficiency is lower at a higher solar cell temperature, the opaque PVT-TEM air heater outperforms the semi-transparent PVT-TEM air heater from the electrical energy viewpoint. This result differs from the general observation wherein a semi-transparent PVT air heater (without TEM) achieves a higher electrical energy gain when compared to an opaque PVT air heater. It is mainly because the absorptivity of TEM material is quite higher than tedlar. Therefore, the product of transmissivity of glass (~ 95%) and absorptivity of TEM is greater than the absorptivity of tedlar. This in turn, causes the temperature of the top-side of TEM to be higher in the case of a semi-transparent PVT-TEM air heater.

From the results stated above, we may state that the choice of the bottom layer material (*i.e.*, glass or tedlar) depends on the respective importance of electrical and thermal energies in a specific application. If the key aim is to maximize the thermal gain, then a semi-transparent PVT-TEM air heater should be favored. Whereas, when an acceptable thermal gain with relatively higher electricity production is desired, an opaque PVT-TEM air heater may be more suitable.

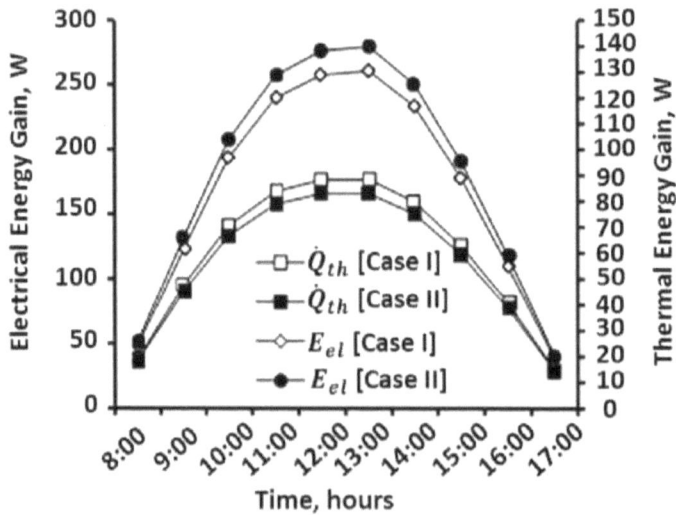

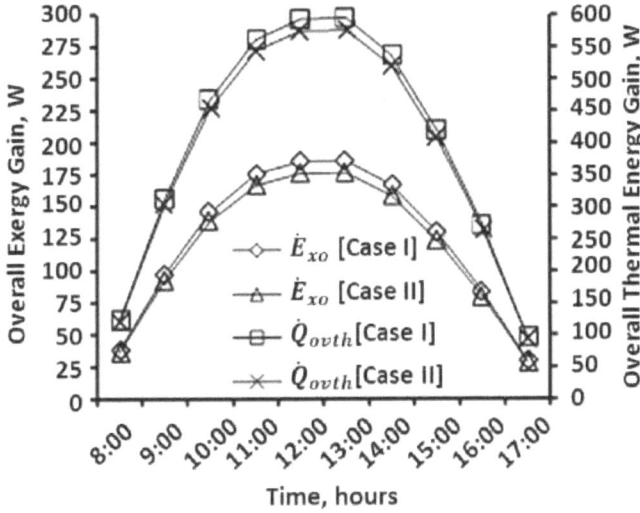

Fig. (7). Rate of useful thermal energy, electrical energy, overall thermal energy and overall exergy of opaque. [Case I] and semi-transparent [Case II] PVT-TEM air heaters [10].

Furthermore, the PVT-TEM air heater with tedlar as the bottom layer (*i.e.*, opaque) performs better from both the overall thermal energy and overall exergy perspectives when compared to the semi-transparent air heater (Fig. 7). This is observed due to a higher magnitude of electrical energy, *i.e.* high-grade energy, in the case of opaque PVT-TEM air heater, which dominates the magnitude of thermal energy counterparts while calculating the overall thermal energy and overall exergy gains. Hence, from the thermal energy point of view alone, a semi-transparent PVT-TEM air heater performs better. While, the opaque PVT-TEM air heater achieves a better overall performance as per overall thermal energy, overall exergy and electrical energy gains.

5. PACKING FACTOR OF TEM

As described in Section 2, a PVT-TEM air heater consists of TEM employed underneath a PV module. In the considered design, the TEM may cover the area under the PV module either fully (as depicted in Fig. **1**) or partially (Fig. **8**). Thus, an important parameter to be discussed in such a context is the packing factor of TEM (β_{tem}). The packing factor of TEM is defined as per the following equation [6]:

$$\beta_{tem} = \frac{Area\ occupied\ by\ TEM}{Total\ area\ of\ PV\ module} \tag{23}$$

The value of β_{tem} is the maximum, *i.e.* one, when the entire area under PV module is covered by TEM as in Fig. (**1**). Whereas, if the PVT-TEM air heater is partially covered with TEM as depicted in Fig. (**8**), then $0 < \beta_{tem} < 1$. An important point to note is the case when β_{tem} equals to zero, denoting the absence of TEM wherein the air heater design becomes equivalent to a conventional PVT air heater.

The energy balance equations can be re-written following Section 3, considering the packing factor of TEM (β_{tem}) in order to derive the respective expressions for different temperatures of the PVT-TEM air heater.

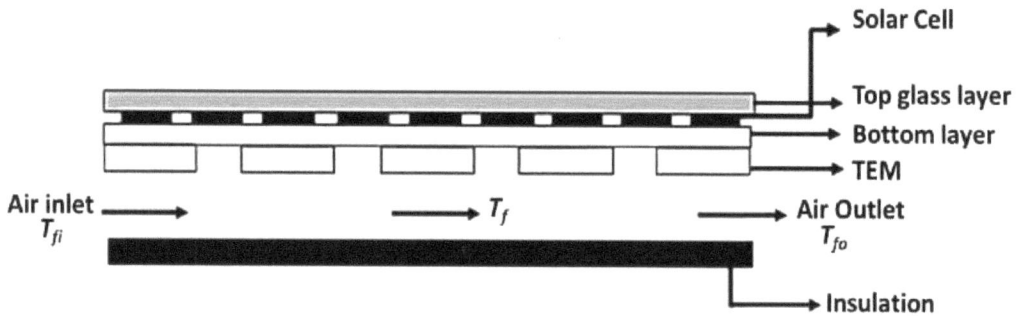

Fig. (8). Cross-sectional view of a PVT-TEM air heater partially covered with TEM.

When the packing factor of TEM (β_{tem}) is less than one, TEM partially covers the total area under the PV module as portrayed in Fig. (**8**). In this case, the air flowing through the duct receives thermal energy from both the bottom-side of TEM, the over packing area of TEM, and the bottom layer of the PV module, along the non-packing area of TEM. Thermal energy flows from the bottom of the solar cells to the top-side of TEM, through glass/tedlar (bottom layer), over the packing area of TEM ($\beta_{tem}A_m$). In addition, there is a transfer of thermal energy from the bottom layer (glass/tedlar) to the air flowing in the duct, over the non-packing area of TEM, *i.e.* $(1 - \beta_{tem})A_m$. When the bottom layer is glass, *i.e.* semi-transparent PVT-TEM air heater, this heat transfer may either correspond to a direct solar gain, if the non-packing area of TEM coincides with the non-packing area of PV or indirect solar gain, if the non-packing area of TEM overlaps with the packing area of PV. This depends on the specific arrangement of TEM under the PV module in the air heater design. However, if tedlar forms the bottom layer of the PV module in Fig. (**8**), *i.e.* opaque PVT-TEM air heater, the air flowing through the duct always receives an indirect solar gain corresponding to either the portion of solar radiation absorbed by tedlar or the thermal energy transferred from the bottom of the solar cells.

CONCLUSION

This chapter deals with the Tem employing PVT air heater. One of the key advantages of a photovoltaic thermal (PVT) air heater over a traditional solar thermal air heater is that it produces both electrical and thermal energy, *i.e.* higher overall energy. The electrical energy could be utilized to drive the pump used for the circulation of air through the duct thereby making the system self-sustained. When a thermoelectric module (TEM) is incorporated within a PVT air heater, the electrical efficiency and, thus, the total energy gain of the air heater improves. Hence, in recent years the research on PVT-TEM design is gaining attention.

NOMENCLATURE

A_m	Area of PV module, m^2	h_{p3}	Tertiary penalty factor resulting from the TEM
l	Length of air heater, m		
b	Breadth of air heater, m	T_{sc}	Temperature of solar cell, °C
c_f	Specific heat of the working fluid *i.e.* Air, J/kgK	$T_{tec,top}$	Temperature of top-side of TEM, °C
		$T_{tec,bottom}$	Temperature of bottom-side of TEM, °C
dx	Elemental length along the air column, m	T_f	Temperature of fluid *i.e.* Air flowing beneath TEM, °C
$I(t)$	Global solar irradiation, W/m^2		
$\dot{m}_f$	Mass flow rate of the working fluid, *i.e.* Air, kg/s	T_a	Temperature of ambient air, °C
		T_{fi}	Temperature of fluid *i.e.* Air flowing beneath TEM at inlet, °c
K			

	Thermal conductivity of the component, W/mK	T_{fo}	Temperature of fluid *i.e.* Air flowing beneath TEM at outlet, °c
L	Thickness of the component, m		
h_t	Coefficient of heat transfer from bottom of solar cells to the top-side of TEM along the bottom tedlar layer, W/m^{2}k	E	Electrical energy gain from the component or in total, W
		$\dot{E}_{xth}$	Exergy of thermal energy, W
h_g	Coefficient of heat transfer from bottom of solar cells to the top-side of TEM along the bottom glass layer, W/m^{2}k	$\dot{E}_{xo}$	Overall exergy, W
		$\dot{E}_{xi}$	Input exergy, *i.e.* Exergy of solar radiation, w
h_{tf}	Coefficient of heat transfer from the bottom-side of TEM to the working fluid, W/m^{2}k	η_o	Solar cell efficiency at standard test condition, with $I(t) = 1000$ W/m^2 and $T_o = 25$ °C
h_o	Coefficient of heat transfer for a flat plate exposed to ambient air/wind through convection and radiation both, W/m^{2}k		
		η_{tem}	Thermal to electrical conversion efficiency of TEM
h_i	Coefficient of heat transfer for a flat plate exposed to ambient air/wind through convection alone, W/m^{2}k	β_o	Temperature coefficient of solar cell efficiency, K^{-1}
R_c	Contact thermal resistance at the interface of TEM, *i.e.* Thermoelectric leg-electrode interface, m^2 K/W	**Greek Letters**	
		α	Absorptivity
		β	Packing factor
$U_{t,c-a}$	Coefficient of total heat transfer from top of solar cells to ambient along the top glass layer, W/m^{2}k	τ	Transmissivity
		η	Energy efficiency
		ε	Overall exergy efficiency
U_b	Coefficient of total heat transfer from bottom to ambient along the insulation, W/m^{2}k		Products of effective absorptivity and transmissivity

U_{tem}	Coefficient of total heat transfer from the top-side of TEM to the bottom-side of TEM, W/m² K	$(\alpha\tau)_{eff}$ $(\alpha\tau)'_{eff}$ $(\alpha\tau)''_{eff}$	
		Subscripts	
		a	Ambient
$U_{tem,top-}$	Coefficient of total heat transfer from top-side of TEM to ambient along the bottom layer *i.e.* Glass/tedlar, W/m² K	eff	Effective
		f	Fluid
		fi	Fluid inlet
$U_{tem,bottd}$	Coefficient of total heat transfer from bottom-side of TEM to ambient, W/m² K	fo	Fluid outlet
		g	Glass
U_{fa}	Coefficient of total heat transfer from the working fluid (air) to ambient, W/m² K	m	Module
		PV	Photovoltaic module
h_{p1}	Primary penalty factor resulting from the top glass layer	sc	Solar cell
		tem	Thermoelectric module
h_{p2}	Secondary penalty factor resulting from the bottom layer (glass/tedlar)	t	Tedlar
		i	Insulation

CONSENT FOR PUBLICATION

Not applicable.

CONFLICT OF INTEREST

The authors declare no conflict of interest, financial or otherwise.

ACKNOWLEDGEMENTS

Declared none.

REFERENCES

[1] A. Braunstein, and A. Kornfeld, "On the development of the solar photovoltaic and thermal (PVT) collector", *IEEE Trans. Energ. Convers.,* vol. EC-1, no. 4, pp. 31-33, 1986.
http://dx.doi.org/10.1109/TEC.1986.4765770

[2] P. Huen, and W.A. Daoud, "Advances in hybrid solar photovoltaic and thermoelectric generators", *Renew. Sustain. Energy Rev.,* vol. 72, pp. 1295-1302, 2017.
http://dx.doi.org/10.1016/j.rser.2016.10.042

[3] D.L. Evans, "Simplified method for predicting photovoltaic array output", *Sol. Energy,* vol. 27, no. 6, pp. 555-560, 1981.
http://dx.doi.org/10.1016/0038-092X(81)90051-7

[4] F.J. DiSalvo, "Thermoelectric cooling and power generation", *Science,* vol. 285, no. 5428, pp. 703-706, 1999.
http://dx.doi.org/10.1126/science.285.5428.703 PMID: 10426986

[5] N. Dimri, A. Tiwari, and G.N. Tiwari, "Thermal modelling of semitransparent photovoltaic thermal (PVT) with thermoelectric cooler (TEC) collector", *Energy Convers. Manage.,* vol. 146, pp. 68-77, 2017.
http://dx.doi.org/10.1016/j.enconman.2017.05.017

[6] N. Dimri, A. Tiwari, and G.N. Tiwari, "Effect of thermoelectric cooler (TEC) integrated at the base of opaque photovoltaic (PV) module to enhance an overall electrical efficiency", *Sol. Energy,* vol. 166, pp. 159-170, 2018.
http://dx.doi.org/10.1016/j.solener.2018.03.030

[7] N. Dimri, A. Tiwari, and G.N. Tiwari, "Comparative study of photovoltaic thermal (PVT) integrated thermoelectric cooler (TEC) fluid collectors", *Renew. Energy,* vol. 134, pp. 343-356, 2019.
http://dx.doi.org/10.1016/j.renene.2018.10.105

[8] B.Y. Ohara, and H. Lee, "Exergetic analysis of a solar thermoelectric generator", *Energy,* vol. 91, pp. 84-90, 2015.
http://dx.doi.org/10.1016/j.energy.2015.08.030

[9] Z. Ouyang, and D. Li, "Modelling of segmented high-performance thermoelectric generators with effects of thermal radiation, electrical and thermal contact resistances", *Sci. Rep.,* vol. 6, no. 1, p. 24123, 2016.
http://dx.doi.org/10.1038/srep24123 PMID: 27052592

[10] N. Dimri, *Artificial neural network (ANN) modelling of photovoltaic thermal (PVT) integrated thermoelectric cooler (TEC) fluid collectors.* Indian Institute of Technology (IIT): Delhi, 2019.

[11] E. Saloux, M. Sorin, and A. Teyssedou, "Exergo-economic analyses of two building integrated energy systems using an exergy diagram", *Sol. Energy,* vol. 189, pp. 333-343, 2019.
http://dx.doi.org/10.1016/j.solener.2019.07.070

[12] K. Sartor, and P. Dewallef, "Exergy analysis applied to performance of buildings in Europe", *Energy Build.,* vol. 148, pp. 348-354, 2017.
http://dx.doi.org/10.1016/j.enbuild.2017.05.026

[13] M.J. Moran, *Fundamentals of engineering thermodynamics,* 7th ed Wiley: Hoboken, N.J., 2011.

[14] A. Bejan, G. Tsatsaronis, and M.J. Moran, *Thermal design and optimization.* Wiley: New York, 1996.

[15] R. Petela, "Exergy of undiluted thermal radiation", *Sol. Energy,* vol. 74, no. 6, pp. 469-488, 2003.
http://dx.doi.org/10.1016/S0038-092X(03)00226-3

[16] J.T. Szargut, "Anthropogenic and natural exergy losses (exergy balance of the Earth's surface and atmosphere)", *Energy,* vol. 28, no. 11, pp. 1047-1054, 2003.
http://dx.doi.org/10.1016/S0360-5442(03)00089-6

[17] A. Bejan, "General criterion for rating heat-exchanger performance", *Int. J. Heat Mass Transf.,* vol. 21, no. 5, pp. 655-658, 1978.
http://dx.doi.org/10.1016/0017-9310(78)90064-9

[18] Y.A. Cengel, and M.A. Boles, *Thermodynamics: An engineering approach.* McGraw-Hill Education: New York, 2015.

CHAPTER 14

Applications and Development of Solar Systems in Buildings

Rishika Shah[1*], R.K. Pandit[1] and M.K. Gaur[2]

[1]*Department of Architecture, Madhav Institute of Technology and Science, Gwalior, India*

[2]*Department of Mechanical Engineering, Madhav Institute of Technology and Science, Gwalior, India*

Abstract: Many harmful effects on the environment can be observed over the past decades due to the extensive usage of non-renewable energy. Most discussed and harmful are the ever-changing global climate change scenarios and their aftermath. As a point of fact, a major part of the world's energy consumption is dependent on non-renewable energy sources, such as petroleum, oil, coal, and gas. Unquestionably, these fossil fuels contribute a great deal to greenhouse gas emissions, carbon dioxide, methane, *etc.*, which further leads to global health issues, global warming, and climate change. With the emergence of sustainable development as a holistic concept since the late 1980s, the issue of global warming has been given prominent attention. It is evident that failure to curb global warming has led to slower progress in achieving sustainable development. About 30% of energy demand is from the built environment sector, which is also responsible for contributing 28% of carbon emissions and continues to add an estimated 1% every year, according to reports by UN Environment [1]. Therefore, the fossil fuel-based energy systems are antagonistic with the goals of sustainable development agendas. Hence, using renewable sources in harnessing clean energy for the built environment has not remained a choice but a fundamental need. Solar energy is one of the cleanest renewable energy sources that provide solutions to climate change and global warming. Often termed as the alternative energy source against oil and coal-based energy sources, solar energy has the potential for abundant availability and is an economical way with a lower ecological and environmental footprint, leading to a better quality of life. Thus, there is a massive amount of global interest in harnessing solar energy for its application and development in building systems.

Keywords: Application, Buildings, Development, Solar systems, Solar in buildings.

*Corresponding author Rishika Shah:** Department of Architecture, Madhav Institute of Technology and Science, Gwalior, India; E-mail: shahrishika24@gmail.com

Manoj Kumar Gaur, Brian Norton & Gopal Tiwari (Eds.)

1. SOLAR ENERGY RESOURCE IN WORLD

Solar energy has become the third most substantial renewable energy source with more than 486GW of installed capacity, with photovoltaic (PV) being the leading technology (Fig. **1**). By the end of 2021, the global concentrating solar power (CSP) installed capacity will reach about 5.5 GW, hinting at the rise of CSP technology. At present, the giant solar PV capacity is being possessed by India, China, US, Germany, Italy, and Japan in the world, whereas 42% of the global CSP capacity is dominated by Spain. During the last five years, the annual growth rate of cumulative solar energy capacity has averaged about 25%, which makes solar the fastest growing renewable power source. Of the total of 94 GW of global solar power expansions, Asia accounted for approximately 70%, while Germany, the United States, and Australia added 3.6 GW, 8.4 GW, and 3.8 GW in recent solar-powered projects during the past year. At present, the largest single-site solar power plant in the world is United Arab Emirates' 1.17GW Noor Abu Dhabi solar project.

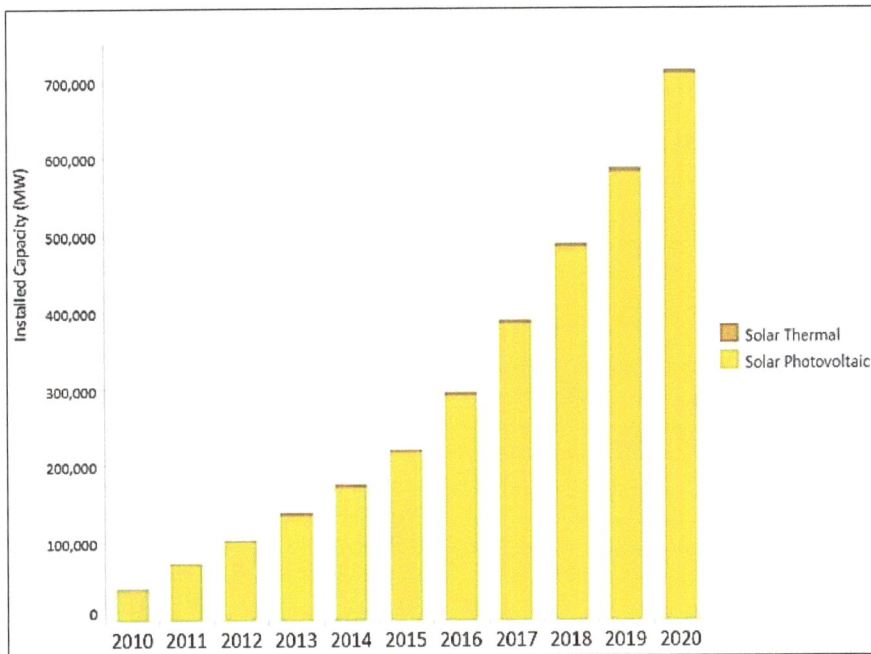

Fig. (1). Global trend of installed capacity in the past decade.

2. APPLICATIONS AND DEVELOPMENT OF SOLAR ENERGY IN THE BUILDING SECTOR

2.1. Solar Water Heating Systems

Since the early 1980s, solar energy has been utilized to arrange hot water at domestic levels and has since evolved technologically as well as at building scales. In this section, an overview of solar water heating developments in the building sector has been discussed. A solar water heating (SWH) system comprises typically of following units:

(i) Solar collectors

(ii) Thermal storagez

(iii) System controller unit

The application and development of solar water heating systems in the building sector are discussed below:

2.1.1. Passive Solar Water Heating Systems

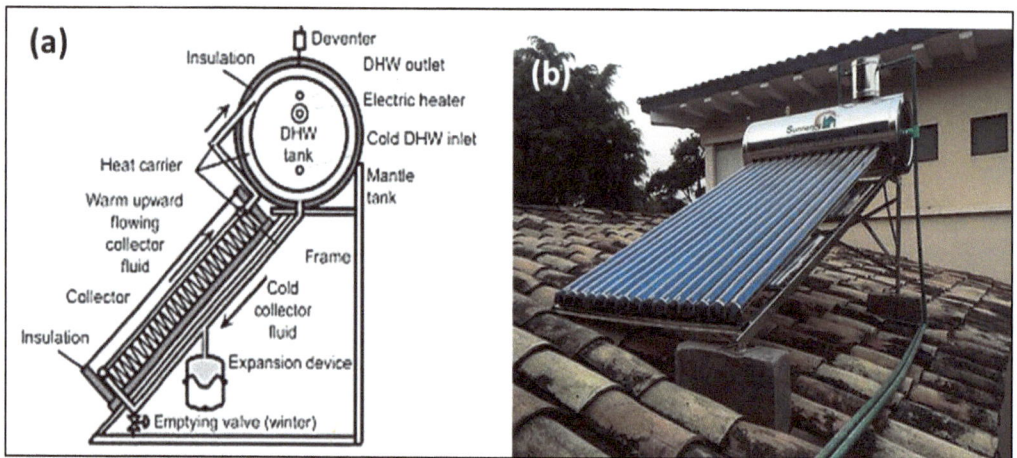

Fig. (2). (a) Schematic diagram of the passive solar water heater **(b)** Installed Thermosyphon SWH on roof.

This is also known as thermosyphon solar water heaters (Fig. **2**), and this type of system does not depend upon electricity and works on the principle of gravity (gravity-driven circulation) in a way that it uses the gravitational differences between hot water and cold water to circulate heat transfer from tube collectors to water collector tanks. The solar collector is placed below the water collector with tubes which permits a natural circulation of heat transfer fluid into the water collector tanks.

2.1.2. Active Solar Water Heating Systems

(a) Forced Circulation SWH Systems

This system makes use of a pump to circulate hot water. In this type of SWH, the solar collector is embedded in the roof profile rather than installed on top of the roof. This system is also known as a direct circulation system (Fig. **3**).

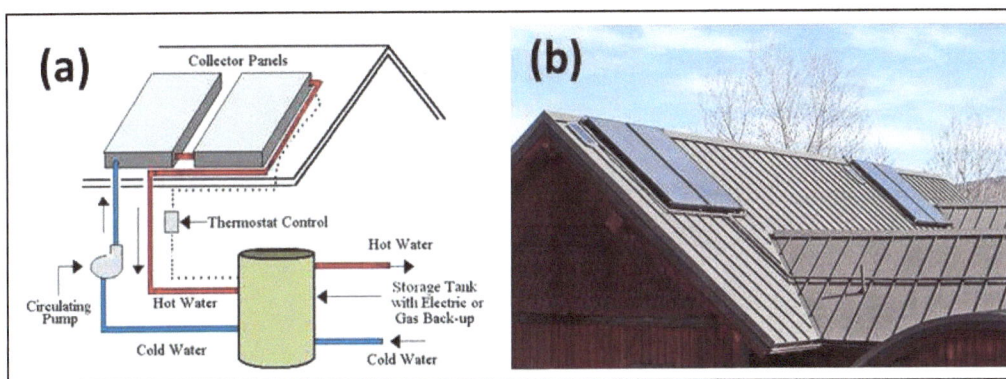

Fig. (3). (a) Schematic diagram of forced circulation solar thermal system **(b)** Installed forced SWH system on the roof.

(b) Drain Back SWH Systems

Designed specifically for cold climatic regions aiming towards fail safe operation, there are just a few more components in the drain back systems. Although the solar control system and collector are similar to the forced circulation system, nevertheless, in this system, an antifreeze solution is circulated through the solar collector and back into a heat exchanger in the hot water storage tank. The indirect solar water heating system is less feasible in comparison to direct circulation systems as the cost of the system is increased due to the addition of a heat exchanger; this also results in a possible degree of energy loss caused by heat transfer. In addition to that, these systems can typically be installed at higher latitudes, where incoming solar radiation is reduced.

Temperature from a sensor placed at the solar collector is compared with that from the sensors placed at the bottom of the hot storage tank where the cold water is stored, with the help of an electronic control system. This electronic control system turns on the pump when a high temperature is observed in the solar collector than that of the one where cold water is placed. However, the potable water is separated in the hot water storage tank from the fluid circulating throughout the solar collector. The fluid which circulates through the solar collector is either glycol solution or water. As soon as the pump stops its function, the solar collector is left empty whenever the fluid is not circulating through it. Consequently, the fluid in the solar collector "drains back" into the drain back tank, and potable water is circulated through a heat exchanger from the hot water storage tank by a second circulating pump into the drain back tank (Fig. **4**).

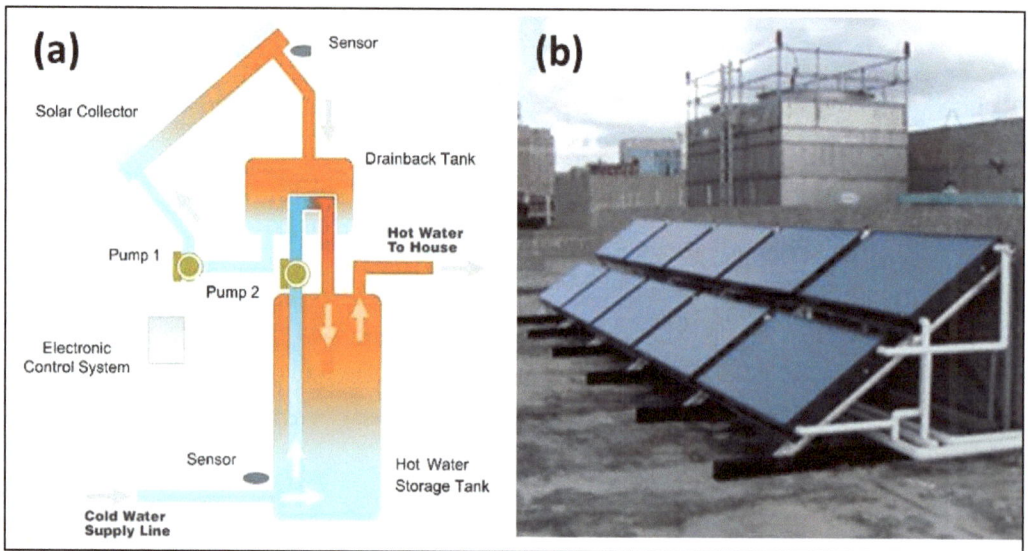

Fig. (4). (a) Schematic diagram of Drain Back SWH systems **(b)** Installed Drain Back SWH systems on the roof [2].

Drain back solar collector loop is an alternative design where only one pump is required. In this setting, the heat exchanger is typically "wrapped around" the hot water storage tank.

2.2. Solar Heating in Buildings

2.2.1. Direct Gain

In this type of system, the actual userspace works as a solar collector, heat absorber, and distribution system. Solar energy enters the living or workable space through windows featuring glass panes and gets absorbed by the buildings component like walls and floors, which have different thermal mass values. South-facing windows admit the most amount of solar energy. The thermal mass building components like floor and walls are enough to store heat, but additionally, water-filled containers can also be placed to store heat from incoming solar energy. However, integrating such water-filled containers into the building system can be economically and functionally a hindrance. The thermal mass property of the building system allows storage of the heat during the daytime and radiating heat during the night. The time taken by thermal mass components to release the heat back is known as thermal lag (Fig. **5**). About 60 -75% of the sun's energy striking the window is utilized through the direct gain system.

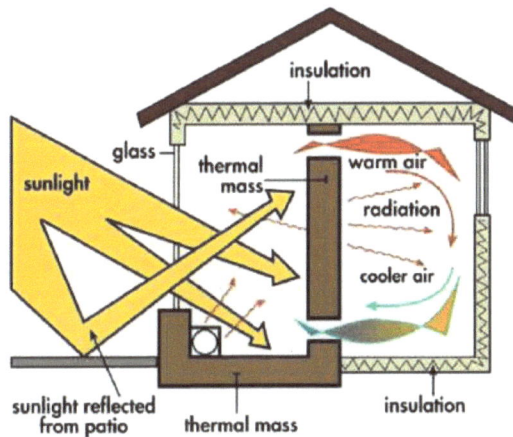

Fig. (5). Components of solar heating of buildings through direct heat gain.

2.2.2. Indirect Gain

In this type of system, buffer space or component having a high thermal mass value is placed between the outside space and the user space. As solar energy does not enter directly into the living space, hence it is known as an indirect gain system. In an indirect gain system, heat is transferred through conduction from thermal mass components to the living space after absorbing solar energy during the day. This thermal mass component acts as a mediator between the sun and living space. About $30 - 45\%$ of the solar energy striking the glass adjoining the thermal mass is utilized in the indirect gain system [3]. The two most common systems of indirect gain are explained below:

(a) Glazed and Unglazed Trombe Wall

Unglazed transpired solar façade is one of the types of building-integrated solar thermal (BIST) system for building facades. It is the tool developed to collect solar energy through a building façade, such that an envelope-like architectural feature is used in the building, and additionally, the setup also accumulates solar energy to increase thermal comfort [4]. In glazed Trombe walls, glass panes are organized on the façade of the building. The cavity and block wall behind the glass wall are heated by the sun, keeping the building warm in winter (Fig. **6**).

Fig. (6). Passive solar homes in Ladakh, India [5].

(b) Photovoltaic Trombe Wall

Over the past few years, the Trombe wall has emerged as a constructive tool for using solar energy for thermal comfort in buildings. It has evolved over the past decade in regard to its energy-saving and easy maintenance characteristics. It has been noted as a zero-running cost tool for providing heat in buildings. The traditional Trombe wall is "a concrete or masonry wall, blackened and covered on the exterior by glazing."

One of the drawbacks of the use of the Trombe wall is its unaesthetic appearance. In view of this, a new idea of photovoltaic Trombe wall emerged, which uses heat from solar energy and reduces the temperature of a photovoltaic cell. 80% of energy is converted into thermal energy when solar energy is directly irradiated; due to this, the efficiency of photovoltaic cells is reduced as their temperature increases. In the case of the photovoltaic Trombe wall, the photovoltaic cells are connected at the back of the glass panel; this established unit is known as the photovoltaic Trombe wall (Fig. **7**). The advantage of this setup is that the temperature of photovoltaic cells is reduced because of the airflow movement between the wall and the PV cells attached to the glass panel [6]. This leads to passive heating and also the production of electrical energy.

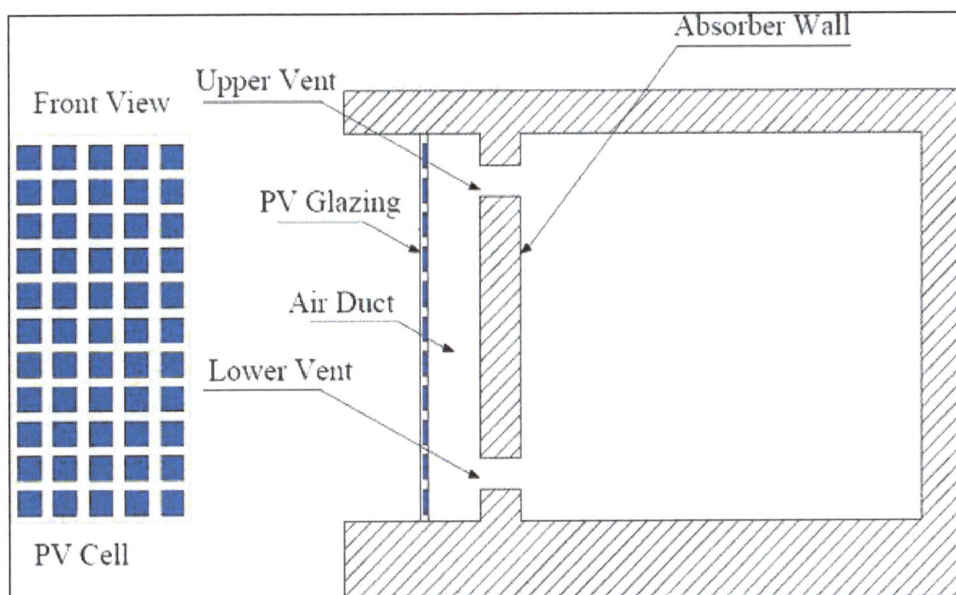

Fig. (7). Photovoltaic-Trombe wall.

2.2.3. Isolated Gain

The isolated gain system has its constitutive region separate from the central living area of a building. A typical example of an isolated gain system is a sunroom (Fig. **8**). The primary agenda of this type of system is isolation from the chief functional areas of a building. About 15 – 30% of the solar radiation is utilized through this system after coming in contact with the glazing, thereby heating the adjacent spaces of the isolated space. A certain amount of solar energy is also conserved in the sunroom itself because of the greenhouse effect.

2.3. Building Integration PV Systems

Apart from being utilized for cooling and heating purposes for the buildings, solar energy is also used to provide electricity for the adequate function of buildings. For this purpose, photovoltaic cells are used. These convert solar energy into electricity through semiconductors such as silicon. These photovoltaics are mostly integrated into building roof systems as maximum radiation fell on roofs, this idea emerged in the early 1990s, and the concept was named as Building Integrated Photovoltaic (BIPV) system. The increased recognition in the past two decades increases the system's capability to facilitate functional net-zero energy building designs.

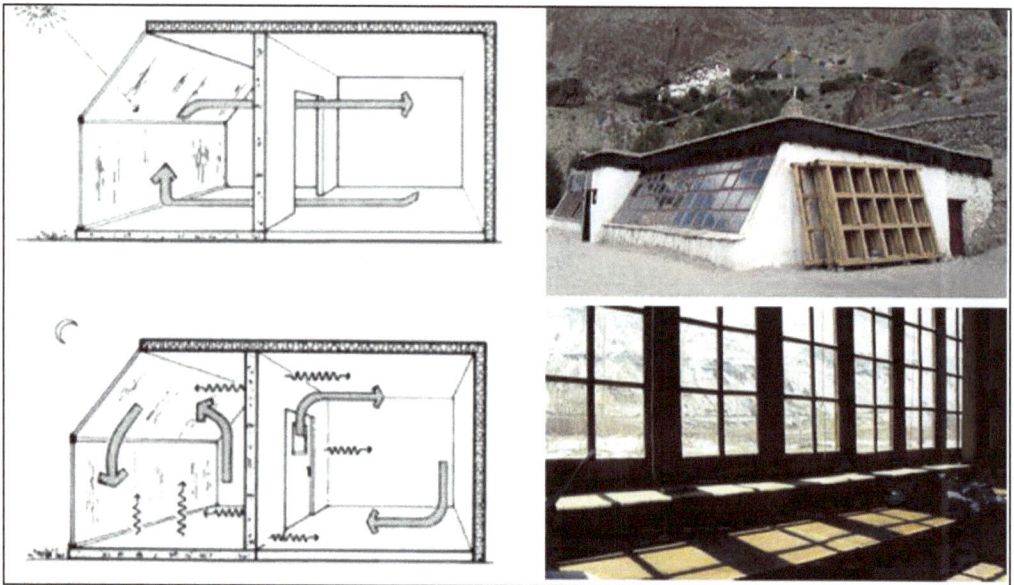

Fig. (8). Schematic overview of a Trombe wall; Solar school in Lingshed, Ladakh, India [7].

Building Integrated Photovoltaic (BIPV) system can be described as when the building envelope is integrated as an element of the structure, replacing traditional high embodied energy materials, rather than being affixed after the construction process [8]. Since this system can be integrated with the building element, as the name clearly suggests, these tools can thus be used in any external building element or facade. These can be easily merged with the design and do not affect the alethic value of the architecture of the building (Fig. **9**). The efficiency and required temperature of photovoltaic cells are enhanced due to the air movement behind these cells.

2.4. Solar Energy Air Conditioning

Any cooling system that uses solar energy for air conditioning in buildings is referred to as solar air conditioning. The most common ways to achieve air conditioning *via* solar energy are – solar thermal energy conversion, passive solar design, and photovoltaic conversion (*i.e.*, solar radiation to electricity).

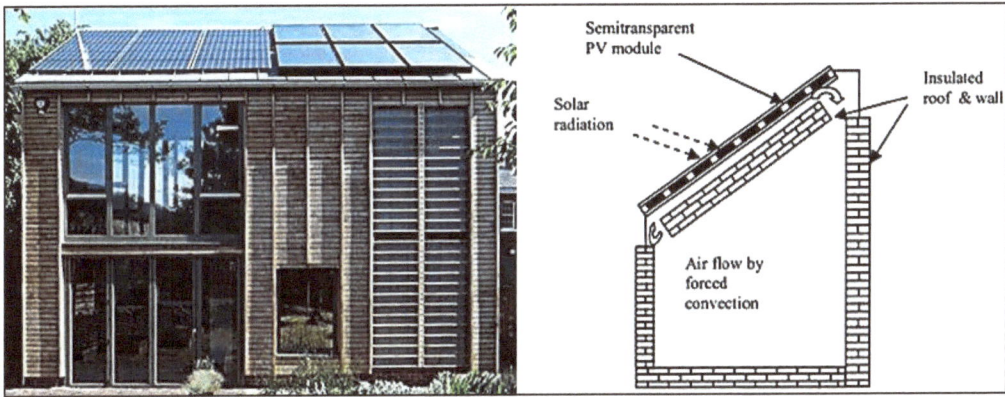

Fig. (9). Building Integrated Photovoltaic (BIPV) system integrated on the roof.

With the rising temperatures providing necessary indoor and outdoor thermal comfort through solar air conditioning in buildings is the most harmonious solution to curb the impact of the urban heat island effect. Another form of active solar conditioning, famously known as Green air conditioning, is also a popular choice, especially in the industrial sector, as no Freon refrigerants are required in the absorption chiller, which utilizes solar energy to protect the ozone layer depletion. The variety of solar air conditioning is described in detail as follows:

(a) Passive Solar Cooling

In passive solar cooling, solar energy is not directly used to provide a cooler thermal environment; instead, the buildings are designed in such a manner so as to reduce the rate of heat transfer from outside to inside to maintain indoor lower temperatures in summers. This method requires fundamental knowledge of heat transfer mechanisms, namely – heat transfer through conduction, convective heat transfer, and short wave thermal radiation from the sun.

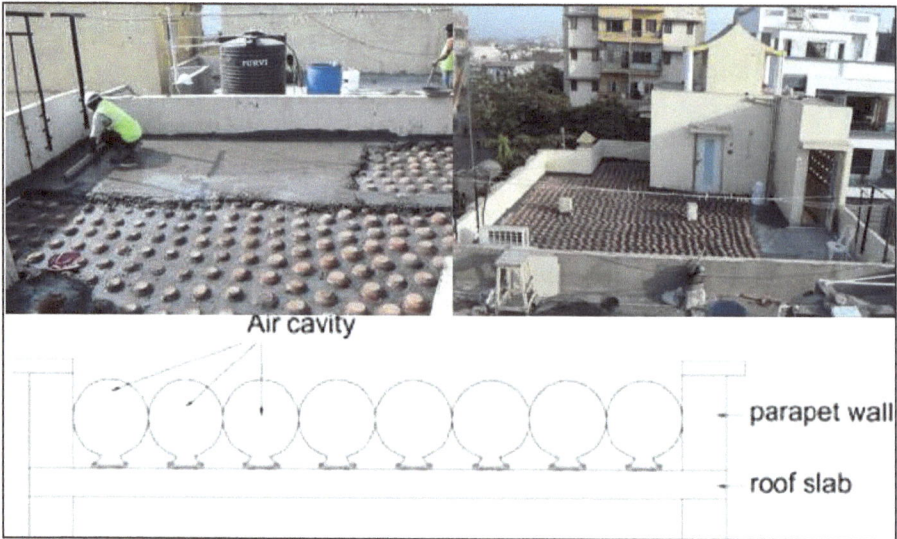

Fig. (10). Under construction insulation layer using inverted earthen pots over RCC slab [9].

Historically, in hot and dry regions of Iran, Afghanistan, India, and Pakistan, earthen pots were installed on roofs to provide a heat buffering system (Fig. **10**). Presently, green roofs are used to provide this thermal buffer between outside-inside temperatures. Downward radiation from the solar-heated roof surface of about 97% is blocked by a radiant barrier and air gap. Passive solar cooling depends upon site climate and topography and is largely observed in traditional architecture but can be easily incorporated in upcoming construction through innovative construction techniques and keeping in mind the solar path of the site.

(b) Photovoltaic (PV) Solar Cooling

Though the most prevalent application of photovoltaic is with compressors, they are known to provide the power for both traditional compressors based and absorption based electrical powered cooling.

The photovoltaic cooling system was not cost-effective until recently. Its economic value largely depends upon the effectiveness of cooling appliances, which have shown poor productivity through electrical cooling methods. However, the scenario is changing as longer payback schedules are allowed, and more efficient electrical cooling methods are used. For example, for total cooling output on a hot day, 7 kW of electric power is required for a 29 kW (100,000 BTU/h) U.S. Energy Star-rated air conditioner with a seasonal energy efficiency ratio (SEER) of 14. A setup like this would require over a 20 kW solar photovoltaic electricity generation system with storage. A smaller less-expensive photovoltaic system would be required for a more efficient air conditioning system. Indirect evaporative coolers can attain a SEER above 20 and also up to 40 in climates that are hot and dry with less than 45% relative humidity. An indirect evaporative cooler of 29 kW (100,000 BTU/h) would only need enough photovoltaic power for the circulation fan plus a water supply. The monthly amount of electricity purchased from the power grid for air conditioning can only be reduced but not completely eliminated using an economic partial-power photovoltaic system.

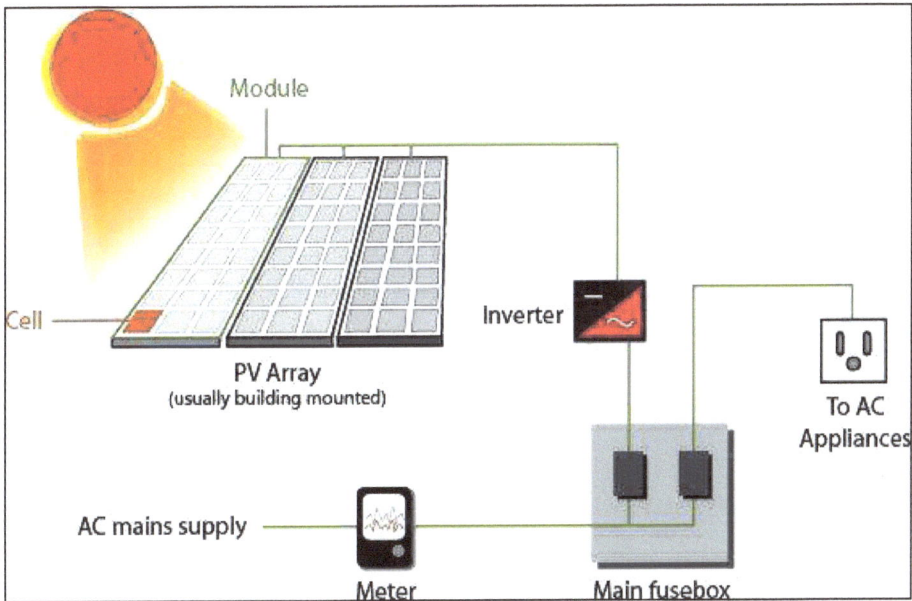

Fig. (11). Block diagram of the PV-AC system [10].

(c) Solar Open-loop Air Conditioning Using Desiccants

In this system, air can be streamed over solid desiccants; common examples are zeolite or silica gel or liquid desiccants – common examples are lithium chloride/bromide. This step allows extracting moisture from the air for an effective mechanical or evaporative cooling cycle. Next is the regeneration of the chosen desiccant. This step involves the use of solar thermal energy to dehumidify in a continuously repeating cycle and is a cost-effective, low-energy-consumption method. A low-energy air circulation fan and a motor to slowly rotate a large disk filled with desiccant are powered by a photovoltaic system.

While reducing the energy loss, energy recovery ventilation systems allow for a controlled way of ventilating a building. In the winter season, heat is transferred from interior warm air being exhausted to the fresh (but cold) supply air by passing air through an "enthalpy wheel" to minimize the ventilation heating cost.

Similarly, in the summer season, the interior air cools the warmer incoming supply air to minimize the cost of cooling the ventilation cooling air. Photovoltaic powers this low-energy fan-and-motor ventilation system in a cost-effective manner, with enhanced natural convection exhaust up a solar chimney - the downward incoming airflow would be forced convection (advection).

A recirculating waterfall is created for dehumidifying a room by mixing a desiccant with water where a photovoltaic-powered low-rate water pump circulates the liquid, and solar thermal energy is used to regenerate the liquid (Fig. **12**). From the many commercially available tools to pass air over a desiccant-infused medium for both the dehumidification and the regeneration cycle, the regeneration cycle can be powered using heat from solar energy. Desiccant regeneration can be greatly enhanced by pre-heating the air.

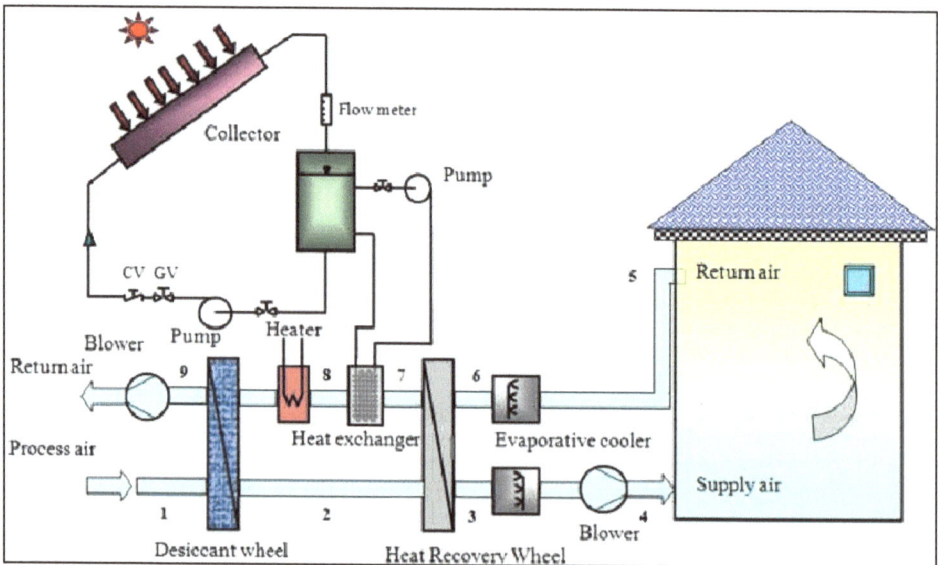

Fig. (12). Schematic diagram of solar open-loop air conditioning using desiccants [11].

(d) Solar Thermal Closed-loop Air Conditioning

The conventional technologies in practice for solar closed-loop absorption cooling are as follows:

- Adsorption: Water/Silica Gel or Water/Zeolite

- Absorption: Water/Lithium Chloride

- Adsorption: Methanol/Activated Carbon

- Absorption: Water/Lithium Bromide

- Absorption: NH_3/H_2O or Ammonia/Water

In an active solar cooling system, absorption chillers or thermally driven chillers receive solar energy from solar thermal collectors. The generator of an absorption chiller is heated from the fluid, which gets its heat from solar energy and which is then recirculated back to the collectors. A cooling cycle is driven by the heat of the generator resulting in the production of chilled water, which is used for industrial and commercial cooling applications (Fig. **13**). Different types of solar cooling systems are categorized by a number of frequentative absorption cooling cycles – single, double, or triple. The efficiency is directly proportionate to this number of cycles. Even though there is less vibration and noise in the case of absorption chillers, their capital price is high than that of compressor-based chillers.

Evacuated tube collectors or flat plate collectors prove to be more productive installations for this arrangement as general inexpensive flat plate collectors can only yield 71 °C water temperature as compared to a required nominal water temperature of 88 °C for enhanced efficiency of absorption chillers. Absorption chillers have been utilized for over 150 years to manufacture ice. Now many industries and commercial agencies have tested the practicality and installed solar thermal cooling systems. Masdar City in the United Arab Emirates is also testing a double-effect absorption cooling plant using a parabolic trough collector, Fresnel array, and high-vacuum solar thermal panels [12].

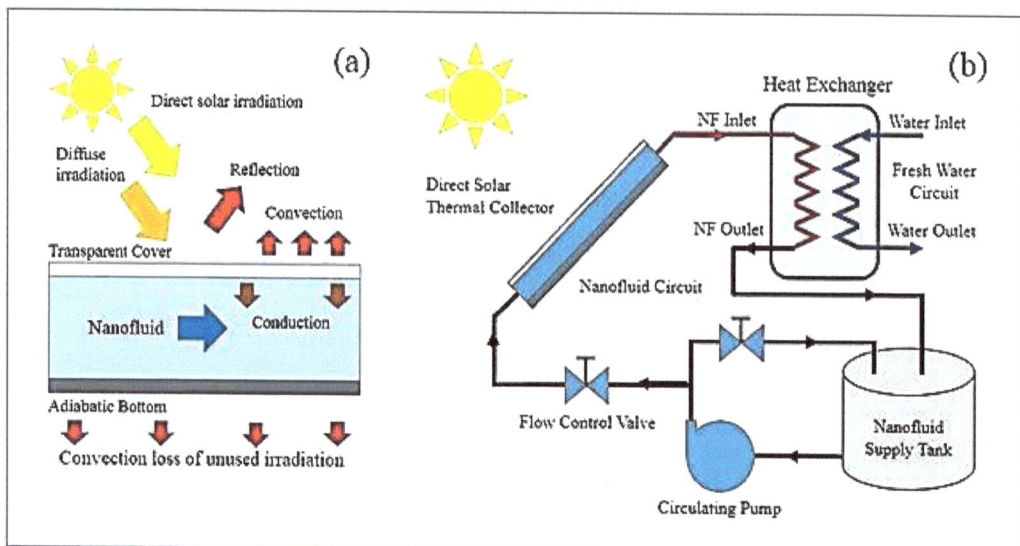

Fig. (13). (a) Schematic diagram of a representative collector showing sources of irradiation and major sources of heat loss **(b)** Schematic diagram showing heat transfer through a simplified closed-loop system from the nanofluid circuit to the water circuit *via* a heat exchanger [14].

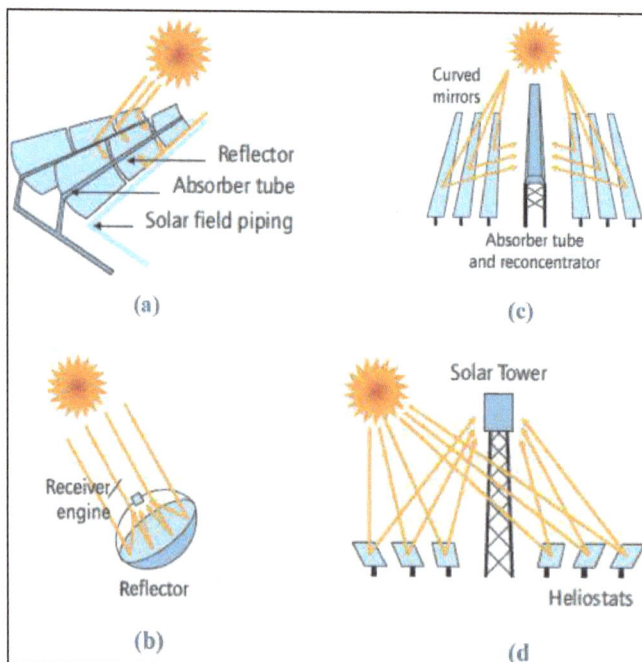

Fig. (14). Four configurations for CST technologies – **(a)** parabolic trough **(b)** linear Fresnel **(c)** disk/engine and **(d)** central receivers [15].

(e) Concentrating Solar Thermal (CST) Technology

High productive air conditioning systems coupled with double/triple effect chillers and solar refrigeration in combination with process heat and steam providing for industrial end-users is the primary reason for installing concentrating collectors in a solar cooling arrangement.

Diverse research in recent times has emphasized the high potential for refrigeration in different areas of the world in the context of industrial relevance. Although this can be accomplished by ammonia/ water absorption chillers, they require high-temperature heat input at the generator, in a range of 120°C – 180 °C. This can only be fulfilled by concentrating solar collectors. Additionally, both cooling and steam are required for a number of industrial processes, and concentrating solar thermal (CST) technology (Fig. **14**) can be very profitable for maximum use.

CONCLUSION

The application and developments of solar energy in buildings are presented in this chapter. With the fast-track economic expansion and budgetary housing policy enforcements, there has been experienced high growth in the building industry globally, and it is foreseen that the public building and urban housing areas will be further developed owing to the rapid global urbanization and industrialization in the future. On that account, it is estimated that the energy consumption rate from the building sector is rising every year in accordance with the total energy consumption trends. Besides, non-renewable energy sources are the main source consumed for building operations. The environment has suffered fatal impacts because of such incompetent energy utilization methods. The implementation and development of renewable energy sources in the building industry are primary affairs for global sustainable policy goals.

Solar energy is one such clean and infinite source and is found to be compatible with various applications in the building industry for different geographic locations. Many countries featuring hot and dry climatic conditions have inexhaustible solar energy, which is expansively utilized in the building sector. The major application includes solar heating and cooling of buildings, solar water heater, solar refrigeration, and photovoltaic power production. Through innovative technologies, countries with low temperatures have also explored and evolved their building systems and shown massive potential for solar energy utilization. These fundamental technologies for heating, cooling, and electricity from solar energy generation for buildings have been discussed.

In many countries, substandard urban policies pose an obstacle to the popularization of solar energy application and development in the building sector; even the governments have shown support for solar energy application in buildings in lieu of their membership with various international sustainable goals platforms. It is often realized that it is not the regressed solar energy technology or the dearth of environment positive intention but the scarcity of economic incentive schemes that have become the bottleneck of energy-efficient buildings. The advancement in the application and development of solar energy in buildings can be brought upon by establishing such economic incentive mechanisms as financial subsidies. Moreover, the steady and consistent technology standards and criteria can aid the endorsement of application and development in the building sector.

CONSENT FOR PUBLICATION

Not applicable.

CONFLICT OF INTEREST

The authors declare no conflict of interest, financial or otherwise.

ACKNOWLEDGEMENTS

Declared none.

REFERENCES

[1]   United Nations, Global status report 2017, Available from: Available from: https://www.worldgbc.org/news-media/global-status-report-2017.

[2]   R. Botpaev, Y. Louvet, B. Perers, S. Furbo, and K. Vajen, "Drainback solar thermal systems: A review", *Sol. Energy,* vol. 128, pp. 41-60, 2016.
      http://dx.doi.org/10.1016/j.solener.2015.10.050

[3]   P. Sharma, "Passive solar technique using trombe Wall - A sustainable approach", *IOSR J. Mech. Civ. Eng.,* vol. 1, no. 1, pp. 77-82, 2016.
      http://dx.doi.org/10.9790/1684-15010010177-82

[4]   G. Quesada, D. Rousse, Y. Dutil, M. Badache, and S. Hallé, "A comprehensive review of solar facades. Opaque solar facades", *Renew. Sustain. Energy Rev.,* vol. 16, no. 5, pp. 2820-2832, 2012.
      http://dx.doi.org/10.1016/j.rser.2012.01.078

[5]   Ladakh Ecological Development Group, Passive solar housing project in western Himalayas, Available from: http://www.ledeg.org/2019/09/17/passive-solar-housing-project-in-western-himalayas/

[6]   J. Jie, Y. Hua, H. Wei, P. Gang, L. Jianping, and J. Bin, "Modeling of a novel Trombe wall with PV cells", *Build. Environ.,* vol. 42, no. 3, pp. 1544-1552, 2007.
      http://dx.doi.org/10.1016/j.buildenv.2006.01.005

[7] C. Hlade, "Passive Solar Mountain Schools", *Smart Shelter Res,* 2017. Available from: https://www.smartshelterresearch.com/architecture/passive-solar-schools/

[8] E. Biyik, M. Araz, A. Hepbasli, M. Shahrestani, R. Yao, L. Shao, E. Essah, A.C. Oliveira, T. del Caño, E. Rico, J.L. Lechón, L. Andrade, A. Mendes, and Y.B. Atlı, "A key review of building integrated photovoltaic (BIPV) systems", *Eng. Sci. Technol. an Int. J.,* vol. 20, pp. 833-858, 2017.
http://dx.doi.org/10.1016/j.jestch.2017.01.009

[9] A. Mozumder, and A.K. Singh, "Solar heat flux reduction through roof using porous insulation layer", *Energy Procedia,* vol. 30, pp. 446-451, 2012.
http://dx.doi.org/10.1016/j.egypro.2012.11.053

[10] X. Li, and V. Strezov, "Energy and greenhouse gas emission assessment of conventional and solar assisted air conditioning systems", *Sustainability (Basel),* vol. 7, no. 11, pp. 14710-14728, 2015.
http://dx.doi.org/10.3390/su71114710

[11] A. Rachman, K. Sopian, S. Mat, and M. Yahya, "Feasibility study and performance analysis of solar assisted desiccant cooling technology in hot and humid climate", *Am. J. Environ. Sci.,* vol. 7, no. 3, pp. 207-211, 2011.
http://dx.doi.org/10.3844/ajessp.2011.207.211

[12] P. Abbate, "Masdar city testing TVP solar's high-vacuum flat solar thermal panels for air-conditioning swiss company to announce preliminary results of pilot project at WFES 2012", Available from: https://www.tvpsolar.com/files/press/TVP-Masdar SAC Test Press Release 16 Jan 2012.pdf

[13] G. Foley, R.C. DeVault, and R. Sweetser, "The Future of Absorption Technology in America", *Adv. Build. Syst.,* vol. 12, 2000.

[14] W. Chamsa-ard, S. Brundavanam, C. Fung, D. Fawcett, and G. Poinern, "Nanofluid types, their synthesis, properties and incorporation in direct solar thermal collectors", *Nanomaterials (Basel),* vol. 7, no. 6, p. 131, 2017.
http://dx.doi.org/10.3390/nano7060131 PMID: 28561802

[15] V. Drosou, P. Kosmopoulos, and A. Papadopoulos, "Solar cooling system using concentrating collectors for office buildings: A case study for Greece", *Renew. Energy,* vol. 97, pp. 697-708, 2016.
http://dx.doi.org/10.1016/j.renene.2016.06.027

The CO_2 Mitigation and Exergo and Environ-Economics Analysis of Bio-gas Integrated Semi-Transparent Photo-voltaic Thermal (Bi-iSPVT) System for Indian Composite Climate

Gopal Nath Tiwari[1], Praveen Kumar Srivastava[2]*, Akhoury Sudhir Kumar Sinha[3], Arvind Tiwari[4]

[1]*Department of Physical Sciences and Humanities, Rajiv Gandhi Institute of Petroleum Technology, Jais (UP), India*
[2]*Institute of Bio-sciences and Technology, Shri Ramswaroop Memorial University, Barabanki, (UP), India*
[3]*Rajiv Gandhi Institute of Petroleum Technology, Jais, Amethi (UP), India*
[4]*BERS Public School, Jawahar Nagar (Margupur), Chikhar-221701, Ballia (UP), India*

Abstract: It is to be noted that biogas production is drastically reduced in cold climatic conditions, especially in winter, due to a drop in ambient air temperature, which is much below an optimum temperature of about 37°C for fermentation of slurry. Many methods, such as hot charging, passive/active for slurry heating, have been tested, and it has been found that the passive heating method is neither practical nor self-sustained. In order to make bio-gas heating self-sustained, economical, and friendly to ecology and the environment, a new approach of Bi-iSPVT has been adopted. Based on the finding, we have made an attempt to analyze the system in terms of CO_2 mitigation, energy matrices, and environ- and exergo-economics to have a clean environment and sustainable climate. An analysis has been performed by using embodied energy, the annual overall thermal exergy of the system for ecological balance for the good health of human beings. It has been found that an energy payback time (EPBT) for a sustainable Bi-iSPVT system is about 1.67years, along with an exergo-economic parameter (*Rex*) of 0.1016 kWh/₹$0.1016\ kWh/₹$.

Keywords: Active heating, Bi-iSPVT, CO_2 mitigation, Energy matrices, Energy payback time (EPBT).

---

*Corresponding author Praveen Kumar Srivastava: Institute of Bio-sciences and Technology, Shri Ramswaroop Memorial University, Barabanki, (UP), India; E-mail: praveen2ku28@gmail.com

Manoj Kumar Gaur, Brian Norton & Gopal Tiwari (Eds.)
All rights reserved-© 2022 Bentham Science Publishers

1. INTRODUCTION

Energy conservation is the way to conserve fossil fuel using either efficient fossil-based tools or renewable energy sources. One of the renewable energy sources is the use of wasted organic products, including cattle manure in biogas plants, by fermentation at an optimum temperature of 37°C. The product from the biogas plant is an excellent renewable fuel providing gas for clean energy applications and can be used for cooking, lighting, and water heating along with farming by using its waste [1-3]. This will help in the reduction of the use of fossil fuels. Energy conservation has become an important topic of research due to the increase in CO_2 levels in the atmosphere to 415 ppm from 270 ppm after world war-II. The main reasons behind this are the fast growth of industrialization and population density. In order to have a clean environment, energy conservation plays an important role. Normally, climatic conditions in the world are classified into six zones, namely (i) cold and humid, (ii) composite, (iii) harsh cold and sunny,(iv) moderate, (v) warm and dry, and (vi) warm humid. It is hard to design any renewable energy system for composite climatic conditions. For such climatic conditions, one needs to know the number of heating and cooling day requirements in a year. During the cold period, ambient air temperature is significantly dropped much below the optimum temperature for fermentation of slurry in the digester of biogas plants [4-6].

The following methods have been adopted in the past for slurry heating in cold climatic conditions:

(a) Passive method: In this case, marginal heating of slurry can be achieved by using a greenhouse over dome [7-13], glazed dome [14], covering of dome at night by an insulating cover, SSP water heater [15].

(b) Active method: Flat plate collector [13, 16,17],Tubular collector [18]

(c) Self-sustained active method: PVT-CPC collector [19,20]

In this paper, we have made an attempt to analyze the CO_2 mitigation, energy matrices, and environ- and exergo-economics of bio-gas integrated photovoltaic thermal (Bi-ISPVT) systems to balance between ecology and environment. The Bi-iSPVT can also be referred to as the active heating of biogas plants. It is observed that the energy payback time (EBPT) of such a system is 1.67 years with an energy production factor of 14.92, which is most economical from an ecological and environmental point of view. An exergo-economic parameter (*Rex*) of Bi-iSPVT has been found to be $0.1016\ kWh/₹$.

2. DESIGN OF BI-ISPVT SYSTEM

Fig. (1) shows that the digester of the biogas plant is integrated with semi-transparent photo-voltaic thermal (SPVT) collectors through heat exchangers to increase the slurry temperature during the winter period to maintain bio-gas production. In the present case, we have only considered a series connection of SPVT [20, 21]. The outlet of the SPVT collector is connected to the lower end of coil type heat copper exchanger for faster heat transfer between the fluid of the heat exchanger and the slurry of the digester. Furthermore, the outlet of the heat exchanger is connected to the inlet of the SPVT collector. The SPVT collector provides both thermal and electrical energy with an efficiency of 45% and 10%, respectively. The electrical energy is used to pump off the water of a capacity of 0.5 to circulate the fluid between the SPVT collector and digester under the forced mode of operation (6 hours/ day). Solar radiation incident of SPVT collector is partially converted into electrical power and the remaining into thermal energy to be used for slurry heating. The average annual solar radiation has been considered 450 Wm^{-2} with sunshine hours of 6 hours. The capacity of digester and production of biogas plants is 1000 kg and 35 m^3 per day [22].

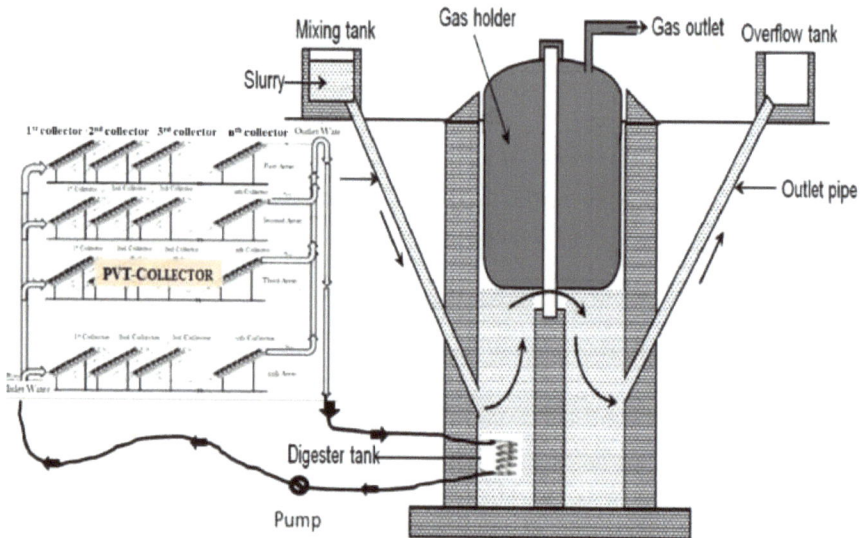

Fig. (1). Cross-sectional view of floating biogas integrated SPVT collectors connected in series and parallel.

3. FORMULATION OF BI-ISPVT MODELLING

In this section, basic formulas of CO_2 emission, mitigation, energy matrices, and exergo economic will be discussed before numerical computations [21].

3.1. CO_2 Emission

It is well known that there is 0.98 kg of CO_2 emissions at a coal-based thermal power plant for 1 kWh of electrical energy generation. Hence, if E_{in} is embodied energy (kWh) of biogas integrated semi-transparent photo-voltaic thermal (Bi-iSPVT), then the CO_2 emission per year by embodied energy at a power plant can be evaluated as follows:

$$CO_2 \ emission \ per \ year, Kg = \frac{E_{in}}{LifeTime} \times 0.98 \tag{1}$$

In India, losses due to the use of poor appliances and wiring and transmission and distribution are 20% and 40%, respectively, the factor of 0.98 increases to 2.04, and the CO_2 emission per year will be:

$$
\begin{aligned}
CO_2 \ emission \ per \ year, Kg &= \frac{E_{in}}{LifeTime} \times 0.98 \times \frac{1}{1-0.2} \times \frac{i}{1-0.4} \\
&= \frac{E_{in}}{LifeTime} \times 2.04
\end{aligned}
\tag{2}
$$

3.2. CO_2 Mitigation

The CO_2 mitigation of the GiSPVT system should be analyzed from the second law of thermodynamics due to the involvement of high-grade energy in embodied energy. If the system works for 300 days per year, then the $CO_2 \ mitigation \ per \ year$ to the environment can be evaluated as,

$$CO_2 \ mitigation \ per \ year, Kg = Ex_{annual,Bi-iSPVT,300} \times 2.04 \tag{3}$$

Furthermore, CO_2 mitigation in whole life will be expressed as,

$$CO_{2,mitigation,life}, Kg = Ex_{annual,Bi-iSPVT,300} \times 2.04 \times Life \ time \tag{4}$$

The net CO_2 mitigation over the lifetime of the GiSPVT system should also consider the effect of CO_2 emission by embodied energy (E_{in}) which has not been considered by earlier researchers.

By considering CO_2 emission by embodied energy (E_{in}), the net CO_2 mitigation over the lifetime of the Bi-iSPVT system in tons (ψ_{co2}) will be as follows:

$$\psi_{co2}(tones) = \left(Ex_{annual,Bi-iSPVT,300} \times Total\ life - E_{in}\right) \times 2.04 \times 10^{-3} \qquad \textbf{(5)}$$

Where, $Ex_{annual,Bi-iSPVT,300}$ is the overall exergy gain for 300 biogas working days, which is the sum of the thermal exergy, electrical energy of SPVT, and exergy of biogas produced from the system.

3.3. Energy Matrices

Energy matrices of any system determine the feasibility of the success of technology [24]. For this, the following parameters have been computed for 300 clear days using Eq. 8b for climatic conditions in New Delhi.

3.3.1. Energy Payback Time (EPBT)

This determines the period to recover invested energy to develop an active biogas plant. It should be as minimum as possible, and it is evaluated from the following equation.

$$EPBT = \frac{E_{in}}{Annual\ energy} \qquad \textbf{(6a)}$$

Here, annual energy can be either thermal or exergy. EPBT should be as minimum as possible in comparison with the life of an active biogas plant.

3.3.2. Energy Production Factor (EPF)

It is the factor that determines how many times energy is obtained from any system in comparison with embodied energy, and it can be expressed as:

$$EPF = \frac{annual\ energy \times lifetime}{E_{in}} > 1 \qquad \textbf{(6b)}$$

The EPF should be as maximum as possible.

3.3.3. Life Cycle Conversion Efficiency (LCCE)

It is the ratio of effective energy obtained and available solar energy in the lifetime of the system, and it is defined as:

$LCCE_{exergy}$ *in fraction*

$$= \frac{Annualenergy \times Lifeofsystem - E_{in}}{Annualsolarradiation \times areaofcollector \times Lifeofsystem} \qquad (6c)$$

The life cycle conversion efficiency is generally less than one, but if it is more than one, then the system is most economical from an energy point of view.

3.4. Enviro-Economic Analysis

Environ-economic analysis of any renewable energy system is one of the ways to control the air pollutants present in the global environment to minimize the emissions of air pollutants by developing and under-developing Nations by giving them economic incentives. This encourages developing and under-developing Nations to use renewable energy technologies at a faster rate. This reduces the emission of carbon (CO_2) to the environment. The analysis is based on environ-economic parameters by considering the price of CO_2 emission [23, 24].

If C_{co2} represents the rate of emission reduction in tones of CO_2 (tCO_2) and ψ_{co2} is a reduction in CO_2 emission in tons for the whole life of the renewable energy system, then the cost of an environ-economic parameter (C_{co2}) can be obtained by using the following expression:

$$C_{co2} = \psi_{co2} \times c_{co2} \quad \text{(unit cost)} \qquad (7a)$$

In order to find out the cost of environ-economic parameters (C_{co2}) Eq.7, it is necessary to evaluate CO_2 emission due to embodied energy and mitigation due to the use of solar energy systems. In the present case, we have to evaluate for Bi-iSPVT (active biogas plant). The expression for ψ_{co2} for Bi-iSPVT is given by Eq. 5.

If the carbon credit rate (C_{co2}) =$10/tone , then total carbon credit earned C_{co2} by Bi-iSPVT in T years can be obtained by using Eqs. 16 and 17 as,

$$C_{co2,T} (\$) = \psi_{co2} \times 10 \qquad (7b)$$

Yearly carbon credit earned per year, $R_{C_{co2}}$ for the lifetime of T years, Eqs. 18 are as follows:

$$R_{C_{co2,25}} = \frac{C_{co2,T}}{T} \tag{7c}$$

This yearly earned carbon credit should be considered as a gain in exergo-economic analysis, which has not been considered earlier by anyone.

3.5. Exergo-Economic Analysis

In this case, the concept of the second law of thermodynamics in combination with cost analysis is considered. Many authors have presented exergo-economic analysis as exergy loss per unit cost with the objective of minimizing the exergy loss.

Following Tiwari *et al.* [24], the exergo-economic parameter (*Rex*) can be expressed as,

$$R_{ex} = \frac{L_{ex,annual,eff}}{(UAC)_{Bi-iSPVT}} = \frac{Annual\,exergy\,loss}{(UAC)_{Bi-iSPVT}} \tag{8}$$

Where, *Lex,annual* represents exergo-economic parameters based on exergy loss.

By using the concept of exergy based on the second law of thermodynamics, one has the following expression for annual exergy loss ($L_{ex,annual,exergy}$),

$$L_{ex,annual,eff} = Ex_{in,eff} - Ex_{net,out,exergy\,SPVT} = Exergy\,loss \tag{9a}$$

Where, $Ex_{net,out,exergy\,SPVT}$ is a total sum of annual thermal exergy and electrical energy of SPVT collectors, and

$$Ex_{in,eff} = Ex_{in,solar} + E_{in,annual} \tag{9b}$$

Till now, no one has considered the effect of annual embodied energy, $E_{in,annual}$ in exergo-economic analysis.

It is to be noted that an annual exergy loss, $L_{ex,annual,eff}$, should be as minimum as possible. The minimization of this can be achieved

i) either increasing $Ex_{net,out,energy\,SPVT}$ by considering only thermal energy or
ii) reducing Ex_{in} without considering embodied energy as done by previous researchers.

Since the present study is based on exergy, exergy of SPVT collectors will be considered.

In Eq, 9b, the annual exergy of solar radiation [24] can be evaluated by:

$$Ex_{in,solar} = I_{annual} \times \left[1 - \frac{4T_a}{3T_s} - \frac{T_a^4}{3T_s^4}\right] \tag{9c}$$
$$= I_{annual} \times 0.95$$

Where, I_{annual} is an average annual solar radiation incident on SPVT collectors. T_a and T_s are the annual average ambient air and sun's temperature in K, and the numerical value in the bracket is approximately 0.95.

The uniform annual cost of Bi-iSPVT, $(UAC)_{Bi-iSPVT}$ in Eq.8 represents uniform end-of-year annual cost [24], which can be obtained by using the following expression:

$$\begin{aligned}(UAC)_{Bi-iSPVT} \\ = (Net\ Present\ Value, NPV) \\ \times (Capital\ recovery\ factor)\end{aligned} \tag{10}$$

Where, $\quad CRF = \left[\frac{i(1+i)^n}{(1+i)^n - 1}\right]$

The net present value (NPV) consists of initial investment, P_i ; operation and maintenance cost, $R_{O\&M}$; annual benefits from Bi-iSPVT, yearly earned value by annual exergy, R_{exergy}; carbon credit, $R_{C_{co2}}$, Eq. 7a; salvage value, (S) *etc*. The cash flow diagram of the GiSPVT system is shown in Fig. (**2**) [19, 20].

Following (Fig. **2**), an expression for net present value can be expressed as:

$$\begin{aligned}NPV = -P_i + [R_{exergy} + R_{C_{co2}} - R_{O\&M}] \times UAC present value + S \\ \times Present\ value\ factor]\end{aligned} \tag{11}$$

Where, $\quad\quad\quad\begin{aligned}UAC\ present\ value\ factor \\ = \left[\frac{(1+i)^j - 1}{i(1+i)^j}\right]\end{aligned}$ and $\begin{aligned}present\ value\ factor \\ = (1+i)^{-n}\end{aligned}$

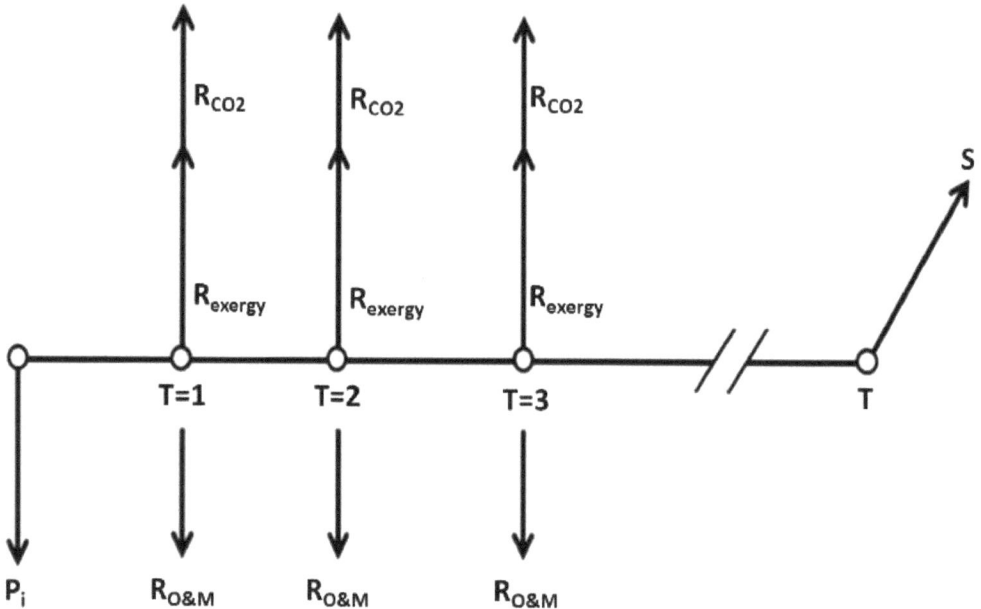

Fig. (2). Cash flow diagram of Bi-iSPVT (RCO$_2$, Yearly earned carbon credit; Rexergy, Yearly income from biogas produced; Pi, Present (Initial) Investment; RO & M, Yearly operation and maintenance; S, Salvage value).

4. NUMERICAL RESULTS AND DISCUSSION FOR BI-ISPVT SYSTEM

4.1. Embodied Energy

The embodied energy, along with the energy density of each component of the PV module, PVT collector, and biogas plant, is given in Tables **1-3**.

Table 1. Embodied energy for 2m^2 PV module in different processes (Tiwari and Mishra, 2012).

Process	Energy Requirement
Silicon purification and processing	
(a) Metallurgical grade silicon (MG-Si) production from silicon dioxide (quartz, sand)	20 kWh per kg of MG-Si
(b) Electronic grade silicon (EG-Si) production from MG-Si	100 kWh per kg of EG-Si
(c) Czochralski Silicon (Cz-Si) production from EG-Si	290 kWh per kg of EG-Si
Solar cell fabrication	120 kWh per m^2 of Silicon cell

(Table 1) cont.....

PV Module assembly	190 kWh per m² of PV module
Rooftop integrated PV system	200 kWh per m² of PV module
Total embodied energy per m²	920 kWh
Embodied energy for 2m² without BOS	1840 kWh

So, the total embodied energy of the active biogas–integrated semi-transparent photovoltaic thermal (Bi-iSPVT) plant can be obtained from Tables **2** and **3**, and it is given as:

E_{in} = 4019.8× 12 +11,212.37= 59,449.97 kWh.

Now, by using Eq. 1, an embodied energy per year for 25 years of a lifetime will be calculated as:

$$E_{in,25} = \frac{59449.97}{25} = 2377.99 \ kWh/year$$

From the above, one can see that an embodied energy per year will decrease with an increase in the life of Bi-iSPVT with proper operating and maintenance.

4.2. Annual Solar Radiation

It is also known that the average number of clear days and solar radiation varies from place to place, between 250 to 300 days and 350-450, respectively. In our present study, we will consider the number of clear days and solar radiation as 300 and 450 W/m².

$$Annual \ Solar \ Radiation \ per \ m^2 \ (I_{annual}) = \frac{450 \times 6 \times N}{1000} \ kWh/m^2$$

In the above equation, six-hour has been considered as a yearly average sunshine hour.

Annual solar radiation incident on 12 SPVT collectors (I_{SPVT}) with each one of 2m² for 300 days will be obtained by using the above equation as,

$$I_{SPVT,300} = \frac{450 \times 6 \times N}{1000} \times 12 \times 2 = 19440 \frac{kWh}{year}, \quad For \ N = 300 \ clear \ days$$

Table 2. Embodied energy for one PVT collector with an effective area of 2 m²

	Items	Quantity	Total Weight	Energy Density	Embodied Energy (MJ)	Embodied Energy (kWh)
Flat Plate Collector Quantity 2	Copper riser ½"	20×1.8 = 36m	8.2	81.0	664.2	184.5
	Header 1"	4×1.15 = 4.6m	3.8	81.0	307.8	85.5
	Al box	2	10	199.0	1990.0	552.0
	Cu sheet	2	11	132.7	1460	405.6
	PV module of 2m² including BOS	1	-	3612/ m²	7224	1972
	Glass wool	13 m²	0.064 m³	139 MJ m⁻³	8.89	2.5
	Nuts / bolts/screws	32	1	31.06	31.06	8.6
	Union /elbow	8	1.5	46.8	70.2	19.5
	Nozzle/flange	8	1	62.1	62.1	17.3
	Mild steel stand	1	40	34.2	1368	380
	Paint	1L	1L	90.4	90.4	25.1
	Rubber gasket	18m	4.2	11.83	49.7	13.8
	G I pipes ½"		9.5	44.1	418.9	116.4
	Al frame 1"	12m	2.5	170	425	118
	Al sheet 24 gauge	-	2.5	170	425	118
	Sub Total					**4019.8**

4.3. Annual Thermal Exergy

Based on the literature, it has been observed that the thermal exergy efficiency of any solar thermal collector is much less than thermal efficiency [19]. Its average value varies between 8 to 12% depending upon operating temperature. SPVT collectors are connected in series and can operate at high temperatures; hence, 12% thermal exergy efficiency of 12-SPVT collectors has been considered [25-27]. In this case, annual thermal exergy for 300 for annual solar radiation is calculated as:

$$Q_{th,exergy,300} = 19440 \times 0.12 = 2332.8 \ kWh$$

4.4. Annual Electrical Energy and Total Exergy of SPVT Collectors

The electrical efficiency of semi-transparent C-Si-based PV modules under standard conditions (I=1000W/m² and surrounding temperature of 25°C) has been obtained more than 20% in open conditions. If it is integrated with a conventional flat collector and tested in outdoor conditions, it varies between 8-10%. Hence, annual electrical energy, $E_{el,annual}$, with 10% C-Si based PVT collector efficiency for 300 days will be calculated as follows:

$$E_{el,annual,300} = I_{SPVT,300} \times 0.1 = 19440 \times 0.10 = 1944 \ kWh$$

Therefore, the total annual exergy of 12-SPVT collectors will be the sum of annual thermal exergy (section 4.3) and above obtained annual electrical energy which can be obtained as:

$$\begin{aligned} \text{Total annual exergy} \quad Ex_{SPVT} &= 2332.8 \ kWh + 1944 \ kWh \\ \text{of SPVT,} \qquad\qquad &= 4276.8 \ kWh \end{aligned} \tag{12a}$$

Table 3. Embodied energy of 1000 kg biogas plant (1 MJ = 0.277778 kWh).

Components	Items	Quantity	Total Weight (kg)	Energy Density (MJ/kg)	Total Embodied Energy	
					MJ	kWh
Bio-gas plant	GRP body digester	1	300	92.3	27690	7691.73
	GRP body dome	1	110	92.3	10153	2820.3
	Compressor a.MS steel b.Copper	1	30 2	34.2 110.19	1026 220.38	285 61.23
	PVC pipes	8	50	11.83	221.8	61.61
	Chullah	1	5	34.2	171	47.5
	Gyser	1	20	44.1	882	245
	Total	-	-	-	-	**11,212.37**

4.5. Annual Energy from the Biogas Plant

The capacity of the proposed Bi-iSPVT biogas plant is 1000 kg. It produces 35 m³ biogas per day if the optimum temperature is maintained at 37°C throughout the year. There is leakage of biogas during burning in transportation/distribution and compressor. This leakage depends upon the proper piping and quality of the compressor. It varies between 10 to 50%. For the present analysis, we have considered these losses as 50%. Then LPG saved per year from 1000 kg capacity of digester of biogas plant will be,

$$Ex_{bio,annual,300} = 35 \times 0.45 \, Kg \, \times (1 - 0.50) \, \times \, 300 \times 13.6 \, = 32{,}130 \, kWh \, with \, loss$$

and, $Ex_{bio,annual,300}$
$$= 35 \times 0.45 \, Kg \, \times \, 300 \qquad\qquad \textbf{(12b)}$$
$$\times \, 13.6 = 64{,}260 \, kWh \, without \, loss$$

Here, 1 m³ of biogas = 0.45 kg of LPG and 1 kg = 13.6 kWh

Here, it is important to mention that $Ex_{bio,annual,300}$ with losses and without losses has been used in evaluating energy matrices and exergo-economic analysis. If losses are reduced, then the biogas system will be more economical [28,29]. The above biogas is used for cooking, lighting, and water heating and hence it can be considered as high-grade energy (exergy) for our present analysis.

4.6. An Overall Exergy from Bi-iSPVT

If a 0.5 hp water pump is used for 6 hours to circulate hot water from the 12-SPVT collector and digester through the copper heat exchanger to heat the slurry at an optimum value of 37°C, then the total annual electrical energy consumed by the pump will be approximately 850.45 kWh. So, in this case, effective electrical energy, $E_{annual,,eff}$ available for Bi-iSPVT for other applications for 300 will be calculated as:

$$E_{el,annual,300,eff} = 1944 - 850.45 = 1093.55 \, kWh$$

Now, the net total annual exergy of SPVT (Eq. 12) will be reduced and become:

Net total exergy of $Ex_{net,SPVT} = 2332.8 \, kWh + 1093.55 \, kWh$
SPVT, $= 3429.35 \, kWh$ **(12c)**

From Eqs. 12b and 12c, one can get net total annual exergy from Bi-iSPVT as,

$$Ex_{annual,Bi-iSPVT,300} = 32,130 + 3429.35 = 35559.35 kWh, \quad with \ 50\% \ loss$$

and,
$$\begin{aligned} Ex_{annual,Bi-iSPVT,300} &= 64,260 + 3429.35 \\ &= 67,689.35 \ kWh \quad (Without \ loss) \end{aligned} \tag{13a}$$

The above numerical values with loss will be used to evaluate energy matrices in the next section.

If the rate of electrical power is ₹5per kWh, then annual income from Bi-iSPVT based on exergy without a loss will be,

$$R_{exergy} = 67,689.35 \ kWh \times ₹5per \ kWh = ₹338,446.75 \tag{13b}$$

Numerical value as annual income will be used for exergo-economic analysis.

4.7. Energy Matrices of Bi-iSPVT System

Here, it is important to mention that energy matrices are the most important parameters for deciding the sustainability of any renewable system, as mentioned earlier in Eqs. 6. In this case, we will consider the embodied energy (section 4.1) and overall exergy of Bi-iSPVT (Eq. 13a) to evaluate energy matrices for the lifetime of 25 years. From section 4.1 and Eq. 13, we have:s

Embodied energy, E_{in} = 59,449.97 kWh and $Ex_{annual,Bi-iSPVT,300}$ = 35559.35 kWh

So, the exergy base matrices can be obtained by using Eqs. 6 as follows:

Energy payback time , $\left(EPBT_{exergy}\right) = \dfrac{59,449.97}{35559.35} = 1.67 \ years$

Energy production factor, $\quad \left(EPF_{exergy}\right) = \dfrac{35559.35 \times 25}{59,449.97} = 14.92$

Life cycle conversion efficiency, $\quad \left(LCCE_{exergy}\right) = \dfrac{35559.35 \times 25 - 59,449.97}{19440 \times 25} = 1.70$

From the above finding, one can conclude that the proposed GiSPVT is economical and sustainable from an exergy (high-grade energy, second law of thermodynamics) point of view, as mentioned above. One can also infer that there are 1.14 years energy payback time (EPBT) and energy production factor (EPF) of 14.62, and life cycle conversion efficiency (LCCE) of 1.67 with 50% biogas losses based on overall exergy of Bi-iSPVT system, which is the best results. If these biogas losses are reduced, then EPBT will be further reduced; the value of LCCE is more than one due to more value of coefficient performance (COP) of the biogas system.

4.8. Environmental Analysis

For detailed environmental analysis, renewable energy systems should include CO_2 emission, mitigation and economics which will be discussed.

4.8.1. CO_2 Emissions

The CO_2 emission mainly depends on embodied energy involved in any renewable energy system. For low embodied energy, there is low CO_2 emission. In the present case, the embodied energy of the Bi-iSPVT system (E_{in}) is 91512 kWh (section 4.1)[30]; the life of the system depends upon the material used for the fabrication of the system. In this case, it has been considered as 25 years.

The CO_2 emission over 25 years of the lifetime of the Bi-iSPVT system by using Eq. 2 can be obtained as:

$$CO_{2,emmison,life} = E_{in} \times 2.04\ kg = 59449.97 \times 2.04 = 121{,}278\ kg = 121.278\ tonnes$$

4.8.2. CO_2 Mitigation

By using Eq. 5 and data of embodied energy as 59,449.97 kWh (section 4.1) and annual exergy of Bi-iSPVT (35559.35 kWh), the CO_2 mitigation of the GiSPVT system should be analyzed from the second law of thermodynamics due to the involvement of high-grade energy in embodied energy of Bi-iSPVT, Tables **1-3**.

The net CO_2 mitigation over the lifetime of 25 years from Bi-iSPVT will be as follows:

$$\psi_{co2,25} = (35559.35 \times 25 - 59{,}449.97) \times 2.04 \times 10^{-3}\ tonnes = 1692.25\ tonnes$$

If the life of Bi-iSPVT is increased due to proper operating and maintenance, then CO_2 mitigation is also increased.

4.8.3. Enviro-economic Analysis

Earlier, it was decided through the Tokyo protocol by the international community to give intensive attention to users of renewable energy in developing and under-developing countries by saving fossil fuel energy at certain rates varying from 25 to 10 dollars per tonnes of CO_2 mitigation, known as carbon credit to sustain a clean environment and climate.

If the carbon credit rate $C_{co2} = \$10/tonne$, then total carbon credit earned C_{co2} by Bi-iSPVT in 25 years can be obtained by using the numerical values of section 4.8.2 as,

$$C_{co2,25} = 1,692.25 \text{tones} \times \$10/tone = \$ 16922.59$$

Yearly carbon credit earned per year, $R_{C_{co2}}$ for the lifetime of 25 years is as follows:

$$R_{C_{co2,25}} = \frac{\$16564.06}{25} = \$ 662.56 = 662.56 \times 73.32 \qquad (14)$$
$$= ₹48,579.06$$

Here, 1\$ = ₹73.32.

This yearly earned carbon credit should be considered as a gain in exergo-economic analysis, which has not been considered earlier by anyone. This will be also be reduced with the increase in the life of the system

4.8.4. Exergo-economic Analysis

The following assumptions have been made for exergo-economic analysis:

i) The Bi-iSPVT will uniformly perform over the life of the system (T=25 years).

ii) The operation and maintenance of the system are about 10% of the initial investment (P$_i$).

iii) Salvage value after the end of life is 20% of the initial investment (P$_i$).

iv) There is not much change in the average value of climatic conditions and

v) 4% interest has been considered.

Refer to sections 3.5, 4.2, and Eq. 9; one has the following:

$$Ex_{in,solar} = I_{annual} \times 0.95 = 19440\frac{kWh}{year} \times 0.95 = 18468 \, kWh/year$$

With annual embodied energy (section 4.1) of $2377.99 \, kWh$ /year, one gets

$$Ex_{in,eff} = Ex_{in,solar} + E_{in,annual} = 18489 + 2377.99 = 20,866.99 \, kWh/year$$

Hence, an annual exergy loss, $L_{ex,annual,exergy}$ can be obtained with the help of the above value and exergy of SPVT collectors, Eq.12a as:

$$L_{ex,annual,exergy} = 20,866.99 - 4276.8 = 16,590.19 \, kWh \qquad (15)$$

Till now, no one has considered the effect of annual embodied energy, $E_{in,annual}$ in exergo-economic analysis. It is to be noted that an annual exergy loss, $L_{ex,annual,eff}$, should be as minimum as possible. The minimization of this can be achieved.

From Eq. 11b and with the help of Eq. 13b and 14, the net present value will be obtained as:

$$\begin{aligned} NPV = &-20,00,000.00 \\ &+ [338,446.75 + 48,579.06 - 1,00,000.00] \times 15.622 \\ &- [2,00,000.00 \times 0.3751] \end{aligned}$$

$$= -20,00,000.00 + 4,483,917.20 + 75020 = ₹2,558,937.20$$

Here,

$$(UAC)present \, value \, factor = \left[\frac{(1+i)^j - 1}{i(1+i)^j}\right] = \frac{2.6658 - 1}{0.04 \times 2.6658} = 15.622, \qquad and$$

$$present \, value \, factor = (1+i)^{-25} = 0.3751 \, for \, T_{life} = 25 \, years,$$

It is to be noted that from NPV that R_{exergy}, the uniform annual income (first term) plays an important role in comparison with R_{CO_2}, carbon credit (second term). Furthermore, carbon credit has a significant impact on the long life of Bi-iSPVT, as can be seen in section 4.8.2, due to an increase in its value. It can be further increased by reducing embodied energy.

The uniform annual cost $(UAC)_{Bi-iSPVT}$ in Eq.10 represents uniform end-of-year annual cost, which can be expressed as follows:

$$(UAC)_{Bi-iSPVT} = 2,558,937.20 \times 0.0638 = ₹163,260.19 \tag{16}$$

Where, $$CRF = \left[\frac{i(1+i)^n}{(1+i)^n - 1}\right] = \frac{0.04(1.04)^{25}}{(1.04)^{25} - 1} = \frac{0.04 \times 2.6658}{2.6658 - 1} = 0.0638$$

From Eqs. 15 and 16, an exergo-economic parameter (Rex) can be obtained as,

$$R_{ex} = \frac{L_{ex,annual}}{(UAC)_{Bi-iSPVT}} = \frac{16,590.19}{₹163,260.19} = 0.1016 kW \tag{17}$$

From the above equation, one can see that the numerical value of R_{ex} should be as maximum as possible to make the system more economical from an exergy point of view. This can be obtained either by:

(i) Reducing exergy loss/ destruction or

(ii) Increasing in uniform annual cost, $(UAC)_{Bi-iSPVT}$

In order to minimize exergy/destruction, one has to increase the system's performance and decrease embodied energy. On the other hand, to increase the uniform annual cost, $(UAC)_{Bi-iSPVT}$, one has to increase carbon credit, R_{exergy}, R_{CO_2} and a salvage value of the system in the calculation of net present value, as shown in Eq.11.

CONCLUSION AND RECOMMENDATIONS

Based on the present studies, the following conclusion and recommendations have been made:

i) The materials with low energy density (Tables **1-3**) should be used for the fabrication of the Bi-iSPVT system to have minimum embodied energy and exergy loss.

ii) The proposed system will be more economical and sustainable for cold and clear climatic conditions. It should be used as a community kitchen for a large number of people in a remote area (section 4.2).

iii) With decreased embodied energy, CO_2 mitigation will be increased (section 4.8.2).

iv) With the decrease in exergy loss, Eq.17, the Bi-iSPVT system will be more economical. In this case, the value of R_{ex} will decrease. This means the rate of exergy gain will decrease.

v) The government should provide a loan with a lower interest rate to encourage the rural population in under-developing and developing countries.

CONSENT FOR PUBLICATION

Not applicable.

CONFLICT OF INTEREST

The authors declare no conflict of interest, financial or otherwise.

ACKNOWLEDGEMENTS

The authors would like to acknowledge the support of the following departments: Department of Physical Sciences and Humanities, RGIPT, Jais, Amethi (UP), India, Institute of Bio-sciences and Technology, Shri Ramswaroop Memorial University, Barabanki, Uttar Pradesh-225003, India and BERS Public School, Jawahar Nagar (Margupur), Chikhar-221701, Ballia (UP), India.

REFERENCES

[1]　B. Jha, R.M. Kapoor, V. Vijay, V.K. Vijay, and R. Chandra, "Biogas: A sustainable and potential fuel for transport application", *J. Biofuels and Bioener.,* vol. 1, no. 1, pp. 28-33, 2015.

　　　http://dx.doi.org/10.5958/2454-8618.2015.00004.8

[2]　V. Vijay, P.M.V. Subbarao, and R. Chandra, "Study on interchange ability of seed based feed stocks for biogas production in multi-feeding mode", *Cleaner Eng. Technol,* pp. 100-11, 2020.

[3]　R. Kapoor, P. Ghosh, B. Tyagi, V.K. Vijay, V. Vijay, I.S. Thakur, H. Kamyab, D.D. Nguyen, and A. Kumar, "Advances in biogas valorization and utilization systems: A comprehensive review", *J. Clean. Prod.,* vol. 273, p. 123052, 2020.

　　　http://dx.doi.org/10.1016/j.jclepro.2020.123052

[4]　G.N. Tiwari, and A. Chandra, "A solar-assisted biogas system: A new approach", *Energy Convers. Manage.,* vol. 26, no. 2, pp. 147-150, 1986.

　　　http://dx.doi.org/10.1016/0196-8904(86)90047-6

[5] G.N. Tiwari, S.K. Singh, and K. Thakur, "Design criteria for an active biogas plant", *Energy,* vol. 17, no. 10, pp. 955-958, 1992.

http://dx.doi.org/10.1016/0360-5442(92)90044-Z

[6] G.N. Tiwari, J.A. Usmani, and A. Chandra, "Determination of period for biogas production", *Energy Convers. Manage.,* vol. 37, no. 2, pp. 199-203, 1996.

http://dx.doi.org/10.1016/0196-8904(95)00167-C

[7] M.S. Sodha, S. Ram, N.K. Bansal, and P.K. Bansal, "Effect of PVC greenhouse in increasing the biogas production in temperate cold climatic conditions", *Energy Convers. Manage.,* vol. 27, no. 1, pp. 83-90, 1987.

http://dx.doi.org/10.1016/0196-8904(87)90057-4

[8] M.S. Sodha, A. Kumar, and J. Kishor, "Solar assisted biogas plant IIA: Experimental validation of the mathematical models for floating drum type biogas plant", *Int. J. Energy Res.,* vol. 12, no. 2, pp. 369-375, 1988.

http://dx.doi.org/10.1002/er.4440120216

[9] J.A. Usmani, G.N. Tiwari, and A. Chandra, "Performance characteristic of a greenhouse integrated biogas system", *Energy Convers. Manage.,* vol. 37, no. 9, pp. 1423-1433, 1996.

http://dx.doi.org/10.1016/0196-8904(95)00228-6

[10] K. Vinoth Kumar, and R. Kasturi Bai, "Solar greenhouse assisted biogas plant in hilly region – A field study", *Sol. Energy,* vol. 82, no. 10, pp. 911-917, 2008.

http://dx.doi.org/10.1016/j.solener.2008.03.005

[11] A.A.M. Hassanein, L. Qiu, P. Junting, G. Yihong, F. Witarsa, and A.A. Hassanain, "Simulation and validation of a model for heating underground biogas digesters by solar energy", *Ecol. Eng.,* vol. 82, pp. 336-344, 2015.

http://dx.doi.org/10.1016/j.ecoleng.2015.05.010

[12] P. Piché, D. Haillot, S. Gibout, C. Arrabie, M.A. Lamontagne, V. Gilbert, and J.P. Bédécarrats, "Design, construction and analysis of a thermal energy storage system adapted to greenhouse cultivation in isolated northern communities", *Sol. Energy,* vol. 204, pp. 90-105, 2020.

http://dx.doi.org/10.1016/j.solener.2020.04.008

[13] H.M. Mahmudul, M.G. Rasul, D. Akbar, R. Narayanan, and M. Mofijur, "A comprehensive review of the recent development and challenges of a solar-assisted biodigester system", *Sci. Total Environ.,* vol. 753, p. 141920, 2021.

http://dx.doi.org/10.1016/j.scitotenv.2020.141920 PMID: 32889316

[14] M.S. Sodha, I.C. Goyal, J. Kishor, B.C. Jayashankar, and M. Dayal, "Solar assisted biogas plants IV A: Experimental validation of a numerical model for slurry temperature in a glazed fixed-dome biogas plant", *Int. J. Energy Res.,* vol. 13, no. 5, pp. 621-625, 1989.

http://dx.doi.org/10.1002/er.4440130513

[15] R. Feng, J. Li, T. Dong, and X. Li, "Performance of a novel household solar heating thermostatic biogas system", *Appl. Therm. Eng.,* vol. 96, pp. 519-526, 2016.

http://dx.doi.org/10.1016/j.applthermaleng.2015.12.003

[16] G.N. Tiwari, A. Chandra, K.K. Singh, S. Sucheta, and Y.P. Yadav, "Studies of KVIC biogas system coupled with flat plate collector", *Energy Convers. Manage.,* vol. 29, no. 4, pp. 253-257, 1989.

http://dx.doi.org/10.1016/0196-8904(89)90029-0

[17] G.N. Tiwari, A.K. Dubey, and R.K. Goyal, "Analytical study of an active winter greenhouse", *Energy,* vol. 22, no. 4, pp. 389-392, 1997.

http://dx.doi.org/10.1016/S0360-5442(96)00118-1

[18] E.S. Gaballah, T.K. Abdelkader, S. Luo, Q. Yuan, and A. El-Fatah Abomohra, "Enhancement of biogas production by integrated solar heating system: A pilot study using tubular digester", *Energy,* vol. 193, p. 116758, 2020.

http://dx.doi.org/10.1016/j.energy.2019.116758

[19] A. Kazemian, A. Parcheforosh, A. Salari, and T. Ma, "Optimization of a novel photovoltaic thermal module in series with a solar collector using Taguchi based grey relational analysis", *Sol. Energy,* vol. 215, pp. 492-507, 2021.

http://dx.doi.org/10.1016/j.solener.2021.01.006

[20] A.K. Singh, R.G. Singh, and G.N. Tiwari, "Thermal and electrical performance evaluation of PVT-CPC integrated fixed dome bio-gas plant", *Renewable energy sources,* vol. 154, pp. 614-624, 2020.

[21] D.B. Singh, "Exergo-economic, enviro-economic and productivity analyses of N identical evacuated tubular collectors integrated double slope solar still", *Appl. Therm. Eng.,* vol. 148, pp. 96-104, 2019.

http://dx.doi.org/10.1016/j.applthermaleng.2018.10.127

[22] H. Dhar, S. Kumar, and R. Kumar, "A review on organic waste to energy systems in India", *Bioresour. Technol.,* vol. 245, no. Pt A, pp. 1229-1237, 2017.

http://dx.doi.org/10.1016/j.biortech.2017.08.159 PMID: 28893504

[23] G.N. Tiwari, and R. Mishra, *Advanced renewable energy sources.* Royal Society of Chemistry: Cambridge, U.K., 2012.

[24] G.N. Tiwari, and A. Tiwari, *Handbook of solar energy.* Springer, 2016.

[25] A.S. Joshi, and A. Tiwari, "Energy and exergy efficiencies of a hybrid photovoltaic–thermal (PV/T) air collector", *Renew. Energy,* vol. 32, no. 13, pp. 2223-2241, 2007.

http://dx.doi.org/10.1016/j.renene.2006.11.013

[26] A.S. Joshi, I. Dincer, and B.V. Reddy, "Analysis of energy and exergy efficiencies for hybrid PV/T systems", *Int. J. Low Carbon Technol.,* vol. 6, no. 1, pp. 64-69, 2011.

http://dx.doi.org/10.1093/ijlct/ctq045

[27] C. Kandilli, "A comparative study on the energetic- exergetic and economical performance of a photovoltaic thermal system (PVT)", *Research on Engineering Structures and Materials,* vol. 5, pp. 75-89, 2019.

http://dx.doi.org/10.17515/resm2019.90en0117

[28] G.N. Tiwari, D.K. Rawat, and A. Chandra, "A simple analysis of conventional biogas plant", *Energy Convers. Manage.,* vol. 28, no. 1, pp. 1-4, 1988.

http://dx.doi.org/10.1016/0196-8904(88)90002-7

[29] G.N. Tiwari, S.B. Sharma, and S.P. Gupta, "Transient performance of a horizontal floating gas holder type biogas plant", *Energy Convers. Manage.,* vol. 28, no. 3, pp. 235-239, 1988.

http://dx.doi.org/10.1016/0196-8904(88)90028-3

[30] N. Gupta, and G.N. Tiwari, "Energy matrices of building integrated photovoltaic thermal systems: Case study", *J. Archit. Eng.,* vol. 23, no. 4, 2017.

http://dx.doi.org/10.1061/(ASCE)AE.1943-5568.0000270

APPENDIX

APPENDIX A

Expressions for various terms used in Eq. 50 and Eq. 66 are as follows:

$$U_{tca} = \left[\frac{1}{h_o} + \frac{L_g}{K_g}\right]^{-1} \quad ; \quad U_{tcp} = \left[\frac{1}{h_i} + \frac{L_g}{K_g}\right]^{-1} \quad ; \quad h_o = 5.7 + 3.8V, \, Wm^{-2}K^{-1} \quad ; \quad h_i = 5.7 \, W/m^2 K$$

$$U_{tpa} = \left[\frac{1}{U_{tca}} + \frac{1}{U_{tcp}}\right]^{-1} + \left[\frac{1}{h_i'} + \frac{1}{h_{pf}} + \frac{L_i}{K_i}\right]^{-1} \quad ; \quad h_i' = 2.8 + 3V \, Wm^{-2}K^{-1} \quad ; \quad U_{L1} = \frac{U_{tcp}U_{tca}}{U_{tcp} + U_{tca}}$$

$$U_{L2} = U_{L1} + U_{tpa} \quad ; \quad U_{Lm} = \frac{h_{pf}U_{L2}}{F'h_{pf} + U_{L2}} \quad ; \quad U_{Lc} = \frac{h_{pf}U_{tpa}}{F'h_{pf} + U_{tpa}} \quad ; \quad PF_1 = \frac{U_{tcp}}{U_{tcp} + U_{tca}}$$

$$PF_c = \frac{h_{pf}}{F'h_{pf} + U_{tpa}} \quad ; \quad PF_2 = \frac{h_{pf}}{F'h_{pf} + U_{L2}} \quad ; \quad (\alpha\tau)_{1eff} = (\alpha_c - \eta_c)\tau_g\beta_c \quad ; \quad (\alpha\tau)_{ceff} = PF_c.\alpha_p\tau_g$$

$$(\alpha\tau)_{meff} = \left[(\alpha\tau)_{2eff} + PF_1(\alpha\tau)_{1eff}\right] \quad ; \quad (\alpha\tau)_{2eff} = \alpha_p\tau_g^2(1 - \beta_c)' \quad ; \quad A_m = WL_m$$

$$A_cF_{Rc} = \frac{\dot{m}_fc_f}{U_{Lc}}\left[1 - exp\left(\frac{-F'U_{Lc}A_c}{\dot{m}_fc_f}\right)\right] \quad ; \quad A_mF_{Rm} = \frac{\dot{m}_fc_f}{U_{Lm}}\left[1 - exp\left(\frac{-F'U_{Lm}A_m}{\dot{m}_fc_f}\right)\right]$$

$$(AF_R(\alpha\tau))_1 = \left[A_cF_{Rc}(\alpha\tau)_{ceff} + PF_2(\alpha\tau)_{meff}A_mF_{Rm}(1 - \frac{A_cF_{Rc}U_{Lc}}{\dot{m}_fc_f})\right] \quad ; \quad A_c = WL_c$$

$$(AF_RU_L)_1 = \left[A_cF_{Rc}U_{Lc} + A_mF_{Rm}U_{Lm}(1 - \frac{A_cF_{Rc}U_{Lc}}{\dot{m}_fc_f})\right]$$

$$(AF_R(\alpha\tau))_{m1} = PF_2(\alpha\tau)_{meff}A_mF_{Rm} \quad ; \quad K_K = \left(1 - \frac{(AF_RU_L)_1}{\dot{m}_fc_f}\right)$$

$$(\alpha\tau)_{effN} = \frac{(AF_R(\alpha\tau))_1}{(A_c + A_m)}\left[\frac{1 - (K_K)^N}{N(1 - K_K)}\right] \quad ; \quad K_m = \left(1 - \frac{A_mF_{Rm}U_{Lm}}{\dot{m}_fc_f}\right)$$

$$U_{LN} = \frac{(AF_RU_L)_1}{(A_c + A_m)}\left[\frac{1 - (K_K)^N}{N(1 - K_K)}\right] \quad ; \quad (AF_RU_L)_{m1} = A_mF_{Rm}U_{Lm}$$

Expressions for a and $f(t)$ used in Eq. 68 are as follows:

$$a = \frac{1}{M_wC_w}\left[\dot{m}_fC_f(1 - K_k^N) + U_sA_b\right] \quad ; \quad \alpha'_{eff} = \alpha'_w + h_1\alpha'_b + h_1'\alpha'_g \quad ; \quad h_1 = \frac{h_{bw}}{h_{bw} + h_{ba}}$$

$$f(t) = \frac{1}{M_wC_w}\left[\alpha_{eff}\,'A_b\underline{I_s}(t) + \frac{(1 - K_k^N)}{(1 - K_k)}(A\,F_R(\alpha\tau))_1\underline{I}(t) + \left(\frac{(1 - K_k^N)}{(1 - K_k)}(A\,F_RU_L)_1 + U_sA_b\right)\underline{T_a}\right]$$

$$h_1' = \frac{h_{1w}A_g}{U_{c,ga}A_g + h_{1w}A_b} \quad ; \quad h_{1w} = h_{rwg} + h_{cwg} + h_{ewg} \quad ; \quad U_s = U_t + U_b$$

$$U_s = U_t + U_b \quad ; \quad U_b = \frac{h_{ba}h_{bw}}{h_{bw} + h_{ba}} \quad ; \quad U_t = \frac{h_{1w}U_{c,ga}A_g}{U_{c,ga}A_g + h_{1w}A_b} \quad ; \quad U_{cga} = \frac{\frac{K_g}{L_g}h_{1g}}{\frac{K_g}{L_g} + h_{1g}}$$

$$h_{ba} = \left[\frac{L_i}{K_i} + \frac{1}{h_{cb} + h_{rb}}\right]^{-1} \quad ; \quad h_{cb} + h_{rb} = 5.7 \ Wm^{-2}K^{-1} \quad ; \quad \begin{aligned} h_{bw} \\ = 100 \ W/m^2K \end{aligned}$$

Expressions for different terms used in Eq. 73 and 82 are as follows:

$$U_{tca} = \left[\frac{1}{h_o} + \frac{L_g}{K_g}\right]^{-1} \quad ; \quad U_{tcp} = \left[\frac{1}{h_i} + \frac{L_g}{K_g}\right]^{-1} \quad ; \quad h_o = 5.7 + 3.8V \ Wm^{-2}K^{-1} \quad ; \quad h_i = 5.7 \ W/m^2K$$

$$U_{tpa} = \left[\frac{1}{U_{tca}} + \frac{1}{U_{tcp}}\right]^{-1} + \left[\frac{1}{h_i'} + \frac{1}{h_{pf}} + \frac{L_i}{K_i}\right]^{-1} \quad ; \quad h_i' = 2.8 + 3V'W/m^2K \quad ; \quad \begin{aligned} U_{L1} \\ = \frac{U_{tcp}U_{tca}}{U_{tcp} + U_{tca}} \end{aligned}$$

$$U_{L2} = U_{L1} + U_{tpa} \quad ; \quad U_{Lm} = \frac{h_{pf}U_{L2}}{F'h_{pf} + U_{L2}} \quad ; \quad U_{Lc} = \frac{h_{pf}U_{tpa}}{F'h_{pf} + U_{tpa}} \quad ; \quad \begin{aligned} PF_1 \\ = \frac{U_{tcp}}{U_{tcp} + U_{tca}} \end{aligned}$$

$$PF_2 = \frac{h_{pf}}{F'h_{pf} + U_{L2}} \quad ; \quad PF_c = \frac{h_{pf}}{F'h_{pf} + U_{tpa}} \quad ; \quad (\alpha\tau)_{1eff} = \rho(\alpha_c - \eta_c)\tau_g\beta_c\frac{A_{am}}{A_{rm}}$$

$$(\alpha\tau)_{2eff} = \rho\alpha_p\tau_g^2(1 - \beta_c)\frac{A_{am}}{A_{rm}} \quad ; \quad (\alpha\tau)_{meff} = \left[(\alpha\tau)_{1eff} + PF_1(\alpha\tau)_{2eff}\right]$$

$$(\alpha\tau)_{ceff} = PF_c.\rho\alpha_p\tau_g\frac{A_{ac}}{A_{rc}} \quad ; \quad A_{rm} = b_rL_{rm} \quad ; \quad A_{am} = b_oL_{am}$$

$$A_cF_{Rc} = \frac{\dot{m}_fc_f}{U_{Lc}}\left[1 - exp(\frac{-F'U_{Lc}A_c}{\dot{m}_fc_f})\right] \quad ; \quad A_mF_{Rm} = \frac{\dot{m}_fc_f}{U_{Lm}}\left[1 - exp(\frac{-F'U_{Lm}A_m}{\dot{m}_fc_f})\right]$$

$$(AF_R(\alpha\tau))_1 = \left[A_cF_{Rc}(\alpha\tau)_{ceff} + PF_2(\alpha\tau)_{meff}A_mF_{Rm}(1 - \frac{A_cF_{Rc}U_{Lc}}{\dot{m}_fc_f})\right] \quad ; \quad \begin{aligned} K_K \\ = \left(1 \right. \\ \left. - \frac{(AF_RU_L)_1}{\dot{m}_fc_f}\right) \end{aligned}$$

$$(AF_RU_L)_1 = \left[A_cF_{Rc}U_{Lc} + A_mF_{Rm}U_{Lm} + A_mF_{Rm}U_{Lm}(1 - \frac{A_cF_{Rc}U_{Lc}}{\dot{m}_fc_f})\right] \quad ; \quad \begin{aligned} K_m \\ = \left(1 \right. \\ \left. - \frac{A_mF_{Rm}U_{Lm}}{\dot{m}_fc_f}\right) \end{aligned}$$

$$(AF_R(\alpha\tau))_{m1} = PF_2(\alpha\tau)_{meff}A_mF_{Rm} \quad ; \quad (AF_RU_L)_{m1} = A_mF_{Rm}U_{Lm} \quad ;$$

$$a = \frac{1}{M_wC_w}\left[\dot{m}_fC_f(1 - K_k^N) + U_sA_b\right]$$

Expressions for a and $\underline{f}(t)$ used in Eq. (68) are as follows:

$$f(t) = \frac{1}{M_w C_w}\left[\alpha_{eff}{}'A_b I_s(t) + \frac{(1-K_k{}^N)}{(1-K_k)}(A\,F_R(\alpha\tau))_1 I_b(t) + \left(\frac{(1-K_k{}^N)}{(1-K_k)}(A\,F_R U_L)_1 + U_s A_b\right)T_a\right]$$

$$\alpha'_{eff} = \alpha'_w + h_1\alpha'_b + h'_1\alpha'_g \qquad ; \qquad h_1 = \frac{h_{bw}}{h_{bw}+h_{ba}} \qquad ; \qquad h'_1 = \frac{h_{1w}A_g}{U_{c,ga}A_g + h_{1w}A_b}$$

$$h_{e,wg} = 16.273\times 10^{-3}h_{c,wg}\left[\frac{P_w - P_{gi}}{T_w - T_{gi}}\right] \qquad ; \qquad h_{1w} = h_{rwg} + h_{cwg} + h_{ewg}$$

$$h_{c,wg} = 0.884\left[(T_w - T_{gi}) + \frac{(P_w - P_{gi})(T_w + 273)}{268.9\times 10^3 - P_w}\right] \qquad ;$$

$$P_w = exp\left[25.317 - \frac{5144}{T_w + 273}\right] \qquad ; \qquad P_{gi} = exp\left[25.317 - \frac{5144}{T_{gi}+273}\right]$$

$$h_{rwg} = (0.82\times 5.67\times 10^{-8})\left[(T_w+273)^2 + (T_{gi}+273)^2\right]\left[T_w + T_{gi}+546\right]$$

$$U_s = U_t + U_b \qquad ; \qquad U_b = \frac{h_{ba}h_{bw}}{h_{bw}+h_{ba}} \qquad ; \qquad U_t = \frac{h_{1w}U_{c,ga}A_g}{U_{c,ga}A_g + h_{1w}A_b} \qquad ; \qquad U_{c,ga} = \frac{\frac{K_g}{T_g}h_{1g}}{\frac{K_g}{T_g}+h_{1g}}$$

$$h_{ba} = \left[\frac{L_i}{K_i} + \frac{1}{h_{cb}+h_{rb}}\right]^{-1} \qquad ; \qquad h_{cb}+h_{rb} = 5.7 \ W/m^2K \qquad ; \qquad h_{bw} = 250 \ W/m^2K$$

Different terms used in Eq. 89-93:

$$(AF_R(\alpha\tau))_1 = PF_1\alpha\tau^2 A_R F_R \qquad ; \qquad (A\,F_R U_L)_1 = (1-K_k)\dot{m}_f c_f \qquad ; \qquad PF_1 = \frac{h_{pf}}{F'h_{pf}+U_{tpa}}$$

$$U_L = \frac{U_{t,pa}\cdot h_{pf}}{F'h_{pf}+U_{t,pa}} \qquad ; \qquad F_R = \frac{\dot{m}_f C_f}{U_L A_R}\left[1 - exp(-\frac{2\pi r'L'U_L}{\dot{m}_f c_f})\right] \qquad ; \qquad K_K = \left(1 - \frac{A_R F_R U_L}{\dot{m}_f c_f}\right)$$

$$h_{pf} = 100 W/m^2K \qquad ; \qquad U_{t,pa} = \left[\frac{Ro_2}{Ro_1 h_i} + \frac{Ro_2 ln(\frac{Ri_2}{Ri_1})}{K_g} + \frac{1}{C_{ev}} + \frac{Ro_2 ln(\frac{Ro_2}{Ro_1})}{K_g} + \frac{1}{h_o}\right]^{-1}$$

Different terms used in Eq. 94 to 98 are as follows:

$$\alpha'_{eff} = \alpha'_w + h_1\alpha'_b + h'_1\alpha'_g \qquad ; \qquad h'_1 = \frac{h_{1w}A_g}{U_{c,ga}A_g + h_{1w}A_b} \qquad ; \qquad h_1 = \frac{h_{bw}}{h_{bw}+h_{ba}}$$

$$h_{1w} = h_{rwg} + h_{cwg} + h_{ewg} \qquad ; \qquad h_{ewg} = 16.273\times 10^{-3}h_{cwg}\left[\frac{P_w - P_{gi}}{T_w - T_{gi}}\right]$$

$$h_{cwg} = 0.884\left[(T_w - T_{gi}) + \frac{(P_w - P_{gi})(T_w + 273)}{268.9\times 10^3 - P_w}\right]^{\frac{1}{3}}$$

$$P_w = exp\left[25.317 - \frac{5144}{T_w + 273}\right] \qquad ; \qquad P_{gi} = exp\left[25.317 - \frac{5144}{T_{gi} + 273}\right]$$

$$h_{rwg} = (0.82 \times 5.67 \times 10^{-8})\left[(T_w + 273)^2 + (T_{gi} + 273)^2\right]\left[T_w + T_{gi} + 546\right] \qquad ; \qquad U_s = U_t + U_b$$

$$U_b = \frac{h_{ba}h_{bw}}{h_{bw} + h_{ba}} \qquad ; \qquad U_t = \frac{h_{1w}U_{c,ga}A_g}{U_{c,ga}A_g + h_{1w}A_b} \qquad ; \qquad U_{c,ga} = \frac{\frac{K_g}{T_g}h_{1g}}{\frac{K_g}{T_g} + h_{1g}}$$

$$h_{ba} = \left[\frac{L_i}{K_i} + \frac{1}{h_{cb} + h_{rb}}\right]^{-1} \qquad ; \qquad h_{cb} + h_{rb} = 5.7\ W/m^2\mathrm{K} \qquad ; \qquad \begin{array}{l} h_{bw} \\ = 250\ W/m^2\mathrm{K} \end{array}$$

Different unknown terms used in Eq. 100 are as follows:

$$a = \frac{1}{M_wC_w}\left[\dot{m}_fC_f\left(1 - K_k{}^N\right) + U_bA_b + \frac{h_{1wE}(P - A_2)A_b}{2P} + \frac{h_{1ww}(P - B_2)A_b}{2P}\right]$$

$$\underline{f}(t) = \frac{1}{M_wC_w}\left[\left(\frac{\alpha'_w}{2} + h_1\alpha'_b\right)A_b\left(\underline{I}_{SE}(t) + \underline{I}_{SW}(t)\right) + \frac{(1 - K_k{}^N)}{(1 - K_k)}(A\ F_R(\alpha\tau))_1\underline{I}_c(t) + \left(\frac{(1 - K_k{}^N)}{(1 - K_k)}(A\ F_RU_L)_1 + U_bA_b\right)\underline{T}_a + \right.$$
$$\left.\left(\frac{h_{1wE}A_1 + h_{1ww}B_1}{P}\right)\frac{A_b}{2}\right]$$

$$A_1 = R_1U_1A_{gE} + R_2h_{EW}A_{gw} \qquad ; \qquad A_2 = h_{1wE}U_2\frac{A_b}{2} + h_{EW}h_{1ww}\frac{A_b}{2}$$

$$P = \left(U_1U_2 - \frac{h_{EW}^2}{A_{gE}}h_{1ww}\frac{A_b}{2}\right)A_{gw} \qquad ; \qquad U_1 = \frac{h_{1wE}\frac{A_b}{2} + h_{EW}A_{gE} + U_{c,gaE}A_{gE}}{A_{gw}}$$

$$U_2 = \frac{h_{1ww}\frac{A_b}{2} + h_{EW}A_{gw} + U_{c,gaw}A_{gw}}{A_{gE}} \qquad ; \qquad B_1 = \frac{(R_2P + A_1h_{EW})A_{gw}}{U_2A_{gE}}$$

$$B_2 = \frac{Ph_{1ww}\frac{A_b}{2} + h_{EW}A_{gw}A_2}{U_2A_{gE}} \qquad ; \qquad R_1 = \alpha'_gI_{SE}(t) + U_{c,gaE}T_a \qquad ; \qquad \begin{array}{l} h_{1gE} \\ = 5.7 + 3.8V \end{array}$$

$$h_{EW} = 0.034 \times 5.67 \times 10^{-8}\left[\left(T_{giE} + 273\right)^2 + \left(T_{giw} + 273\right)^2\right]\left[T_{giE} + T_{giw} + 546\right]$$

$$U_{c,gaE} = \frac{\frac{K_g}{T_g}h_{1gE}}{\frac{K_g}{T_g} + h_{1gE}} \qquad ; \qquad U_{c,gaw} = \frac{\frac{K_g}{T_g}h_{1gw}}{\frac{K_g}{T_g} + h_{1gw}} \qquad ; \qquad R_2 = \alpha'_gI_{SW}(t) + U_{c,gaw}T_a \qquad ; \qquad \begin{array}{l} h_{1gw} \\ = 5.7 + 3.8V \end{array}$$

$$h_{1wE} = h_{rwgE} + h_{cwgE} + h_{ewgE} \qquad ; \qquad h_{1ww} = h_{rwgw} + h_{cwgw} + h_{ewgw}$$

$$h_{ewgE} = 16.273 \times 10^{-3}h_{c,wgE}\left[\frac{P_w - P_{giE}}{T_w - T_{giE}}\right] \qquad ; \qquad h_{cwgE} = 0.884\left[\left(T_w - T_{giE}\right) + \frac{(P_w - P_{gi}E)(T_w + 273)}{268.9 \times 10^3 - P_w}\right]^{\frac{1}{3}}$$

$$h_{ewgw} = 16.273 \times 10^{-3}h_{c,wgw}\left[\frac{P_w - P_{giw}}{T_w - T_{giw}}\right] \qquad ; \qquad h_{cwgw} = 0.884\left[\left(T_w - T_{giw}\right) + \frac{(P_w - P_{giw})(T_w + 273)}{268.9 \times 10^3 - P_w}\right]^{\frac{1}{3}}$$

$$P_w = exp\left[25.317 - \frac{5144}{T_w + 273}\right] \qquad ; \qquad P_{giE} = exp\left[25.317 - \frac{5144}{T_{gi}E + 273}\right]$$

$$P_{giW} = exp\left[25.317 - \frac{5144}{T_{giW} + 273}\right]$$

$$h_{rwgE} = (0.82 \times 5.67 \times 10^{-8})\left[(T_w + 273)^2 + (T_{giE} + 273)^2\right]\left[T_w + T_{giE} + 546\right]$$

$$h_{rwgW} = (0.82 \times 5.67 \times 10^{-8})\left[(T_w + 273)^2 + (T_{giW} + 273)^2\right]\left[T_w + T_{giW} + 546\right]$$

APPENDIX B

$$U_{gaE}$$
$$= \left[\frac{h_{1gE}\left(\frac{K_g}{L_g}\right)}{\left(\frac{K_g}{L_g}\right) + h_{1gE}}\right] \quad ; \quad U_{gaW} = \left[\frac{h_{1gW}\left(\frac{K_g}{L_g}\right)}{\left(\frac{K_g}{L_g}\right) + h_{1gW}}\right] \quad ; \quad U_{ga} = \frac{h_{b,bf}}{2(h_{b,bf} + h_{ba})} \quad ; \quad \frac{h_1}{= \frac{h_{b,f}}{2(h_{b,f} + h_{ba})}}$$

$$U_b = \frac{h_{b,f}h_{ba}}{(h_{b,f} + h_{ba})} \quad ; \quad A = C_1 U_2 + C_2 \quad ; \quad \begin{aligned} C_1 \\ &= \alpha_g I_{SE} A_{gE} \\ &+ U_{gaE} T_a A_{gE} \end{aligned} \quad ; \quad A' = C_1' U_1 + C_2'$$

$$\begin{aligned} C_2 &= \alpha_g I_{SW} h_{EW} A_{gE} A_{gW} \\ &+ U_{gaW} T_a h_{EW} A_{gE} A_{gW} \end{aligned} \quad ; \quad C_1' = \alpha_g I_{SW} A_{gE} + U_{gaW} T_a A_{gW}$$

$$\begin{aligned} C_2' &= \alpha_g I_{SE} h_{EW} A_{gE} A_{gW} \\ &+ U_{gaE} T_a h_{EW} A_{gE} A_{gW} \end{aligned} \quad ; \quad U_1 = h_{1fE}\left(\frac{A_B}{2}\right) + h_{EW} A_{gE} + U_{gaE} A_{gE}$$

$$U_2 = h_{1fW}\left(\frac{A_B}{2}\right) + h_{EW} A_{gW} + U_{gaW} A_{gW} \quad ; \quad B = \left(h_{1fE} U_2 + h_{1fW} h_{EW} A_{gE}\right)\left(\frac{A_B}{2}\right)$$

$$H = U_1 U_2 - h_{EW}^2 A_{gE} A_{gW} \quad ; \quad \begin{aligned} \alpha_{fb} \\ &= \left(\frac{A_b}{2}\right)\left(\alpha_f + 2\alpha_b U_{ga}\right) \end{aligned}$$

$$\left(AF_R(\alpha\tau)\right)_1 = \left[A_c F_{Rc}(\alpha\tau)_{c,eff} + PF_2(\alpha\tau)_{m,eff} A_m F_{Rm}\left(1 - \frac{A_c F_{Rc} U_{L,c}}{\dot{m}_f C_f}\right)\right]$$

$$\left(AF_R U_L\right)_1 = (A_c F_{Rc})U_{L,c} + (A_m F_{Rm})U_{L,m}\left(1 - \frac{A_c F_{Rc} U_{L,c}}{\dot{m}_f C_f}\right) \quad ; \quad \begin{aligned} K_k \\ &= 1 - \frac{(AF_R U_L)_1}{\dot{m}_f C_f} \end{aligned}$$

$$A_c F_{Rc} = \frac{\dot{m}_f C_f}{U_{L,c}}\left[1 - exp\left(-\frac{F' A_c U_{L,c}}{\dot{m}_f C_f}\right)\right] \quad ; \quad A_m F_{Rm} = \frac{\dot{m}_f C_f}{U_{L,m}}\left[1 - exp\left(-\frac{F' A_c U_{L,m}}{\dot{m}_f C_f}\right)\right]$$

$$(\alpha\tau)_{c,eff} = PF_c(\alpha_p\tau_g) \quad ; \quad (\alpha\tau)_{m,eff} = \left[(\alpha\tau)_{2,eff} + PF_1(\alpha\tau)_{1,eff}\right]$$

$$(\alpha\tau)_{1,eff} = \alpha_c\tau_g\beta - \eta_c\tau_g\beta \quad ; \quad (\alpha\tau)_{2,eff} = \alpha_p(1-\beta)\tau_g^2 \quad ; \quad PF_1 = \frac{U_{tc,p}}{U_{tc,p} + U_{tc,a}}$$

$$PF_2 = \frac{h_{pf}}{h_{pf} + U_{L2}} \quad ; \quad \frac{PF_c}{= \frac{h_{pf}}{F'h_{pf} + U_{tp,a}}} \quad ; \quad U_{L,m} = \frac{h_{pf}U_{L2}}{F'h_{pf} + U_{L2}} \quad ; \quad U_{L2} = U_{L1} + U_{tp,a}$$

$$\frac{U_{L,c}}{= \frac{h_{pf}U_{tp,a}}{F'h_{pf} + U_{tp,a}}} \quad ; \quad U_{L1} = \frac{U_{tc,a}U_{tc,p}}{U_{tc,a} + U_{tc,p}} \quad ; \quad U_{tc,p} = \left[\frac{1}{h_i} + \frac{L_g}{K_g}\right]^{-1} \quad ; \quad h_o = 5.7 + 3.8V$$

$$h_i = 2.8 + 3V \quad ; \quad z = \frac{2\pi r_{11}UL}{\dot{m}_f C_f}$$

APPENDIX C

$$h_o = 5.7 + 3.8V \qquad h_g = \frac{K_g}{L_g} \qquad h_i = 2.8 + 3V \qquad V = 1\,m/s$$

$$h_t = \frac{K_t}{L_t} \qquad U_b = \left[\frac{L_i}{K_i} + \frac{1}{h_i}\right]^{-1} \qquad h_{tf} = 2.8 + 3V \qquad U_{t,c-a} = \left[\frac{L_g}{K_g} + \frac{1}{h_o}\right]^{-1}$$

Constants used in Eqs. (6) – (10):

$$(\alpha\tau)_{eff} = \tau_g\beta_{sc}(\alpha_{sc} - \eta_{sc})$$

$$(\alpha\tau)'_{eff} = \tau_g^2\alpha_{tem}(1 - \beta_{sc}) + (1 - \eta_{tem})h_{p1}h_{p2}(\alpha\tau)_{eff}$$

$$h_{p1} = \frac{h_g}{U_{t,c-a} + h_g} \qquad\qquad h_{p2} = \frac{U_{tem}}{U_{tem,top-a} + U_{tem}}$$

$$h_{p3} = \frac{h_{tf}}{(1-\eta_{tem})U_{tem,bottom-a} + h_{tf}} \qquad\qquad U_{fa} = \frac{h_{tf}(1-\eta_{tem})U_{tem,bottom-a}}{(1-\eta_{tem})U_{tem,bottom-a} + h_{tf}}$$

$$U_{tem,top-a} = \frac{U_{t,c-a}h_g}{U_{t,c-a} + h_g} \qquad\qquad U_{tem,bottom-a} = \frac{U_{tem}U_{tem,top-a}}{U_{tem,top-a} + U_{tem}}$$

Constants used in Eq. (11):

$$W = \frac{\tau_g \beta_{sc}}{U_{t,c-a} + h_g}$$

$$X = \frac{\tau_g \beta_{sc} \alpha_{sc}}{(U_{t,c-a} + h_g)} + \frac{h_g h_{p1} (\alpha\tau)_{eff}}{(U_{t,c-a} + h_g)(U_{tem,top-a} + U_{tem})} + \frac{h_g U_{tem} (\alpha\tau)'_{eff}}{(U_{t,c-a} + h_g)(U_{tem,top-a} + U_{tem})[(1-\eta_{tem})U_{tem,bottom-a} + h_{tf}]} +$$

$$\frac{h_g U_{tem} (\alpha\tau)'_{eff} h_{tf} h_{p3} \left[1 - \frac{\left\{ 1 - exp\left(\frac{-(U_{fa}+U_b)A_m}{\dot{m}_f c_f} \right) \right\}}{\frac{(U_{fa}+U_b)A_m}{\dot{m}_f c_f}} \right]}{(U_{t,c-a} + h_g)(U_{tem,top-a} + U_{ec})[(1-\eta_{tem})U_{tem,bottom-a} + h_{tf}](U_{fa}+U_b)b}$$

$$Y = \frac{U_{t,c-a}}{(U_{t,c-a} + h_g)} + \frac{h_g U_{tem,top-a}}{(U_{t,c-a} + h_g)(U_{tem,top-a} + U_{tem})} + \frac{h_g U_{tem} (1-\eta_{tem})U_{tem,bottom-a}}{(U_{t,c-a} + h_g)(U_{tem,top-a} + U_{tem})[(1-\eta_{tem})U_{tem,bottom-a} + h_{tf}]} +$$

$$\frac{h_g U_{tem} h_{tf} \left[1 - \frac{\left\{ 1 - exp\left(\frac{-(U_{fa}+U_b)A_m}{\dot{m}_f c_f} \right) \right\}}{\frac{(U_{fa}+U_b)A_m}{\dot{m}_f c_f}} \right]}{(U_{t,c-a} + h_g)(U_{tem,top-a} + U_{tem})[(1-\eta_{tem})U_{tem,bottom-a} + h_{tf}]}$$

$$Z = \frac{h_g U_{tem} h_{tf} \left[1 - exp\left(\frac{-(U_{fa}+U_b)A_m}{\dot{m}_f c_f} \right) \right]}{(U_{t,c-a} + h_g)(U_{tem,top-a} + U_{tem})[(1-\eta_{tem})U_{tem,bottom-a} + h_{tf}] \frac{(U_{fa}+U_b)A_m}{\dot{m}_f c_f}}$$

SUBJECT INDEX

www.ingramcontent.com/pod-product-compliance
Lightning Source LLC
Chambersburg PA
CBHW050759220326
41598CB00006B/67